110 — C

AF531591

A HANDBOOK OF
BIOTECHNOLOGY

Part - II

A HANDBOOK OF BIOTECHNOLOGY

Part - II

Edited by

Neelam Yadav

ANMOL PUBLICATIONS PVT. LTD.
NEW DELHI - 110002 (INDIA)

ANMOL PUBLICATIONS PVT. LTD.
4374/4B, Ansari Road, Daryaganj,
New Delhi - 110 002
Phone : 23278000, 23261597, 23255577

A Handbook of Biotechnology — Part II

© Reserved

First Edition, 2003

ISBN 81-261-1483-5

[Responsibility for the facts stated, opinions expressed, conclusions reached and plagiarism, if any, in these volumes is entirely that of the Editor. The publishers bear no responsibility for them, whatsoever.]

PRINTED IN INDIA

Published by J.L. Kumar for Anmol Publications Pvt. Ltd., New Delhi - 110002 and Printed at Tarun Offset Press, Delhi.

Contents

SECTION 3—ROLE IN INDUSTRY AND AGRICULTURE

Preface

Biotechnology is a branch of science that combines the study and inter-action of biology and technology. The complex relationship between science and technology has benefited the mankind, since prehistoric times. Those were times when Biotechnology was used and practised in processes like fermentation by common people, without knowing their complex mechanisms. However, it was in the twentieth century and particularly its latter half that Biotechnology witnessed a revolutionary change. Many researches were conducted and newer horizons reached during this period. Newer products and drugs are being produced on a large scale now. This has benifited a huge section of society in particular, and the whole mankind in general.

This work is an effort to provide the students of Biotechnology with its latest elements, uses and applications. Tissue culture techniques, gene technology, involving recombinant DNA, gene cloning and PCR, hybridoma, protein engineering and enzymes etc. are discussed in some detail in this book. One great use of Biotechnology is in identifying criminals, with the help of the samples of semen, blood, urine, tears, saliva, perspiration and hair roots etc. This aspect also forms part of this book.

Metabolic engineering, agricultural research industrial microbiology, intellectual property rights and patents, environmental applications and implications are some other topics taken up here. Intellectual property and environmental aspects are in particular of great relevance to India, today. Towards the end of twentieth century, India has seen hectic activity in this field. Efforts are on to create better infrastructure and trained manpower.

Good literature on any subject helps a lot to all those, involved with the subject, be they students, researchers industrialists or others in any way connected with the discipline. This effort aims at providing exactly that to those, associated with the discipline of Biotechnology. It requires a lot of devotion, hard work, research and compilation on the part of the individual or individuals involved in creation of such a work.

This book has been written in simple language, still being a book on technology, it is full of technical terms. The text matter of the book is supported by relevant data in the form of tables. Several figures are also given to enrich the subject matter.

The subject of Biotechnology has gradually become very important. Its importance can be gauged from the fact that a National Biotechnology Board was constituted by the Government under the Department of Science and Technology on the recommendations of Scientific Advisory Committee of the Parliament in 1982. Biotechnology plays an important role in employment, productivity, trade, economics and quality of life. As such, literature on the subject is of great importance and relevance to lots of people.

Hopefully, this effort would prove to be purposeful for and acceptable by those, concerned.

— Editor

SECTION - 1

MICROBIAL TECHNOLOGY

ONE

Sources of Energy

What is Energy?

Energy is the capacity to do work. The product of force multiplied by the distance moved in the direction of the force is the physicists definition of work. It is often useful to speak of energy as existing in one of two states, Kinetic and Potential. Kinetic energy is the energy of motion. It is the energy involved in the process of doing work. In addition to the Potential energy and Kinetic energy, it is also useful to recognize five other forms of energy: chemical, electrical, mechanical, radiant and atomic. Atomic energy has little direct relationship to the normal functioning of the individual organism, but the others are all directly related. In recent years, there has been a lot of interest in using the physical and biological systems together for exploitation of energy source. An attempt has been made to describe some crucial bioenergy sources of current biotechnological importance.

Biogas and Methanogenic Microbes

A living cell is a potential store house of energy. This energy is generated and released by multiple biological methods. One such important group of organisms are methanotrophs. These organisms including *Methanobacterium* and *Methanococcus* species which utilize CO_2 as an electron acceptor and methane is formed. Many people suggest that methane formation is an indication of the fact that these bacteria were present on the earth before the atmosphere became filled with oxygen, at which point aerobic bacteria developed. In fact, biomethanogenesis is an ancient process for the conversion of biomass into energy. It was discovered in 1776 by Volta, who demonstrated the presence of methane in marsh gas. This marsh gas is also known as biogas or 'Gobar gas'. There are other names such as Klar-gas, refused derived fuel, will-of-the -wisp given to biogas. It is odourless, smokeless and burns with a blue flame. Biogas production occurs in many microbial ecosystems such as organic sediments of aquatic systems, marshes and soil and in the rumen and large intestines especially in herbivorous animals. It involves a complex mixture of anaerobic bacteria which convert upto 90 per cent of the combustible energy of the degradable organic matter to methane and carbon dioxide.

The gas which comes out of a bio-gas plant contains 65 to 85 per cent methane, 15 to 35 per cent carbon dioxide, some hydrogen, nitrogen and sometimes traces of hydrogen sulphide or other gases. A typical biogas sample containing 68 per cent methane, 31 per cent CO_2 and 1 per cent N_2 gives a calorific value of around 5871 $Kcal/m^3$.

As a fuel, bio-gas is similar to natural gas which contains around 80 per cent methane and has a calorific value of 8660 $Kcal/m^3$. Thus, the calorific value of bio-gas is around 60 to 80 per cent that of natural gas, depending on its CO_2 content. This heat value of the biogas can be further improved by about 30 per cent by reducing its CO2 content.

Bio-gas forms combustible mixtures in the range of 6 per cent to 15 per cent concentration in air. It takes nearly 350 Kg/cm^2 pressure to liquefy it for easy storage (for comparison it may be noted that propane liquefies at 10.5 Kg/cm^2).

As compared to hydrocarbon fuels, bio-gas has a slower rate of flame propagation and larger delay period. It boils at a temperature lower than 157°C and, therefore, has to be kept under considerable pressure for transportation purposes. It is lighter than air and has a density of around 1.1 Kg/m^3 at N.T.P. Its ignition temperature is around 750°C.

The presence of CO_2 in bio-gas lowers its calorific value. It also interferes with complete combustion, consequent resulting in higher carbon deposits. The carbon dioxide content of bio-gas can be reduced by passing the gas through a lime solution. Upto thirty per cent reduction in CO_2 concentration has been reported to have been achieved by a group at the I.I.T., New Delhi using this method.

Traces of Hydrogensulphide (H_2S) in bio-gas should be removed, as it can cause corrosion of certain components of the diesel engine. H_2S gas is generally removed by passing the gas through ferric oxide and iron filings.

The production of biogas takes place in three steps (i) solubilization and hydrolysis of the organic components (ii) acidogenesis (iii) methanogenesis. By enzymatic hydrolysis, the complex substances are solubilized and serve as the substrates for microorganisms in the second stage. This first stage, in general, is never complete because it would be too long. The conversion of energy concerns only half of the organic matter, but the residues or sludge of the methanė fermentation are useful in agriculture because they can serve as fertilizers. In the second stage, these soluble organic compounds are converted to organic acids. Three groups of bacteria are involved in this process. The first converts complex organic substrates into butyric, propionic and lactic acid; the second, the acetogenic bacteria converts these organic acids into acetic acid, hydrogen and carbon dioxide. The acids serve as substrates for the final stage of decomposition accomplished by methanogenic bacteria. The methanogenic bacteria reduce carbon dioxide to methal by several types of anaerobic bacteria, serves as electron donor for these methanobacteria. With an efficiently operated digestor and substrates such as cattle waste, as much as 4.5 liters of methane can be produced per liter of reactor material a day. With some vegetable materials or forages as much as 65-70 per cent of the matter can be converted to methane. Methane can also be derived from aromatic compounds under conditions of strict anaerob'rosis. This is achieved with organisms such as Methanobacterium formicicum and Methanospirillum hungati. This process is quite widespread in nature, particularly in the treatment of sewage and effluents as well as in the conversion of certain biocides. The aromatic ring is broken down to acetate which is degraded by the methanobacteria as follows:

$4\ C_6H_5COOH + 24H_2O \rightarrow 12\ CH_3COOH + 4HCOOH + 8H_20$

Benzoate Acetate

12 CH3COOH $\rightarrow 12\ CH_4 + 12CO_2$

Acetate

4 HCOOH $\rightarrow 4CO_2 + 4H_2$

Formate

$3CO_2 + 12H_2 \rightarrow 3CH_4 + 6H_2O$

$4\ C_6H_5COOH + 18H_2O\ \ 15\ CH_4 + 13CO_2$

In nature, aromatic compounds are derived from the slow decomposition of tannins and of lignin, generally, carried out by extracellular microbial enzymes. Methanogenesis of lignins and tannins, which are an important component of the soil organic matter, is very important in the carbon cycle.

The methanoganic bacteria are highly sensitive to oxygen and grow very well in the presence of hydrogen and carbon dioxide. Recently, Matsushita Electric Industrial Co. in Japan has discovered a new methanogenic bacterium *Methanobacterium kadomensis* St. 23 which could produce methane much faster than produced by normal strains.

In India and China, biogas production has become very common particularly in rural areas. Besides, other countries including the Philippines, Guatemala, Israel, and France have now undertaken extensive biogas programme. The raw material including household refuse (with water 1 : 1), agricultural refuse such as banana, cassava, citrus fruits, sugarcane are also being used. The major raw material is cow dung in India and pig dung in China. In China, however, animal dung is not utilized for any other purpose except biogas in contrast to India where cow dung is still used as fuel in several villages. As a matter of fact, the use of cow dung as a direct fuel source in India has led to considerable loss in its use as fertilizers, as well as environmental pollution as a result of excessive cow dung burning as fuel. On a national scale, the most significant contribution to country's energy budget is in China, where over seven million small family scale digesters have been built. Biogas in its crude form is primarily used for cooking or lighting. It may also be used to power stationary engines for generation of electricity or be compressed into cylinders for use in cars or tractors or fed into piped gas distribution system.

Biogas production is carried out in water-tight cylindrical vats with a side opening as inlet for the substrate material. Gas is collected in a steel container and the cylinder is fully airtight because the entire process is strictly anaerobic. The digester is generally made up of clay bricks or steel. The substrate is generally seeded with acetogenic, or methanogenic bacteria or with sludge from another digester. The optimum digestion takes place between PH 6.0 and 8.0. A higher C/N ratio in the digester stimulates the process, and an excess of nitrogen is avoided. This C/N ratio can be modified by the addition of straw or bagasse. The suspension of fermentative substances is stirred to prevent stratification, which inhibits fermentation. The cow dung plant (particularly in India) generally requires 2-3 weeks for complete digestion. In Scotland, however, where liquid manure from pigsties is used, the digestion is complete within ten days. The Rowett Research Institute in Scotland has developed a fast digestion system even for poultry and cattle dung which takes about 15-20 days for complete digestion. Digestion is generally carried out at temperatures between 30 and 40°C or 50 and 60°C to kill all pathogenic enteric bacteria.

The manorial value of the dung is not diminished in the process of biogas production, it is enhanced. Thus, a biogas plant helps in obtaining both fuels and manure from the same quantity of cattle dung. A large scale programme for the promotion of biogas in India has been taken up as the National Project for Biogas Development

(Table). Over 6.5 lakh family-based biogas plants have already been installed. These are generating more than 7200 lakh cubic metres gas annually equivalent to about 25 tonnes of firewood (valued at Rs. 100 crores per annum). Besides, these plants also produce 120 lakh tonnes of high quantity organic fertilizers (about Rs. 100 crores per annum).

According to the NPBD, by the year 2000/2001, about 12 million biogas plants will be installed at a total estimated outlay of about Rs. 10,600 crores. These plants will save a minimum of 42 million tonnes of fuel wood annually.

Interest in biogas production is world wide and includes programmes in both the developed and the developing countries. Village and community scale production has many obvious advantages in developing countries and developments in China and, to a lesser extent, in India bear witness to this. In both Europe and North America, very extensive biogas research programmes have been established. Such programmes have been directed towards both farm-scale production with on-site use and commercial scale gas production coupled to a distribution net work (Fig.). Clearly, such a process will comprise much more than the bioreactor, unlike the simple systems employed in developing countries where even biogas storage is commonly within the bioreactor. It involves a thorough evaluation of cost of biogas production projects.

Table

The Number of Biogas Plants Set Up in Different States/Union Territories Under the National Project for Biogas Development

State/UT	*No. of plants set up during NPBD period i.e. from 1981-82 to 1984-85*
1. Andhra Pradesh	31393
2. Uttar Pradesh	71166
3. Haryana	7837
4. Maharashtra	88211
5. Bihar	21093
6. Karnataka	20772
7. Punjab	4899
8. Tamil Nadu	31905
9. Gujarat	22949
10. Madhya Pradesh	16399

State/UT	No. of plants set up during NPBD period i.e. from 1981-82 to 1984-85
11. Madhya Pradesh	6273
12. Kerala	6488
13. Orissa	5900
14. Rajasthan	14304
15. Delhi	182
16. Assam	1101
17. Other States/UTs.	5015
TOTAL	355887

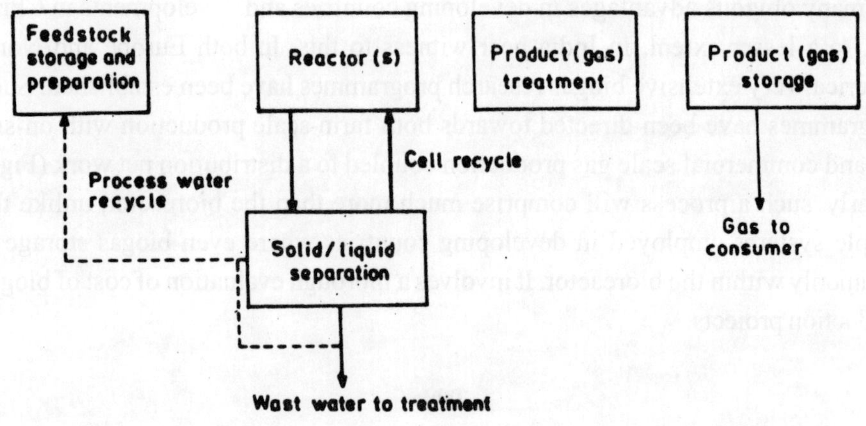

Fig. Biogas production flowsheet

Ethanol Production and Utilization as an Energy Source

Ethanol is generally produced from petroleum products such as ethylene. Ethylene could be converted to ethanol at a high temperature in the presence of water and catalyst. In India, Ethanol production is gaining importance. By 1980, India became almost self sufficient in Ethanol production. However, the growing demand for ethanol in industries such as organic chemicals and pesticides, Pharmaceutical, Paints and Varnishes and several other laboratory requirements envisages a need for approximately 1200 million tons of ethanol by 1990. This is almost double the requirement in 1980. In United States also, more than 70 per cent ethanol is produced by synthetic methods. Now, there is greater need for exploiting material sources and a large number of plants such as cassava, maize, potatoes, Jerusalem artichoke and sweet sorghum are being used for ethanol production. Brazil has been producing large quantities of ethanol since 1975 from sugarcane and Cassava. Brazil is the second largest sugarcane producing country in the world. In recent years, the use of ethanol as fuel is gaining importance because of rising cost and diminishing petroleum reserves. It can be used as a fuel in internal combustion engines primarily as anhydrous alcohol (99.8%) or mixed with petrol upto a 20 per cent ratio, or as hydrated ethanol (94%). In fact, ethanol blending has resulted in changing the octone rating and other engine performance characteristics such as starting, caruration and emissions. The economic value of ethanol as a petrol blend is about 20 per cent higher than that of ethanol used as a direct petrol substitute. Currently, several other countries including Costa Rica, Africa, Venezuela and Philippines are also exploiting sugarcane and cassava for production of ethanol. Of all the crops grown in the U.S.A., corn contains the highest amount of starch and is the most obvious choice as a feed stock for ethanol. Around 5% of the gasoline sold in the U.S.A. is now blended with 10% ethanol. France is also now using sugarbeet, Jerusalem artichoke and sugarcane for mass production of alcohol. India has also been producing millions of hectolitres of ethanol from the fermentation of molasses.

Methanol, ethanol or other higher alcohols can be used as transportation fuels. The use of alcohols for transportation purposes in Brazil dates back to 1922. Brazil's gasohol is a mixture of 80 per cent gasoline and 20 per cent pure ethanol.

Methanol has about only half the energy content of gasoline. It burns as clearly as any other liquid fuels and in LC. engines, it burns knockfree. Small amounts of methanol can be mixed with gasoline and used in automobile engines immediately. Engines can be modified to use methanol as fuel. Though methanol is produced mainly from natural gas, it can also be produced from biomass, coal, lignite, peat, etc. by existing technologies.

Ethanol, on the other hand, has about two-thirds the energy content of gasoline and burns as clearly as other liquid fuels in LC engines knockfree. It can be used as transportation fuel in combination with gasoline. Several other countries including Papua, New Guinea, and Australia have already launched massive programmes for extracting alcohol for fuel purposes.

Biologically produced fuels and feedstocks have the apparent advantage of being renewable and raw materials for their production are frequently assumed to be cheap, free or even of negative value. Since 1973, the concept of producing ethanol as a fuel and/or chemical feedstock from agriculturally produced carbohydrates has been extensively investigated, and five or six years ago, enthusiasm for such processes mounted to a climax. Ethanol programmes in the USA and brazil have been widely discussed. Both the reasons for establishing, the programmes and the ultimate outcomes differ markedly in the two countries, in spite of the original stimulus for their establishment being OPEC mediated crude oil price increases.

In the USA, although concern existed about the rate of depletion of fluid fossil fuel resources, it was the over-production of maize (corn) and seriously depressed corn prices for several years in the mid-1970s, that promoted the politically attractive expedient of converting surplus corn into fuel ethanol (gasohol). In Brazil, however, for several other political and economic reasons, a national ethanol production programme designed to significantly increase ethanol production from sugarcane for fuel and as a feedstock for chemical manufacture was developed.

Liquid fuels produced from crops, grown specifically for the purpose, are expensive unless production occurs in countries with climatic conditions that petmit very fast crop growth and where the agricultural labour is cheap. The Brazilian ethanol programme has increased ethanol production from 6 x 10' m3 in 1975-76 to 9.2 x 106 m^3, in 1985. Sugarcane production for ethanol employed a direct labour force of 0.5 million in 1984. Brazil's target production for the crop year 1987-88 is 14.3 x 106 m^3, but with production destined exclusively for the home market.

Ethanol produced from corn is emerging as a leading agent in reducing carbon monoxide pollution. A recent study conducted for the city of Phoenix, Arizona, shows the CO_2 levels could be reduced by about 23 per cent if cars were run on a mixture of 90 per cent gasoline and 10 per cent ethanol. Recent data show that even new cars and the ones with modern emission controls, will run "cleaner" using blends containing oxygen rich fuels like ethanol.

It is currently being debated whether the system and source used in a particular country are worthy of being adapted. Several countries including Sweden, Malawi, Kenya, United States and Zimbabwe have made substantial investment in alcohol production. At present, the biggest programme is in Brazil where the Government spends about 15 billion Rupees per year on subsidizing the production of alcohol from sugarcane. Zimbabwe is another country which has launched a massive alcohol production programme from sugarcane. Higher purity molasses results in a higher fermentation efficiency for the distillery. Currently, all the alcohol produced is blended with gasoline (12% alcohol/gasoline mix) and this is the major fuel available for spark ignition vehicles. It has been now proved that cars fueled with gasohol (gasoline + alcohol) managed to obtain 6.7% more miles to the gallon on an average.

Ethanol from agricultural crops is a viable proposition even in Europe, AC Biotechnics would claim, since the company have just brought on stream a 20,000 litre per day plant operating on the proprietary Biostil principle.

The plant is situated in Skaraborg in central Sweden and the ethanol is being sold to the UK petroleum company for incorporation at 4% into petrol sold at pumps in the Stockholm area. It uses as an input substrate whole wheat grain which is produced by Sweden's farmers in an excess of a million tonnes per year and which would otherwise have to be sold off in the world markets at subsidised prices.

In addition to the advantages gained in reducing oil imports (albeit marginally) and alleviating export subsidies, the Biostil process also generates byproduct credits in the form of carbon dioxide gas, starch, and protein rich animal fodder. The great advantage of the Biostil system, however, is that it is a continuous process which uses feedback of biomass into the fermentor to increase the concentration of ethanol produced. Distillation processes are, therefore, much more economic.

AC Biotechnics sees the Skaraborg plant, which is owned and run by Agroenergi Utveckling AB, as the first in a series of Biostil operations throughout Sweden. These plants could have capacities upto 200,000 liters of ethanol per day.

The lesson here, perhaps, is that it is not substrate cost per se that determines the commercial viability of operations such as ethanol production, but the balance between the costs of producing the fuel and those of the alternative integrated scenario. The savings in consumption of crude oil can be calculated from such results.

Currently, several crops are being exploited as an energy source. Some high yielding energy-cane hybrids have been obtained by crossing a standard variety with an Argentinian wild clone of Saccharum spontaneum (Sugar cane is highly productive with photosynthetic efficiency upto 4%). Another crop plant which has a potential energy source is Sorghum and it can be cultivated rapidly over a range of climates. The genes for sweetness and lignin have already been located in this plant and yields as high as 5,000 L of ethanol per hectare per year have been achieved experimentally (minus the cellulose portion).

Recently, Melvin Calvin has been attempting to produce gasoline directly from plant biomass and *Euphorbia lathyris* is being seriously investigated for the extraction of hydrocarbons which have molecular weight fairly close to that of gasoline. Another important plant is the water hycinth (Eichornia crassipes) which is a major water weed in India as well as in several other countries. It is now being exploited for methane production.

Besides these plants, palm tree tissue culture experiments have been successful and it may be possible to exploit it commercially. At present, more than 14% of the world's vegetable oils are obtained from Palm. It is a perrenial plant and can produce 6 t ha $^{-1}$ of oil.

Biophotolysis and hydrogen Photoproduction

A large percentage of solar energy that reaches the earth goes waste since it is not exploited in total for commercial purposes. Perhaps the most important step would be to produce fuels and biomass directly involving lot of technological operation. Basically, all green plants are able to use this solar energy and reduce CO2 using water as a source of electrons in the presence of the light reactions of photosynthesis. The photosynthetic efficiency of the conversion of solar energy into rich organic compounds is low, averaging only around 1% in good conditions, but this represents a stored energy efficiency on an annual basis. In many parts of the developing world, biomass is being converted into energy faster than it is being replaced. In fact, there is potentially afar greater long-term danger from deforestation and desertification as a consequence of the over-exploitation of biomass than there is from lack of food. Currently, about 13% of the world's primary energy is derived from bio-mass equivalent to 22M barrels oil. It is primarily exploited in. the rural areas of the developing countries (with about half of the world's population). A number of developed countries also depend upon a considerable amount of energy from biomass, including the United States (4%), Sweden (10%), Canada (8%) and Australia (3%). Inspite of this biomass, however, resources are extremely limited and also the production and use of biomass is inefficient with low yields and poor conversion. Therefore, attention is now being focussed on other natural sources of energy, particularly, photosynthetic microbes. Besides, higher plants algae and cyanobacteria contribute substantially towards solar energy bioconversion.

Biophotolysis refers to the production of hydrogen from water with the help of Sunlight. The phenomenon was first documented by Hans Gaffron about four decades ago in algal systems. Later, it was shown to operate in isolated spinach chloroplast, The process is facilitated with the help of enzyme hydrogenase.

$$H_2 2 \xrightarrow{\text{Hydrogenase}} 2H^+ + 2e^-$$

As a matter of fact, it is theoretically possible to produce hydrogen from water using any plant or algae that contains this hydrogenase enzyme.

At room temperature, hydrogen is a colorless and odourless gas. It is highly combustible and thus a source of energy. However, like oxygen, it does not dissolve in water appreciably. At very low temperature, it can exist at liquid or even as solid (Table). On the earth, free or uncombined hydrogen is rare. It is found occasionally in volcanic gases. It is also found in traces in the upper atmosphere. In producing hydrogen for commercial use, the major consideration is usually the cost. For a commercial hydrogen, the primary sources are water and hydrocarbons and it can be made inexpensively by passing steam over hot carbon.

$$C(S) + H_2O \xrightarrow[1000^0C]{} CO + H_2$$

It is hard to get pure hydrogen from this source because carbon monoxide (CO) is difficult to separate completely from hydrogen. The mixture of H2 and CO is an industrial fuel known as *water gas*.

Table

Physical Properties of Hydrogen

Molecular mass	2.016
Bond Length	0.479 A°
Bond Energy	103 Kcal 1^{-1} M
Approx. Molecular diameter	2 A°
Normal Melting point	14.1° K
Normal Boiling point	20.4° K
Critical temperature	33.2° K
Density of STD	0.0899 gl $^{-1}$
Density of Liquid (20°K)	0.07 g ml $^{-1}$

Purer and still relatively inexpensive hydrogen can be made by passing steam over hot iron.

$$3 Fe + 4H_2O \rightarrow Fe_3O_4 + 4H_2$$

The iron can be recovered by reducing Fe_3O_4 with water gas. The purest (99.9 per cent) but most expensive hydrogen available commercially is electrolytic hydrogen made from the electrolysis of water.

$$2H_2O \xrightarrow{\text{electrolysis}} 2H_2 + O_2 \qquad H = +135 \text{ Kcal}$$

The reaction is endothermic and requires energy which must be supplied by the electric current. It is the power consumption and not the raw material that makes electrolytic hydrogen expensive.

In petroleum refineries, where gasoline is made by the catalytic breakdown of hydrocarbons, hydrogen is a by-product. When gaseous hydrocarbons are passed over a hot catalyst, decomposition occurs to form hydrogen and other hydrocarbons.

Under appropriate conditions chemically, H_2 is stable, and combines directly with most elements. With oxygen, H_2 reacts to release large amounts of energy by the change which occurs at an appreciable rate only at high temperature or

$$2\ H_2 + O \rightarrow 2\ H_2O \qquad \Delta H = -116 \text{ Kcal}$$

in the presence of a catalyst. In the oxyhydrogen torch, the above reaction occurs to produce temperatures of about 2800°C and the reaction is self sustaining. Hydrogen also reacts with certain compounds. In some cases, it simply adds on the other molecule, as for instance, in forming methyl alcohol (CH_3OH) from CO.

$$CO + 3\ H_2 \xrightarrow{\text{catalyst}} CH_3OH$$

Such addition reactions are known as hydrogenation reactions and account for much of the industrial consumption of hydrogen.

Hydrogen, when mixed with oxygen, provides 29,000 calories per gram as compared to 11,200 and 7,800 calories per gram of gasoline and coal respectively. It can be substituted for hydrocarbons in a broad range of applications often with increased combustion efficiency. Its burning process is non-polluting and it can be used in fuel cells to produce both electricity and useful heat. Rigorous efforts are needed to solve technical and economic problems related to production, storage, transmission utilization and safety aspects.

Hydrogen is a versatile fuel and can be used for automobiles, space crafts, aeroplanes, helicopters, and engines. Hydrogen driven scooters, buses, cars and aircrafts have been developed and tested in USA, UK, and Japan. If the problems relating to on-board fuel supply, the servicing of hydrogendriven vehicles, and the proper safety measures can be solved, perhaps transport sector would be the largest consumer of hydrogen.

The VSSC/ISRO are keenly interested in liquid hydrogen for future *space programmes in India*. Liquid hydrogen is considered one of the best fuels for subsonic and supersonic aircrafts world over.

The use of several algae as a means of trapping solar energy to expand fuel supplies is attractive for reasons other than its potential as a cheap energy alternative. Hydrogen as a fuel is nonpolluting and both water and sunlight is inexhaustible. Earlier, it had also been observed that spinach chloroplasts and bacterial extracts containing hydrogen would, on irradiation with visible light, evolve hydrogen if ferredoxin was added as an electron carrier. The bacterial extract (from Clostridium) containing hydrogenase is very sensitive to oxygen and the reaction is carried out in the presence of Nitrogen. David Hall from King's College London has been able to produce 50 micromoles of hydrogen every hour per milligram of chlorophyll (and continuously for six hours). Other laboratories have been able to sustain the reaction for longer periods. Hydrogen produced from cell fractions such as the chloroplast has theoretical advantages with a reduced system size for trapping light energy and improved efficiency.

Hydrogen so produced can easily be stored and has a very high calorific value of 20 Kcal g^{-1} (carbohydrates have 3.5 Kcal g^{-1}). Besides, hydrogen production does not produce any toxic intermediates.

A large number of algae are being exploited for hydrogen production. Some of the important algae include *Chlorella, Chlamydomonas, Scenedesmus, Ankistrodesmus, Dunaliella, Corralina and Porphyridium.*

Recently, a group of workers from France, USSR and Switzerland have shown that hydrogen could be produced in the presence of suitable artificial electron donors (other than water) and light absorbing pigments (other than chloroplast membranes).

A lot of work is currently going on photoproduction of hydrogen by cyanobacteria. These organisms can be adapted to produce H_2 photosynthetically by growing them in a medium free from both Oxygen and Nitrogen such as one argon and CO_2. It has been estimated that a stable system (which is extremely difficult to obtain, maximum stability so far has been nineteen days) would be very useful for the production of Hydrogen by water photolysis. While receiving 10 million Joules per square metre per day of solar energy, it would produce about 90 liter of Hydrogen per square metre per year, with an energy content of about 40 million Joules (about 3 kilograms of Hydrogen per square metre per year). In principle, the development of a stable and efficient photochemical system that mimics a functional PSII (O_2 evolution and reduction of electron carrier) of chloroplast will be very helpful in the process of converting solar energy into hydrogen.

Another important source of such energy production is the halophilic bacterium *Halobacterium halobium,* which can utilize solar energy trapped by purple pigment *bacteriorhodopsin* found in the cell membrane. This light sensitive pigment is a derivative of vitamin A and consists of a single protein chain which is attached to a molecule of retinaldehyde. This pigment behaves like a light-driven proton pump and the whole series of changes that take place in the bacteriorhodopsin molecule are completed within about 10 milliseconds. Thereafter, the molecule is once again ready to be activated by light and to pump another electron. These bacteriorhodopsin molecules are arranged in the cell membrane in groups of three and work synergistically. The bacteriorhodopsin can work even in isolation. Several people have been working on introducing bacteriorhodopsin into an artificial system. However, all such systems have so far proved to be relatively inefficient. Several countries are now seriously trying to exploit this system to obtain energy. The organism can be grown in shallow ponds containing common salt and minerals. It has a very feeble Oxygen requirement. The bacterial cell is very rich in the purple pigment. It has been calculated that a 10-litre bacterial culture would yield about 0.5 gram of purple membrane. Techniques are currently being developed to impose these membranes on artificial support in such a way that the bacteriorhodopsin molecule should pump.

Biophotolysis for the production of hydrogen requires large reactors that are transparent to radiant energy and impermeable to hydrogen. These reactors may be less economical. Therefore, the cost of the reactor has to be competitive with the alternative energy source. Although Calvin's laboratory has initiated the research, the techniques for microencapsulation and stabilization of subcellular fractions capable of hydrogen evolution need to be developed further. Besides, improving the output of hydrogen depends upon seeking hydrogenases which are less sensitive to oxygen.

Utilization and Degradation of Cellulose

Cellulose is a very important renewable raw material produced in large amount by plants. It has been estimated that productivity of cellulose, on a global synthesis basis, can be as high as 11.4 x 10^{15} Kg/Year. Cellulose is primarily associated with lignin—the structural material of plants-and the resulting composition is known as lignocellulose. A chemical bond between lignin and cellulose had been suggested. However, recently it has become quite clear that the association is primarily physical in nature; with lignin and amorphous cellulose forming a mutually interpenetrating system of high polymers. Unlignified cellulose (from non woody aquatic plants and natural fabrics, such as cotton) is very useful substrate for microbial processing. In fact, the global availability of the total amount of non-woodyplant cellulose is extremely high and is more than a billion metric ton. Sugarcane bagasse is the highest amongst the growing crops besides having an extremely high solar energy utilization capacity. It also has a very high yield (about 85 tons/ha). Moreover, bagasse has a very high L-cellulose content (about 33%). If agricultural productivity ever exceeds the food requirements of the world, the residual cellulose (which is primarily an agricultural waste, byproduct) may be exploited for petroleum production and, thus, become an important source of energy. In recent years, it has become possible to commercially exploit cellulose for low cost quantitative recovery of hexoses and pentoses. It is also becoming possible to obtain cellulose derived sugar syrups which will have a wider industrial application.

The degradation of cellulose requires a cellulose-cellulase system. Cellulases can be defined simply as a system of enzymes which are able to break down the highly ordered cellulose polymer into sufficiently small sugar molecules which are able to pass through the microbial cell walls. Pretreatment of cellulose is required to enhance cellulose susceptibility to enzyme attack by removing lignin barrier and thus changing the structure of cellulose. Besides, methods such as milling and heat treatment, swelling, dissolving and regeneration have been used to increase cellulase susceptibility to enzymic attack. Lignin is also degraded by microorganisms. However, in vitro degradation of lignin by enzymes has yet to be demonstrated and proved. So far, it has not been possible to characterize and describe any one of the enzyme systems involved in the microbial degradation of lignins. A few years ago, Swedish Forest Products Research Laboratory developed a mutant of a white rot fungus *Sporotrichum pulverulentum* which is a cellulase negative species. It is capable of degrading kraft and wood lignocellulase actively. Using such a mutant, a sharp decrease in the lignin and xylan component of wood has been demonstrated. The laboratory has been able to produce biological pulp without any chemical treatment for delignification.

Many microorganisms are known to produce the enzyme cellulase (Table 4.3). However, the most promising is Trichoderma reesi which produces extracellular cellulase. Cellulases are inducible enzymes and their synthesis is strongly repressed by soluble sugars or other easily metabolizable substrates. The cellulase enzyme derived from *Clostridium thermocellum* (a thermophilic anaerobe) has been reported to be resistant to end product inhibition and repression. Several new mutant strains have been developed in Trichoderma reesi with enzyme activity more than 16 fold over the present strain.

It is important to achieve enhanced production of cellulase and that could be done by several methods as listed.

1. Selection of mutant strains with a higher production of extracellular cellulase.

Table

Process Characteristics for Microorganisms Utilizing Lignocellulose or Cellulose

Microorganism	***Substrate***	***Process Temperature***	***Time Required***
Volvariella sp.	Lignocellulose (straw, etc.)	Ambient (high temperature and humidity)	Weeks
Lentinus edodes	Lignocellulose (wood)	Ambient	Months
Pleurotus sp.	Lignocellulose (straw, wood, etc.)	Ambient	Weeks
Trichoderma reesei	Cellulose or pretreated lignocellulose	Ambient	Days
Thermoactinomycetes sp. and other thermophilic actinomycetes	Cellulose or pretreated fignocellulose	55°-65°C	Days
Phanerochaete chrysaiporium	Lignocellulose (wood)	Ambient to 40°C	Weeks
Clostridium thermocellum	Cellulose or pretreated lignocellulose	55°-65°C	Days

Microorganism	***Substrate***	***Process Temperature***	***Time Required***
Pseudomonas fluorescens var. cellulosae lignocellulose	Cellulose; pretreated lignocellulose;	Ambient	Days
Thermophilic Sporocytophaga	Cellulose or pretreated lignocellulose	55°-65°C	Days

2. Since cellulose is an inducible enzyme, it would be essential to raise constitutive mutant strains for production of cellulose.

3. Production of strains, by genetic manipulation, which are resistant to feed back or end product inhibition.

Enzymatic break down of cellulose requires four major enzymes.

(i) Endoglucanases, which randomly split one cellulose polymer into two smaller molecules.

(ii) Exoglucanases, which remove one glucose unit at the non-reducing end of the molecule already split by endò-glucanases.

(iii) Cellobiosylhydrolase, which resembles exoglucanase in action with the exception of removing one cellobiose unit from the chain.

(iv) Cellobiase, which catalyses the hydrolysis of cellobiose to glucose in aqueous phase.

The enzymes cellobiase and 1-4, ß-glucosidase are quite-similar in structure and action. However, cellobiase catalyses hydrolysis of cellobiose and 1-4,ß glucosidase splits dextrin at the ß 1-4 linkages. The cellulase enzyme obtained from T, rusei lacks xylanase and does not degrade hemicellulose. However, when mixed with cellulose derived from Aspergillus wentii, it breaks down complex compounds in bagasse. The genetics of T. reesi has not been worked out in detail and the genes coding for various enzymes need to be dissected and manipulated to increase enzyme production.

Cellulose is present in excess on this planet and lignocellulose is one of the most important renewable source of energy, chemicals and proteins. There are two main sources (i) agricultural residues and by-products (ii) waste generated by industrial and community activities. For developing countries including India, cellulosic waste from industrial and community activity is very important source and is much higher than the agricultural residues (100 m ton). Exploitation of cellulose forms the most important step towards vitalization of several economic sectors. It might involve problems of transport, as well as management particularly with regard to cellulose extraction from industrial and community activities. However, there are local problems and, keeping in view the potential resource wealth from cellulose, it should not be a difficult planning for any developing country.

Other Nonconventional Bioenergy Sources

Energy is the capacity to do work. The word has been derived from Greek 'Energia' and coined by Thomas Young (1773-1829). Lately, however, energy is considered to be another form of matter which is interconvertible. For long, significant natural sources of energy have been water falls, and wind energy for driving ships. In addition, the first fossil fuels to be exploited were surface deposits of asphalt, Peat and coal, oil from surface seepage and gas venting from underground reservoirs. The first oil well drilled in Modena, Italy in 1640 provided fuel for street lighting. Oil as lamp fuel was used on the island of Sante in the Ionian sea late in 400 B.C. By 1880, crude oil accounted for more than 13% of all mineral fuels consumed. Steam power developed in 18th century led to revolution in energy exploitation. Energy could be categorized as (i) Physical energy (such as Mechanical, Heat and Light and electro-magnetic energy and Nuclear Energy), (ii) Chemical Energy (Tied up in molecules of food or fossil fuel) and (iii) Bioenergy (energy present or derived through biological systems).

In 1980, the United National General Assembly meeting put forward the following conclusion:

"The International community will have to make substantial and rapid progress in the transition from the present international economy based primarily on hydrocarbons. It will have to rely increasingly on new and renewable sources of energy, seeking to reserve hydrocarbons for non-energy and nonsubstitutable uses." The emphasis was further supported by U.N. conference on New and Renewable Sources of Energy (Table) at Nairobi in August 1981. In view of the growing need for an organized "energy

Table

Conventional And Non-conventional Energy Sources

Conventional Energy Sources	*Non-conventional Energy Sources*
Coal	Bio-Energy
Water (Hydro)	Geothermal Energy
Nuclear	Hydrogen Energy
Oil and Natural gas	Solar Energy Tidal Energy
	Wing Energy

budget", the Government of India set up the Department of Nonconventional Energy Sources (DNES) on September 6, 1982. Before that, the Commission for Additional Sources of Energy (CASE) was also set up by the Government of India in March 1981 which is now looked after by DNES. (Fig).

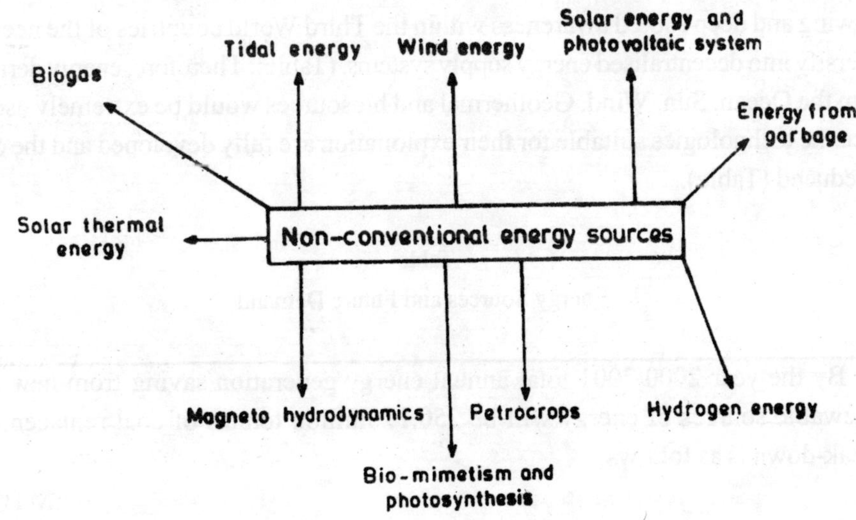

Fig. Spectrum of possible N-C energy sources

Due to growing needs, non-renewable sources of energy such as fossil fuel will be exhausted in due course (Table). Eighty per cent of India's rural population lives on what is known as "Non-commercial" energy sources. It amounts to about 275 m ton of coal equivalent. Therefore, nonconventional sources of energy needs to be tapped for commercial purposes. All fossil fuels including coal, oil and natural gas are generally considered to be the result of decomposition and conversion of plants and other living matter. Coal comprises 3/4 of the total fuels of the world and in India the total reserves amount to about 80,000 million tons. However, the present day technology wastes nearly 50% of the coal recovered. Similarly energy derived from other conventional sources (except nuclear and to some extent Hydro-energy)

Table

World Energy Reserve

	Total Reserve	***Proven Recoverable Reserve***
Gas	46	8
Oil	72	17
Coal	730	47
Biomass	60	Renewable

is decreasing and some of these sources have become inaccessible. There is also a growing and deep rooted awareness within the Third World countries of the need to diversify into decentralised energy supply systems. (Table). Therefore, energy derived from the Ocean, Sun, Wind, Geothermal and biosources would be extremely useful once the technologies suitable for their exploitation are fully developed and the cost is reduced (Table).

Table

Energy Sources and Future Demand

By the year 2000/2001 total annual energy generation/saving from new and renewable sources of energy will be 250.17 million tonnes of coal replacement. Break-down is as follows:

		MTCR
1.	Power from Biomass (6,000 MW)	30.94
2.	Power from Wind (5,000 MW)	9.42
3.	Power from solar systems (2,000 MW)	8.78
4.	Small hydro (2,000 MW, 10.5 bkWh)	8.24
5.	Improved cook stoves (10.00 crores chulhas)	79.80
6.	12 million biogas plants (4 cum average size)	47.88
7.	Energy plantations (2.5 million hectare with average 20 tonnes/ha)	47.5
8.	Sewage dude (50 MW)	0.26

9.	Solar thermal systems	20.00
10.	Photovoltaic pumps (50,000 numbers)	0.08
11.	Wind pumps (50,000 numbers).	0.92
12.	Small battery charges and stand alone sys.	0.04
13.	Energy from distillery wastes (140 proj.)	1.54
14.	Energy from Municipal solid wastes (40 projects)	0.67
	Total	256.07

The above table represents nearly 20% of the level of energy demand by 2000/2001. (Courtesy to DNES, Ministry of Energy, New Delhi).

(i) Solar Energy

The use of solar energy goes back to 212 B.C. when Archemedes used reflecting mirrors to set fire to a Roman fleet besieging syracuse. Records are also available which reveal that Proculus used solar reflectors to drive off the fleet of Vitellius during the siege of Constantinpole (626 A.D.). Basically, all flesh is green and life is dependent on Sun directly or indirectly. As a matter of fact, the source of power since the earth came into existence has been fusion energy from the Sun. Solar flares are a source of X-rays which on

Table

Consumption of Commercial Energy in Different Regions of the World

	(Quantities in Million Metric Tonnes of Coal equivalent in Kilograms per Capita)	
Region	***Total Commercial Energy Consumption***	***Per capita Consumption***
World	8755	2074
Developed (Market Economies)	4953	6360
Developing (Market Economies)	936	449
Centrally Planned Economies	2867	2112

(Quantities in Million Metric Tonnes of Coal equivalent in Kilograms per Capita)

Region	*Total Commercial Energy Consumption*	*Per capita Consumption*
Africa	179	113
North America	2736	11231
Other American Developing Countries	175	917
Caribean American Developing countries	211	1359
Mid East	162	1241
Far East	749	568
Centrally Planned Asia	821	834
Western Europe	1566	4245
Centrally Planned Europe	2046	5477
Oceania	110	4950
Developing Countries of Asia	310	257
(Market Economies) India	114	178

Sources: World Energy Supplies (Statistical Papers series J. No. 22), United Nations, New York, 1979.

absorption produce ionization 75-80 kilometres high in the earth's atmosphere. Although the visible wavelengths (400-700 nm) are the most noticeable, about 60 per cent of the energy of sunlight is in the invisible infrared span, the limit of which is indefinite in sunlight. Solar energy is the major source of earth's heat without which, it would be impossible to live. The emission from the sun in the visible and infrared parts of the spectrum suggests that the sun is a body with a surface temperature of 6000° absolute. The ultimate source of energy is the nuclear reaction which takes place on the surface of the sun.

$$4'\,{}_1H^+ \rightarrow {}^4He^{+2} + 2°\,{}_1e^+ + 3Y\ (\text{gamma}) + 2n\ (\text{neutrino}) + \text{energy}$$

This energy is released by the thermonuclear fusion of two hydrogen atoms (${}_1H^1$) to form an atom of Deuterium (${}_1H^1$). This Deuterium then fuses with another hydrogen atom to form helium (2 He^{+3}) which in turn fuses with another helium isotope to form an ordinary helium atom and two hydrogen atoms. The byproduct of this process is an enormous amount of energy. It has been determined that 0.1 ml of Deuterium produces energy that is one thousand times a litre of petrol. Solar Energy could possibly be utilized as follows:

1. Direct conversion of energy into heat.
2. Conversion to electricity.
3. Conversion to chemical energy as in the formation of carbohydrates in green plants. .
4. Indirect transmission of energy is used for—
   (i) Solar cookers
   (ii) Solar desalination
   (iii) Solar water heaters
   (iv) Solar air heaters (especially for drying of food grains and vegetables).
   (v) Solar steam generators
   (vi) Solar wood Seasoning

Solar Thermal Energy has applications in all the sections and is capable of supplementing the fossiy fuels and other existing sources of energy (Fig.). Using computer simulation models, three solar houses have been constructed at three different places (keeping in view the climatic regions) in India: (i) zone climate of Jodhpur (ii) cool and humid climate of Srinagar and (iii) composite climate of Delhi. Recently, DNES has also installed solar huts and green houses at Leh which represent the cool-desert climate.

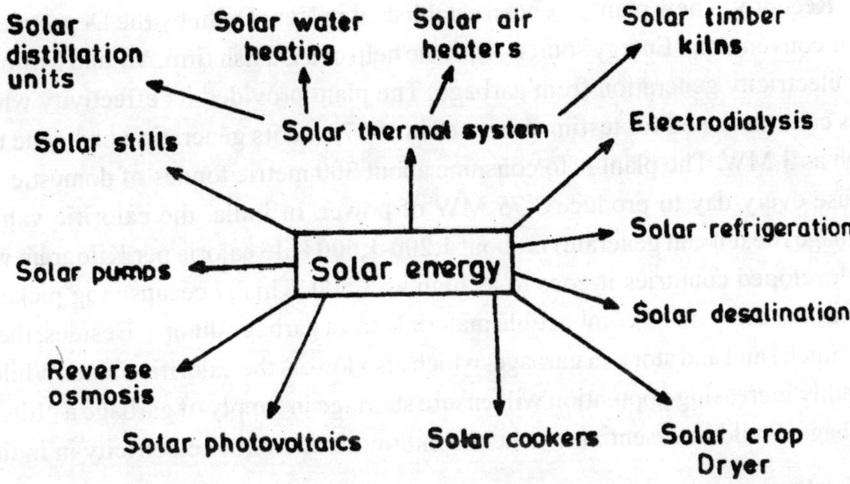

Fig. Spectrum of possible solar energy utility

The conversion of solar energy to electricity by means of photocells (Photovoltaic technology) is gaining momentum. This route of solar energy utilisation is attractive because of favourable solar radiation conditions and large requirements for electrical energy for decentralised application. The case of installation and maintenance, lack of noise and pollution, and long life make photovoltaic systems ideally suited for use in remote and isolated areas, forests, hilly and desert regions. DNES had drawn massive plans for setting up several photovoltaic plants in different parts of the country during 1988-89.

So far, about 300 villages have been electrified through solar photovoltaics street lighting and other community facilities have been provided. Over 600 photovoltaic pumps have also been put up for the purpose of drinking water supply and micro-irrigation.

In arid, semiarid or coastal areas of India, there is abundant sunlight that can be utilized to produce potable water from brackish or seawater. The salination technology is simple and small solar stills can be fabricated locally in rural areas. Over 100 solar desalination systems have already been installed at different places in India. In the years to come, it is proposed to establish solar power plants at high altitude to grab maximum amount of incident sun light. Besides, especially designed space stations harbouring solar power plants are also being envisioned in the foreseeable future.

Power from Garbage

Recently, a new plant has been established in New Delhi by the Department of Non-conventional Energy sources with the help of a Danish firm, Volund Miliotenic for electricity generation from garbage. The plant provided its effectivity when it was commissioned for testing purposes in 1987, and its generation had gone up as high as 3 MW. The plant is to consume about 300 metric tonnes of domestic solid refuse every day to produce 3.75 MW of power. In India, the calorific value of garbage (heat it can generate) is about 1,200-1,600 Kilo calorie per Kilogram while in developed countries it goes to as high as 3,500. This is because rag pickers in India take away many combustible materials from garbage dumps. Besides, there is too much sand and stone in garbage, which also lowers the calorific value. While the steadily increasing population will ensure shortage in supply of garbage a "filtered" garbage would be essential to generate enormous amount of electricity in India.

Tidal Energy

The energy crisis has spurred the search for non-conventional renewable sources of energy throughout the world. Energy from the seas and oceans, which cover about three-fourths of the earth, can make a significant contribution in finding over the crisis.

Considerable improvement in the tidal power technology in many countries in the past two decades has generated interest in India and other countries, which have attractive potential sites.

Tidal energy has been used by coast dwellers in more than 10 countries, particularly in Britain and France. Simple tidal mills on the sea shore used to be installed in the middle ages or even earlier to aid the coast dwellers in their daily work. The tidal mill is a version of the conventional water mill but requires just a little extra skill and care to operate.

The ocean presents thousands of times more energy potential than the current global consumption of around 3.5 million MW. Of the various forms of ocean energy such as tidal power, ocean thermal energy conversion and waves, only tide can be harnessed for largescale development with the present technology.

According to a study conducted by the National Oceanography and Atmospheric Administration of the U.S. Department of Commerce, about 3,000,000 MW power is continuously dissipated through the motion of tide on the surface of earth. Due to physical limitations, only a portion of this potential can be harnessed.

There are many primary advantages of a tidal power scheme: essentially no fuel is required, there is minimal air and water pollution, a predictable power output, high plant reliability, minimum land requirements, minimum public safety hazards related to power plant structures or operations, minimal environmental impacts as there is no extra submergence and no rehabilitation problem, independent of the vagaries of nature like failure of monsoons, etc. several socio-economic benefits to the region.

Due to the well-defined nature of the tidal phenomenon, tidal power can be predicted well in advance. Tidal power development can also be planned to provide a certain amount of dependable capacity and firm energy regardless of the time of the day and tidal conditions. However, additional investment is required on such schemes.

In India, the pace of progress of power development upto 1987 was rather slow.

The total installed capacity was hardly 1,300 MW. Now, it has increased to around 52,000 MW, an average annual growth rate of about 9.5 per cent. Presently, thermal generation constitutes about 66 per cent, hydro 32 per cent and nuclear two per cent. But the power supply position is still not satisfactory. Demand has been increasing at a much faster rate, outstripping growth in capacity.

India has a few potential sites for tidal power development. These are in the Gulf of Kachchh and the Gulf of Khambat (Cambay) on the western coast in Gujarat and delta of the Ganga in Sunderbans in West Bengal. The overall tidal potential is as much as 10,000 MW.

A beginning was made in the tidal power field when the erstwhile Central Water and Power Commission in 1975 hired a British expert's services. Subsequently, the Central Electricity Authority identified an alternative medium scale development in the Gulf of Kachchh for which the technoeconomic feasibility is being considered.

The proposed Kachchh tidal power project-India's first-envisages an installation of 900 MW. The main tidal barrage will be 3.25 Km long. The power house, comprising 36 bulb units of 25 MW each will occupy a length of more than 1.5 kms. The Floated-incaisson construction technique has been proposed for the construction of power house and sluices. DNES has developed Ocean Thermal Energy conversion (OTEC) and tidal energy technology.

Wind Energy

A strong wind is certainly a vast source of energy which could be trapped and utilized. In Indian mythological literature also, "Vayu Devta" or "Pawan" (wind) is supposed to be the most powerful and energetic entity. During 1985-86, (when the seventh plan began), the wind energy programme of DNES expanded rapidly. Attempts are being made to monitor the wind to generate electricity from it. In India, three Wind Farms for generation of electricity have been commissioned on the coast line of Mandavi in Gujarat, Tuticorin in Tamil Nadu and Okha in Gujarat. Two more are expected to start working at Puri (Orissa) and Deogarh (Maharashtra). More than 1700 wind pumps have already been installed in India for lifting water for irrigation purposes. A typical wind mill start lifting water at 12 Km/hr wind speed and yields about 30-35 cubic metres of water per day.

Another area where the company has been doing significant work is in the conversion of wind energy into electricity. Under a government-level project between

West Germany and Greece, a project has been undertaken in a Greek island called the Kythonos Wind Park. Five wind energy converters operate in parallel with the grid. The wind park power output is 100 KW. It is so costeffective that under its fuel-saver operation, 80,000 litres of diesel are saved per annum.

The German firm which has made first generation wind-mills of 30 MW now plans to develop wind-mills with a capacity of 300 MW. It is convinced of the cost-effectiveness of wind energy for a variety of reasons.

Research and application has established that among the various plants for utilisation of renewable energy sources, wind energy converters will reach a broad economic efficiency first; in certain regions, they are already or will become economical.

Wind energy has many advantages over even solar energy. Its utilisation is possible in large areas of the world without any pronounced dependence on latitude, season or the time of day. The specific energy density in windfavoured regions is high compared to solar irradiation in areas of intense sunshine.

Also, in favourable sites, the rate of availability-the equivalent number of hours of rates power per year is higher for wind energy converters than for solar energy plants.

Bio-Energy (and Biomass)

Bioenergy is indirectly the solar energy trapped by biomolecules in living cells through the process of photosynthesis. About 99.4% (equivalent to 1.7 x 1023 Cal/a) of non-nuclear energy available on the earth comes from the sun and some of this is continually being converted to biomass, although with a relatively low efficiency of 1 or 2% (Table 4.8). Biomass, therefore, represents a continuously renewable

Table

Annual World Production of Biomass (Categorized)

	In Tonnes
Total Primary production	2×10^{11}
Total Cereals production	1.5×10^{9}
Total Root crops production	5.7×10^{8}
Total Sugar crops production	1×10^{9}
Total Forest production	9×10^{10}

source of chemical energy which can be burnt or readily converted to convenient liquid or gaseous fuels (such as methane, ethanol or hydrogen) by microorganisms. In short, biomass is referred to the primary product of photosynthesis taking place in cultivable (agricultural land) and noncultivable (forests, lakes, ocean and other natural aquatic systems) areas. There are more than 200 billion tons of matter (C-H compounds) produced every year.

The use of biomass as a source of energy has wide ranging applications. More than 40% of the total energy consumed in India comes from sources such as wood, plant waste and animal waste. India has substantial quantities of biomass available in various forms (equivalent to 275 millions tonnes of coal (Tables).

Table

Area Lost Due to Deforestation: State-wise

State	***Area lost (Thousand ha.)***	***% of area lost (Total 4.3 m.ha)***
West Bengal	324.8	7.55
Karnataka	308.4	7.40
Uttar Pradesh	221.8	5.16
Maharashtra	215.5	5.00
Andhra Pradesh	202.2	4.73
Kerala	189.0	4.39
Madhya Pradesh	180.8	4.20
Gujarat	180.4	4.20
Orissa	109.3	2.54
Jammu & Kashmir	90.8	2.11
Rajasthan	85.8	2.00
Assam	72.6	1.69
Bihar	67.6	1.57
Tamil Nadu	65.6	1.63

Source: Central Forestry Commission, Ministry of Agriculture, 1980.

(Courtesy DNES, New Delhi).

Domestic cooking energy needs of the country are estimated to account for more than 50% of the total energy consumption. Fuel wood is used for 80% of the cooking energy needs. Wood is listed as the fourth largest source of energy after petroleum products, coal and natural gas.

DNES is implementing a programme that consists of activities including "assessment, production, improvement, conversion and utilization of biomass for meeting domestic and other energy requirements and to preserve ecological balance."

DNES has established five Biomass Research Centres to select and adopt fastgrowing species with high biomass potential, high calorific value, nitrogen

Table

State-wise Forest Area

State/Union Territory	*Forest area (000 ha)*	*Forest as % of total area*	*Per capita Forest Area ha*
1. Madhya Pradesh	16,239	30.67	0.34
2. Orissa	6,793	43.61	0.28
3. Maharashtra	6,672	21.68	0.12
4. Andhra Pradesh	6,437	23.25	0.13
5. Arunachal Pradesh	5,154	61.67	9.30
6. Uttar Pradesh	5,075	17.24	0.05
7. Karnataka	3,757	19.59	0.11
8. Rajasthan	3,461	10.11	0.12
9. Assam	3,081	39.24	0.17
10. Bihar	2,928	16.84	0.05
11. Tamil Nadu	2,254	17.33	0.05
12. Himachal Pradesh	2,177	39.11	0.56
13. Jammu & Kashmir	2,104	9.47	0.40
14. Gujarat	1,959	10.00	0.07
15. West Bengal	1,183	13.47	0.02
16. Kerala	1,133	29.15	0.05
17. Andaman & Nicobar	746	39.99	4.91
18. Meghalaya	637	28.32	0.55
19. Tripura	603	57.44	0.33
20. Manipur	602	26.92	0.47
21. Nagaland	289	17.48	0.47
22. Sikkim	282	38.62	1.20
23. Punjab	206	4.09	0.01
24. Haryana	146	3.30	0.01
25. Other	713	33.81	N.A.

Source: Report of the Fuelwood Study Committee, Planning Commission, Government of India New Delhi, 1982. (Courtesy DNES, New Delhi).

fixing species and adaptability to wasteland situations (Fig.). The major research centres are operational at Garhwal University, Srinagar; NBRI Lucknow; M.K.U. Madurai; ML Sukhadia University, Udaipur and Purulia, West Bengal. The main objective of these centres will be to select fast growing species with short rotation cycle, establishment of plantation trials for assessing the yield potential under a variety of agrosilvicultural practices including spacing, pest control, irrigation and cropmixing. The systematic and intercropping integrated study is likely to increase bioproductivity of fuelwood trees and shrub species under adverse soil conditions. According to an estimate made by the Society for Promotion of Wasteland, about 93 million hectares of wasteland is available in India which is not under productive use.

In this regard, the concept of Energy Plantation and Power of "Petrocrops" is already gaining impetus. The work on the gasification of biomass done at the Indian Institute of Petroleum at Dehradun has shown encouraging result.

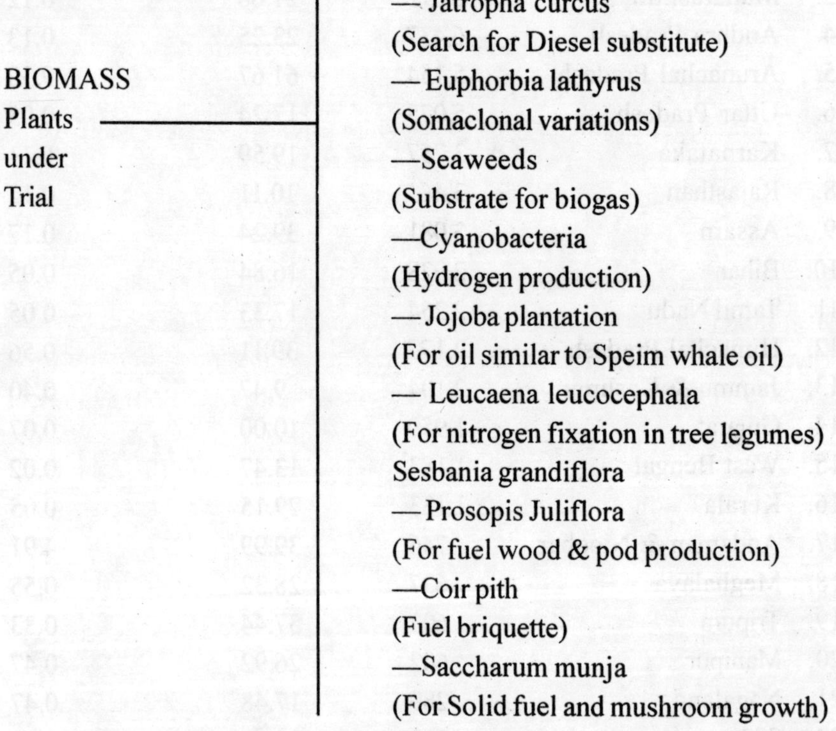

Fig. Selected Biomass Yielding Plants/Microbes for Extraction of Various Products (With Special Reference to India).

Suitable nursery technology for mass selection at the seedling level has been evolved at M.K.U. Madurai with several trials for the most widely used plant-species for biomass production including Eucalyptus, Casuarina Acacia nilotica and Leucanea K-8. The Biomass Research Centre for Coromandal Coast set up at INCOR Vishakhapatnam, aims to optimise the production of woody biomass from selected species for fuel wood under coastal environmental conditions. A lot of work has been done on Sesbania grandiflora in this direction.

TWO

Metabolic Engineering

Proteins can be modified in a desired manner to suit specific needs for their industrial or other specific uses. In these cases the three dimensional structure of the proteins is dependent on the DNA sequence of a gene, which could be suitably modified to get the desired gene product. However, often we are interested in getting metabolites produced in the cell, at an industrial level. However, cell's metabolic networks that evolved in nature are not optimized for practical applications. In such cases, performance of metabolic pathways or bioprocesses can be modified (deregulated) by genetic manipulation of the cell, so that the metabolite is overproduced. The technology involved in this biotechnology is described as metabolic engineering which is defined as follows: "metabolic engineering is the improvement of cellular activities by manipulation of enzymatic; transport and regulatory functions of the cell with the use of recombinat DNA technology".

Although for metabolic overproduction, sometimes the strains of living organisms are improved by mutations and selection, the opportunity to introduce heterologous genes and regulatory elements made 'metabolc engineering', a very fascinating area of research. In this area, cell function can be modified using targeted alterations in normal cellular activities. This may involve not only biosynthesis of a metabolite, but may also involve manipulation of protein processing pathways. Metabolic engineering, at present, is only a collection of examples, which promise not only future technological benefits, but also contribute to basic science, agriculture and medicine. One should, however, recognize that although in a particular case, metabolic engineering might have been shown to be feasible, the strains thus developed may still be unsuitable for practical yield, so that the limitations may still need to be addressed by subsequent genetic manipulations. In this chapter, we discuss the possibilities, achievements and limitations of the approach of 'metabolic engineering' as another subject of biotechnology.

Cloning and Expression of Heterologous Genes

Introduction of foreign genes may lead to expression of a heterologous enzyme or transport system in any organism leading to the production of new compounds. However, these new compounds may undergo further reactions which can not be anticipated, thus putting a limitation on its use for metabolic engineering. Further, the protein thus synthesized may or may not fold properly and may not be suitably localized. Despite these limitations, the number and scope of positive experiments encourage further application of this approach. Following are some of the possible applications of cloning and expression of foreign genes in metabolic engineering:

Table

Hetrologous activities recruited to alter metabolite and protein end products.

Host Organism	*Original metabolite*	*Heterologous enzyme added (source organism)*	*New product*
1. Erwinia herbiola	2, 5-DKG Corynebacterim C	2, 5-DKG-reductase	2-KLG
2. Acremonium chrysogerrum	Cephalosporim C	(i) D-amino acid oxidase (Fusarium salani) (ii) Cephalosporin acylase (Pseudormonas diminuta)	7 ACA
3. CHO cells	Terminal β galactosyl residues in N-acetyllactosamine sequences	P-Galactoside α 2, 6 sialyl-transferase (rat)	Sialyl α 2,6-galactosyl linkages
4. Mouse L cells	Unsubstituted type II, N-acetyllactosamine glycoconjugate end groups	G-Dp-L-fucose : β-D-galactoside 2-α-L-fucosyl-transferase (A431 human cell line)	H fucosyl α 12 galactosyl linkages

Completion of partial pathways giving new products

Natural strains of any organism may be imperfect from an applied perspective and can be improved by extension of native pathways. Some of the examples, include the following : (i) 2-keto-L-gulonic acid (2-KLG) is a precursor of ascorbic acid (vitamin c) and its synthesis involves two steps, first involving conversion of glucose to 2,5-diketo-Dgluconic acid (2,5-DKG) in Erwinia herbicola, and second involving the conversion of 2, 5- DKG to 2-KLG in a species of Corynebacterium. It has been shown that by cloning and transfer of gene for 2,5-DKG reductase from Cotynebacterium to Erwinia, glucose can be converted to 2-KLG in a single fermentation step, (ii) A similar approach (as for 2-ICLG) was used for the synthesis of 7- aminocephalosporanic acid (7ACA), which can be used as the precursor for several semi-synthetic antibiotics in Acremonium chrysogenum. (iii) A glycoprotein like erythropoietin produced by chinese hamster ovary (CHO) cells can be subjected to post translational modification by the transfer of a gene for 3-galactoside a 2,6 sialyltransferase. This enzyme causes sialyl a. 2,6-galactosyl linkage on surface glycoprotein and can convert erythropoietin into a kind more closely resembling human erythropoietin which is rich in these linkages. (iv) Human H blood group antigen could be produced in mouse cells due to transfection with human DNA.

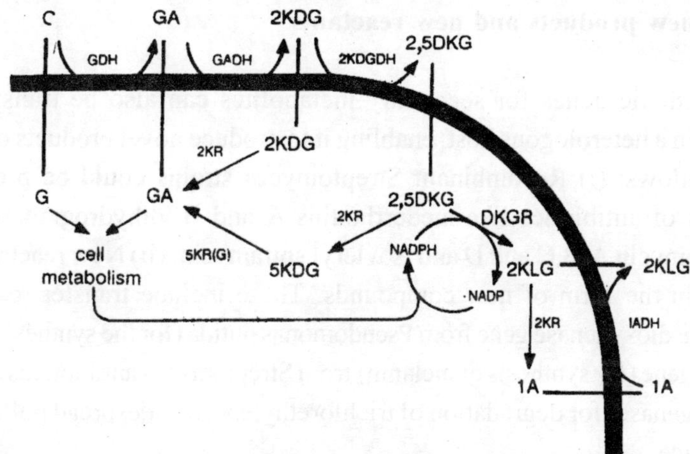

Summary of the enzymes, intermediates and by-products encountered in the synthesis of 2-keto-L-gluconic acid (2-KLG) from glucose in Erwinia herbicola. DKGR (2,5 DKG reductase) is derived from a cloned gene and needs NADPH to direct metabolic flow to 2-KLG; wide dark band is cell membrane; straight bars are transport systems. (Abbreviations: GDH=D-glucose dehydrogenase; GADH=D-gluconate dehydrogenase; 2KDGDH=2-ketoD-gluconate dehydrogenase; DKGR = 2,5 diketo-D-gluconate reductase; IADH = L-iodate dehydrogenase; 5KR(G)=S-keto-D-gluconate reductase; G=D-glucose; GA=D-gluconate; 2, KDG=2 keto-D-gluconate; 2,5 KDG= 2,5 diketo-D-gluconate; IA=L iodate; 5,KDG=Sketo-D-gluconate.)

Transfer of an entire biosynthetic pathway

The transfer of an entire biosynthetic patway is facilitated by the cultstering of genes involved. Following are some examples: (i) Genes for biosynthesis of antibiotics 'actinorhodin' and 'erythromycin A' were tranferred to Streptontyces lividans, separately from S. coelicolor and S. erythreus respectively; similarly genes for the biosynthesis of the antibiotic, 'penicillin V' were transferred from Penicillium citrysogenum to Neurospora crassa and Aspergillus niger. In each case, the new host synthesized the new product using genes for entire biosynthetic pathway transferred from original species; (ii) Genes for biosynthesis of a biodegradable polyester, polyhydroxybutyrate (PHB) were transferred from Alcaligenes eutrophus to E. coli. This recombinant E. coli was shown to produce PHB concentrations approaching 50% of the dry cell weight, when nitrogen source is depleted; (iii) It was possible to construct bacterial strains (having as many as five degradation pathways derived from three organisms), which could simultaneously cause degradation of mixed substrates involving biochemical and metabolic complexities; (iv) Genes for tryptophan and threonine biosynthetic pathways could be transferred to E. coli, which could be used as amino acid producers.

Creating new products and new reactants

Biosynthetic genes for secondary metabolites can also be transferred and expressed in a heterologous host, enabling it to produce novel products or reactants such as follows: (i) Recombinant Streptomyces strains could be produced for production of antibiotics like mederrhodins A and B, dihydrogranatirhodin, 2-nonerythromycin A,B,C and D and isovleryl spiramycin, (ii) New reactants may be produced in the form of new compounds. These include transfer to E. coli, of naphthalene dioxygenase gene from Pseudomonas putida (for the synthesis of indigo), tyrosinase gene (for synthesis of melanin) from Streptomyces antibioticus and toluene mono-oxygenase (for degradation of trichlorethylene, a widespread pollutant) from P. mendocina.

Altering nutrient uptake and metabolite flow

Recombinant DNA has also been used for increasing growth rate, decreasing nutrient demand and higher attainable cell densities in several microbes for a variety of applications. These include the following examples: (i) A **glutamate dehydrogenase** gene, when transferred from E. coli to Methylophilus methylotrophus, increased efficiency of carbon conversion from 4% to 7% and eliminated the requirement for ATP. (M methylotrophus has been developed as an animal feed material.); (ii) Transfer of pyruvate decarboxylase and alcohol dehydrogenase genes from Zymomonas mobilis to E. coli led to a shift from acetate production (which inhibits growth) to ethanol production (ethanol is less inhibitory, and is an important industrial chemical). This led to three fold increase in growth and has a potential application in conversion of several sugary found in biomass;

(iii) Transfer of a -acetolactate decarboxylase (a -ALDC) gene from Klebsiella terrigena or Enterobacter aerogenes to budding yeast (S. cerevisiae) leads to conversion of acetolactate to acetonin rather than to diacetyl, which has an undesirable flavour in beer. Enzymatic reduction of diacetyl (lagering step) requires 5 weeks, while the process for conversion to acetonin (which does not influence flavour at moderate concentration) requires only two weeks (no lagering step required); (iv) Transfer of yeast ornithine decarboxylase gene to Nicotiana rustica, doubled the nicotine content in roots; similarly the transfer of hyoscyamine 6/3-hydroxylase from Hyoscyamus niger to Atropa belladonna increased the scopolamine content three to ten times. In both cases, the recombinant strains had rearrangement or shift of metabolite flux towards higher yield of desired product; (v) Quantity of active protein can also be increased by genetic manipulation of protein processing pathways. Transfer of cyanobacterial rubisco genes to E. coli, and its coexpression with over-expression of E. coli genes for chaperone proteins GroES and GroEL led to ten fold increase in the assembly of rubisco. Similar assembly of rubisco of higher plants in altered E. coli may be possible in future.

Transfer of promising natural motifs

Gene for Vitreoscilla haemoglobin (VHb), which helps in cell growth and product synthesis under limited oxygen supply, has been transferred to E.coli, Streptomyces lividans and S. coelicolor. (Vitreoscilla is a gram negative aerobic bacterium). Under limited supply of oxygen, the expression of VHb gene in E. coli and Streptomyces allowed much more rapid synthesis of total cell protein. A motif in VHb, perhaps increasses efficiency of net ATP synthesis, facilitates oxygen transfer to respiratory centre, and modifies some aspect of cellular redox chemistry. This strategy may be used for production under stress conditions. Genes for resistance against other stresses may also be used in future for production of specific proteins under these other stress conditions (e.g. abnormal temperature and pH).

Redirecting Metabolite Flow

In a branched biosynthetic pathway, there are always several forks, where intermediates can enter alternative pathways. At such forks, resources (substrate, enzyme, transport system, ribosomes, etc.) are distributed equally or unequally to two or more processes. When we are interested in one of the several possible products at a fork, the desired route or process can be made a priority, so that resources are mainly channelized to this route and the traffic to other routes is minimized without decreasing cell viability. This can be achieved by several methods including the following: (i) use of cloned genes inhibiting feedback inhibition by the end product; (ii) use of mutant ribosomal RNA and (iii) use of positive regulatory sequences.

Desensitizing feedback inhibition

Very often, particularly in the biosynthesis of amino acids, the surplus end product acts on one of the first few enzymes of biosynthetic pathway rendering it inactive. In industry, we may like to get the end product in maximum quantity disregarding the need of the cell. This can be achieved if the feedback inhibition by the end product is desensitized. Mutant enzymes have actually been isolated, which do not respond to feedback inhibition. However, this objective can also be achieved by using cloned genes.

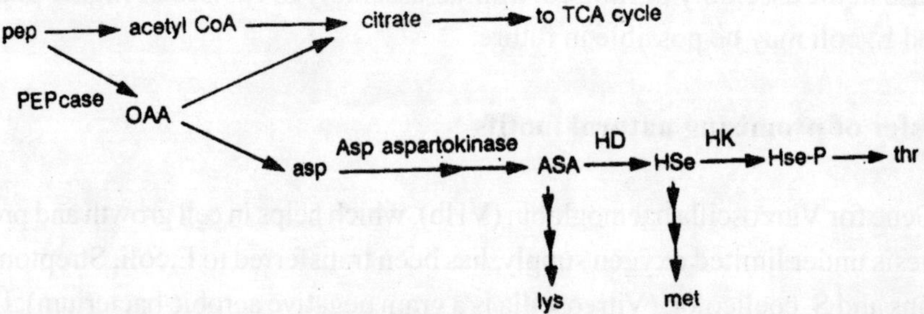

Pathways of the biosynthesis of the aspartate family of amino acids. (Abbreviations: acetyl coA = acetyl coenzyme A; TCA=tricarboxylic acid; Asp =aspartate; ASA =aspartate. semialdehyde; Hse=homoserine; Lys=lysine; Met=methionine; Hse-P=O-phosphohomoserine; Thr=threonine.

One of the popular examples invloving desensitizing feedback inhibition for metabolic engineering includes improved threonine production by Brevibacterium lactoferntentum. A mutant M-15 was isolated, which lacked feedback inhibition of aspartokinase by threonine and lysine, and of homoserine dehydrogenase (HD) by threonine. When cloned HD and HK (homoserine kinase) genes were transferred and their expression amplified in this mutant, the production of threonine increased from 17.5g/ liter to 33g/liter. In the mutant, aspartokinase and HD could not be rendered inactive by lysine or threonine; the amplified HD and HK expression redirected ASA (aspartate semialdehyde) towards synthesis of threonine rather than lysine. Threonine production by M-15 could be increased by yet another method. Expression of cloned phosphoenolpyruvate carboxylase (PEP II case) in M-15 led to 15% increase in threonine production, due to increased flow towards oxaloacetate (OAA) rather than towards acetyl CoA.

In another lysine producing strain, transfer of gene for a mutant HD enzyme (not responding to feedback inhibition) led to shift of lysine concentration from 65g/liter to 4g/liter and that of threonine from zero to 52g/liter.

Elevation of the activity of rate limiting enzyme

Cephalosporium aremonium strains are used for the production of cephalosporin C. However, penicillin N, which is an intermediate product in biosynthesis of cephalosporin C, accumulates. Thus, the activity of enzyme (encoded in gene cef EF) responsible for converting penicillin N limits the rate of overall pathway. Expression of cef EF was elevated through increased gene dosage, and the recombinant C, acremonium exhibited 15-fold reduction in penicillin N concentration and 15% increase in the production of cephalosporin C.

Alteration in protein processing pathway

Genetic manipulation may also be used at the post-translational level (after the protein synthesis is over) to increase the yield of a product. Low yields of a human tissue plasminogen activator (tPA, used for heart diseases) is attributed to its association with GRP78 in rough endoplasmic reticulum of CHO cells. (GRP78 is a 78-kD glucose regulated stress response protein). It has been shown that coexpression of an antisense GRP78 message (due to cloned antisense RNA sequence) resulted in smaller quantities of GRP78 and faster tPA secretion.

Reduction of competition for limited resources

Sometimes, there is a competition for resources between host genes and cloned genes during protein synthesis. There are examples, where 16S rRNA was altered to create ribosomes which will allow protein synthesis on messages containing a corresponding mutation in Shine-Delgarno sequence in mRNA. This will allow translation of native genes on native ribosomes and that of cloned genes on mutant ribosomes, thus eliminating r competition. A 35% increasee in the expression of a cloned galactosidase gene was obtained following this approach.

Modification of metabolic regulation

A metabolically engineered baker's yeast was developed through modification of the regulation of expression of genes for maltose permease and maltase. This recombinant yeast was intended to reduce the time for leavening of sweet doughs. These genes are glucose sensitive, so that glucose represses their expression. Therefore, one would like that glucose uptake and catabolism should be associated with that of maltose. Same constitutive yeast promoters were used for separate genes responbible for uptake and catabolism of glucose and maltose, so that simultaneous uptake (by permease enzyme) and catabolism of both sugars is possible.

In yeast, a futile and accessory metabolic cycle was created, which led to increased CO_2 generation. Increase in CO_2 is favoured, because CO_2 in yeast consumes ATP, which otherwise inhibits the enzymes (phospho fructokinase or PFK, and pyruvate kinase) used in sugar catabolism. Similarly, alkylbenzoate degradation pathway encoded on a Pseudomonas plasmid was modified (through mutation), so that a promoter Pm for the operon (metaoperon) meant for degradation of 4-ethylbenzoate (a pollutant) was activated by the same compound (4- ethylbenzoate). The metabolically engineered Pseudomonas can thus be used for the processing of 4- ethylbenzoate.

Limitations

As discused earlier in this chapter, several steps are involved in metabolic engineering. We discussed the possibilities and achievements, but limitations are experienced at each step, which will be discussed briefly in this section.

Limitations of cloning in industrial strains

In many industrial strains, cloning is difficult due to non- availability of (i) suitable vectors and (ii) methods for introduction of exogenous DNA. Even after successful transfer, stable propagation of cloned genes is problematic as observed in Bacillus subtilis. This is due to rearrangement, deletion and cleavage of cloned sequences and the plasmid vector carrying it. However, several genes for metabolism have been cloned from Clostridium acetobutylicum and the technology for their integration in B. subtilis is being optimized and standardized. C acetobutylicum is used for the biological production of acetone and butanol.

Limitations due to lack of knowledge about concentrations of intracellular proteins and metabolites

Different analytical methods are often used for determining either the concentrations of cellular proteins and metabolites, or the rates of reactions, or the fluxes in specific biochemical pathways. These include the following: (i) two dimensional electrophoresis; (ii) use of isotopically labelled precursors, and (iii) nuclear magnetic resonance (NMR) spectroscopy. Despite the availability of these analytical methods, the time and instrumentation required to evaluate metabolite concentrations limits rational metabolic engineering. Measurements of concentrations of intermediate products also becomes necessary in some cases to find out the possible rare-determining reactions. Once these rate-determining reactions are known, genetic transfer of stress response motifs can be tried. Under these conditions, the use of organisms that can function over wide ranges of temperature and pH would help in the industrial production of a metabolite.

Metabolic control theory and metabolic engineering

Theoretical development of metabolic control theory has been a significant advance in metabolic engineering. It allows calculation of sensitivity which gives estimates of 'flux control coefficients' to be estimated as follows:

$$\text{flux control coeff.} = \frac{\text{expected fractional change of flux}}{\text{fractional change in the amount of enzyme involved}}$$

The sensitivity of flux (through the pathway) to other parameters in kinetic expressions for each enzyme can also be estimated under different growth conditions (temperature, pH, etc.). This will provide guidance for useful metabolic engineering leading to acceleration of the pathway. Since generally several steps are rate-limiting in a pathway, sequential improvement of metabolic pathway through repeated cycles of metabolic engineering may be required in some cases. Unanticipated cell responses to metabolic engineering are also noticed, which need to be corrected before its practical application. This would require mathematical calculations, simulation of responses, modeling and other theoretical exercises using computers. These are the subjects of 'chemical engineering' and are beyond the scope of this book.

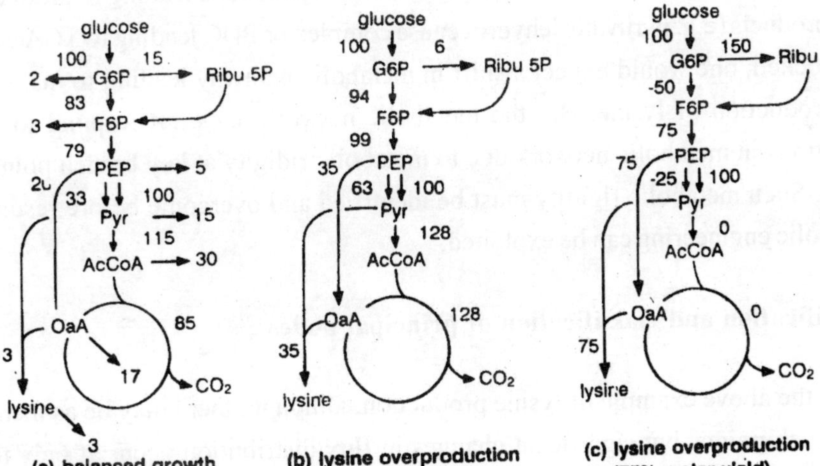

(a) balanced growth **(b) lysine overproduction (35% molar yield)** **(c) lysine overproduction (75% molar yield)**

Flux distributions, normalized by glucose uptake in Corynebactetium glutamicum; a network of 34 reactions are involved; biomass synthesis is indicated by arrows directed towards an arable numeral; the two fluxes between PEP and Pyr represent reactions catalysed by pyruvate kinase (PEP synthetase) and the glucose: PEP phosphotransferase systems (Abbreviations: G6p=glucose-6-phosphate; Ribu5P =ributose-S-phosphate; F6P=fructose-6phosphate; PEP=phosphoenol pyruvate; Pyr=pyruvate; AcCoA=acetyl coenzyme A; OaA=oxaloacetate.

Limitations in Metabolic Engineering due to Network Rigidity

During metabolic engineering involving primary metabolism (e.g. glycolysis, TCA cycle and pentose phosphate pathway, etc.) the carbon flux distributions at key branch points (calld 'nodes') is often radically redirected from the flux distributions that are normally associated with balanced growth. Such flux alterations are often directly opposed by mechanisms that have evolved to maintain original flux distributions for optimal growth. This opposition to flux alterations at key branch points or nodes is often described as 'metabolic rigidity' or 'nodal rigidity'. There are relatively few examples where this nodal rigidity has been overcome through genetic manipulations leading to successful flux alterations.

The concept of metabolic or network rigidity can be illustrated using the example of overproduction of lysine by Coynchncierum ghrnmicum.The metabolic flux distributions (using glucose as a substrate) for balanced growth, for 30-40% molar yield and for maximum yield (75% molar yield) are shown where glucose uptake was used as a standard indicated by 100 units. It will be seen that flux distribution for maximum overproduction represents a significant deviation from that for balanced growth or from that for 30-40% overproduction. If pathways leading to production of byproducts (e.g. pyruvate dehydrogenase complex or PDC leading to TCA cycle) are blocked, one would expect a shift in metabolic pathway leading to maximum overproduction of lysine. But the metabolic network does not respond to such alterations in metabolic network due to metabolic ridigity at key branch points or nodes. Such metabolic rigidity must be identified and overcome before results of metabolic engineering can be exploited.

Identification and classification of principal nodes

In the above example of lysine production, although, there may be as many as 30 branch points, but significant changes in flux distribution occur at only three nodes, when lysine production increases from 35% to 75%. These include glucose-6-phosphate (G6P), phosphoenolpyruvate (PEP), and pyruvate (Py). For yield above 60%, fructose-6 phosphate (F6P) becomes a principal node in place of G6P (for details consult Stephanopoulos and Vallino, 1991). Identification of these nodes in any metabolic pathway is a prerequisite for any exercise involving metabolic engineering.

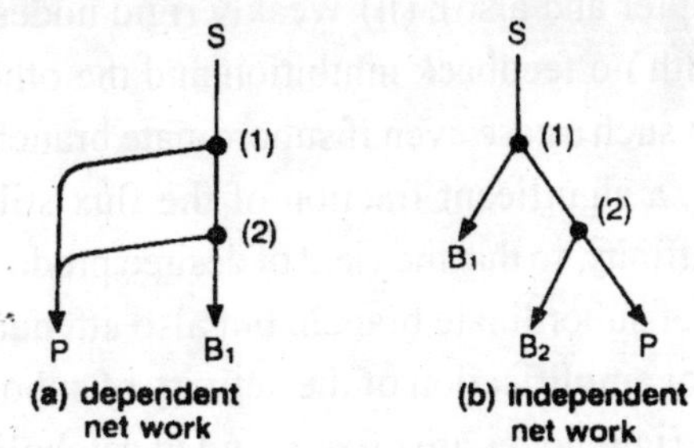

Three types of branch points or nodes. They indicate different levels of rigidity to modifications in flux partitioning (dashed lines represent negative feedback; curved arrow indicates positive feedback.)

Identification of principal nodes, where flux distribution needs to be altered also depends on whether a network or subnetwork architecture is 'dependent' (where each node contributes to a consumed component of final product; e.g. OaA and Py in lysine network) or 'independent' (where metabolites produced from principal nodes do not condense. The metabolic engineering in a 'dependent' pathway is much more difficult than in an 'independent' pathway. The yield of desired product in an 'indepedent' network can be improved simply by altering flux partitioning at any one of the principal nodes. In a 'dependent' network, on the other hand, flux partitioning should be altered at both 'dependent' nodes, so that flux partitioning at these nodes is altered in a co-ordinated manner.

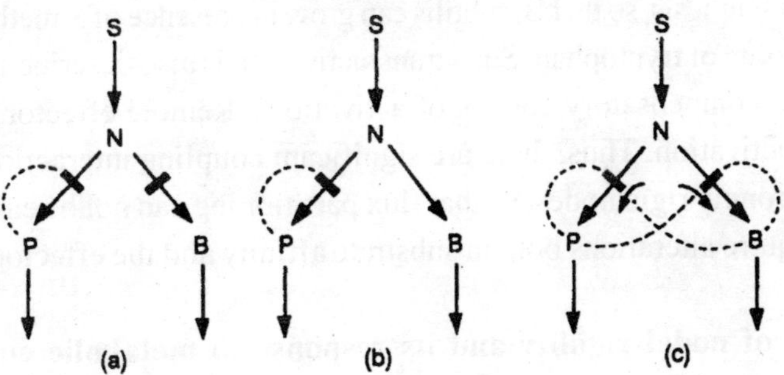

Three types of branch points or nodes. They indicate different levels of rigidity to modification in flux partitioning (dashed lines represent negative feedback; curved arrow indicates positive feedback.

The principal nodes are clssified into three main categories depicted in: (i) Flexible nodes are those, where flux alterations can be made easily to meet the demands of metabolic engineering without any adverse effects. Such flexible nodes are often controlled by feedback inhibition in each branch. One of the examples of flexible nodes is aspartate semialdehyde (ASA), where flux can be redirected towards lysine synthesis by deregulating feedback inhibition of aspartokinase by lysine and threonine (see earlier in this chapter and also); (ii) Weakly rigid nodes are those, where one branch is dominant with no feedback inhibition and the other is subordinate with feedback inhibition. In such a case even if subordinate branch is deregulated, to get a specific product 'P', a significant fraction of the flux still enters the dominant branch due to enzyme affinity, so that the yield of desired product is limited. Therefore, not only deregulation of subordinate branch, but also attenuation of the activity of dominant branch and/or amplification of the activity of subordinate branch will be required. Such weakly rigid nodes are often found in catabolic pathways leading to CO formation. One such example is isocitrate branch point between glyoxylate shunt and TCA cycle in E. coli cultures grown on acetate. When glucose is added to this culture, the flux supported by isocitrate lyase (ICL, which is the first enzyme of glyoxylate shunt) drops to zero (although glyoxylate shunt is fully active) and isocitrate dehydrogenase (ICDH) of TCA cycle exhibits complete dominance over ICL, because Michaelis constant (Km) of ICDH for isocitrate is 1/75 that of ICL. However, if the activity of ICDH is attenuated by 60% or more, a siginificant fraction of isocitrate can be diverted into glyoxylate shunt (ICDH activity can be attenuated by phosphorylation). Therefore, for preferential flux distribution towards glyoxylate shunt, feedback inhibition of ICL needs to be deregulated and ICDH activity needs to be; (iii) Strongly rigid nodes are those, in which feedback inhibition by a metabolite in its own branch is associated with trans- activation of enzyme in an opposite branch. In B. subtilis, histidine activates anthranilate synthetase, an enzyme involved in tryptophan synthesis, so that B. subtilis can grow in presence of 5-methyltryptophan, an antimetabolite of tryptophan. Such trans-activation is also described as 'metabolic interlock' or 'compensatory control of activation'. Remote effectors may also be involved in activation. Thus, there are significant coupling interactions associated with these strongly rigid nodes, so that flux partitioning can not be easily modified. This may require alterations both in substrate affinity and the effectors involved.

Assessment of nodal rigidity and its response to metabolic engineering

Since information on the degree of nodal rigidity is needed for chosing an appropriate technology for metabolic engineering, various methods have been used for assessment of rigidity of a branch point or a node. This information can be obtained using mathematical models of metabolism, 'metabolic control theory (MCT)' or 'biochemical system theory (BST)'. In these approaches, simulations of a network are conducted to assess rigidity. Various technologies of metabolic engineering, when applied to a network may also yield information about nodal rigidity. The types of alterations used to assess rigidity are, however, case specific and can be classified into following four general categories : (i) attenuation of enzyme activity using an inhibitor ; (ii) amplification of attenuation of enzyme activity through genetic modification; (iii) environmental perturbation, such as change in carbon source; (iv) deregulation of a different metabolite to increase metabolic burden. Using these techniques, rigidity of G6P, Pyr and PEP principal nodes in lysine biosynthesis network could be assessed. For more details and examples, readers are advised to consult the review article written by Stephanopoulos and Vallino (1991).

To examine the response of nodal rigidity, let us consider an independent network given earlier where increase in yield of P is the desired goal. Further, the node 1 is flexible and node 2 is strongly rigid, and B_1 does not inhibit its own synthesis. In this case, if B_1 branch is blocked, the yield of P would increase due to flexibility of node 1, but if B_2 branch is blocked, the yield of P will actually decrease, since due to rigidity of node 2, flux will be diverted to B_1 and the flux towards P will be attenuated. If instead, node I is rigid and node 2 is flexible, then blocking B, will improve yield of p, but if B_1 is blocked, although yield of P will not be affected, the rate of the synthesis of *p* will be severely attenuated, due to the effect of flux distribution at S.

Nodal rigidity in 'dependent' networks presents even greater problems for metabolic engineering. If node 1 or node 2 is rigid, then attenuation of B_1synthesis would not affect the yield of P, but its synthesis rate will be correspondingly attenuated thus causing metabolic collapse. Response in 'depetiden' network is independent of the location of rigidity, since alterations are needed at all nodes for coordinated flux partitioning at these nodes.

Conclusions

'Metabolic engineering' is an emerging science in biotechnology and often falls within the jurisdiction of 'chemical engineering'. The lack of significant progress in this area is due to complex interactions of metabolic networks. When the nodes or key branch points are flexible, the metabolic engineering is easy, but greater effort is needed for dealing with rigid principal nodes. Several aspects of the technology and mathematics of the kinetics of chemical reactions need to be understood and used for rational designing of metabolic engineering. In many cases, complete biosynthetic pathway may be transferred from another organism to overcome the problem of nodal rigidity. Several limitations to the successful metabolic engineering are the challenges for future chemical engineers involved in this area of study.

THREE

Process of Fermentation

Technically, speaking fermentation, is the chemical transformation of organic compounds with the help of enzymes is an old process. The ability of yeast to make alcohol-beer was known to the Sumerians and the Babylonians before 6000 B.C. Much later, by about 4000 B.C., the Egyptians discovered that the carbon dioxide generated by the action of brewer's yeast could leaven bread. References to wine, another product of fermentation, can be found in the Book of Genesis, where it is noted that the Noah consumed a bit too much of the beverage. Reference has also been made to wine as "Soma" in Indian mythology as the liquid frequently drunk on several occasions. Religious Indian script like "Ramayan" also indicates that alcohol was consumed by "Asuras", millions of years ago.

By the 14th century A.D., the distillation of alcoholic spirits from fermented grain, a practice thought to have originated in China or the Middle East, was common in many parts of the world. Similarly, some other fermentation processes such as the production of vinegar (and cultivation of acetic acid bacteria) as well as yogurt and cheese were also known. Indian literature is full of examples of "Curd" being utilized by people since millenia.

Although there are reports of microorganism providing food and drink for more than 8000 years, yet their existence was recognized only in the 17th century by Dutch microscopist Anton van Leeuwenhoek. It was, however, not until the second half of the 19th century that Louis Pasteur of France and John Tyndall of Britain demolished the concept of spontaneous generation and proved that existing -microbial life comes from preexisting life.

Fermentation as a Biochemical Process

Virtually all cells possess the capacity to convert glucose to pyruvate with the extraction of energy in the form of two ATP molecules per glucose molecule broken down. This pyruvate occupies a key position in several metabolic positions and the most important factor is the availability of oxygen. If oxygen is not present, or present only in low concentrations, the pyruvic acid molecule may be further broken down by an oxygenic or anaerobic process and end up either as lactic acid (as in some bacteria, fungi and animal cells) or as ethyl alcohol (in yeast and most plant cells). Yeast cells, for example, can live without oxygen. Under such anaerobic conditions, they convert glucose to pyruvic acid by routine glycolytic sequence. The pyruvic acid so formed is then converted to ethyl alcohol.

When the glucose-filled juice of grapes and other fruits is extracted and stored in airtight kegs, the yeast cells, present as a bloom on the skin of the fruit, turn the fruit juice to wine by converting glucose into alcohol. This process is Fermentation. Yeast, like all living things, has a limited tolerance for alcohol and when a certain concentration (about 12 per cent) is reached, the yeast cells die and fermentation ceases.

The complete balanced ec 'on for the fermentation of glucose to ethyl alcohol is:

$$C_6H_{12}O_6 + 2ADP + 2Pi \ -2\ CH3CH2OH + 2CO_2 + 2\ ATP$$

Two molecules of ATP are generated from ADP and Pi during this fermentation. During this process, about 7 per cent of total available energy of the glucose molecule—about 52 kilo calories is released. However, 93 per cent of energy is still locked in the two alcohol molecules. An anaerobic cell must, therefore, consume much larger quantities of glucose (or other fuel) per unit of time to accomplish the same amount of cellular work as an aerobic cell.

It is worth noting how efficiently the anaerobic cell handles the 7 per cent (47 kcal/mol) of energy available to it. Based on the standard free-energy changes (For details see "Biochemistry" by the same Author), the two ATPs with a D, G° of - 7.3 Kcal represent an efficiency of energy conservation of about 31 per cent (14.6/47 x 100).

The fact that the glycolytic sequence does not require oxygen suggests that this series of reactions evolved early, before free oxygen was available and that the primitive heterotrophs used glycolysis or a similar process to extract energy from the organic compounds that they absorbed from their watery surroundings. However, this major biochemical process is centuries old and is now a treasure house of a variety of several important products as described in the following sections:

Microbial Processing and Raw Material

Since time immemorial, microbial processes have been harnessed for the production of food, pharmaceuticals and alcoholic beverages. In general, most raw materials are naturally occurring substances from which more useful materials can be produced. In fact, any abundant carbon source might be employed for microbial processing including cellulose, sugar, starch, lingnocellulose, coal, petroleum, and organic acids. Currently, attempts have been made to use carbon dioxide directly as a raw material, other than photosynthesis. Although coal is basically carbon, it is biologically inert and may contain substances inhibitory or toxic to microbes.

Both plants, and to a lesser extent animal biomass represent a sizeable source of carbon for microbial processes. Agricultural plants, accounting for just 7 percent of the primary photosynthetic productivity, provide structural materials, textiles and paper products besides being a major source of food for man and animals. Agricultural raw materials are the most important source of carbon for microbial conversion products. Most agricultural crops and residues are relatively free from toxic materials and this might have stimulated their use as a raw material for microbial processes.

Traditionally, the bacteria or/and fungi required for the fermentation process would have been present in the raw material. Some people consider using natural flora as ideal since the end product has a complexity of flavor that cannot be obtained in any other way. The realities of this situation are, however, debatable and the disadvantage of relying on a variable inoculum is that the fermentation may be erratic.

Many developed countries are using various raw materials and fermentation products including Ethanol, Whey, Paraffins, Molasses and sulphite waste liquor. These raw materials provide all the necessary requirements such as carbon, trace elements, nitrogen salts and vitamins. Besides, a host of other raw materials (Table) are now currently under potential exploitation for the microbial processes.

Table

Typical Raw Materials and Products in Industrialized Countries

Raw Materials	*Products*
Sulphite Waste Liquor Single-Cell Protein (SCP)	
Ethanol	SCP Acetic Acid
Methanol	SCP

Raw Materials	*Products*
Whey	SCP
	Lactic Acid
Paraffins	SCP
	Citric Acid
	Amino Acids
Molasses	Ethanol
	Glutamic Acid

In a survey made by United Nations Environment Programme (UNEP) in 1978, it was reported that about 1,700 million tons of cereal straw is wasted per year because it has no use in several countries. Indonesia, however, has been using residual waters of molasses to fertilize paddy fields and pressing residues are used as soil fertilizer or even mixed up to form animal fodder.

Among raw materials commonly used for microbial processes, molasses is certainly the most easily available for use as a substrate in India and some other developing countries. It is readily utilized by several microbes since it contains easily assimilable sugars and essential trace elements. However, it is relatively poor in nitrogen content.

Another example is pineapple, of which a major part is wasted. As a matter of fact, more than 75 percent of the pineapple fruit is wasted during canning. The pulp, husks and fresh waste products of pineapple are often crushed to extract more juice and the dried residues are used as cattle fodder.

Out of several crops (Table) that may be good sources of starch, some of these are more important for microbial processes. Cassava is a very good substrate for the production of ethanol and single cell protein. Cassava is relatively less nutritious because of the low protein to calorie ratio. In addition, Taro (Colocasia esculenta) commonly known as 'Kachalu' in India is equally important as a substrate for microbial processes. It is also grown in Africa, Egypt, South East Asia and some Pacific and Caribbean islands.

Table

Estimated Production of Major Food Crops in Developing Countries

Crop	*Metric Tons (in thousands)*
Paddy	186230
Casava	103486
Wheat	95048

Crop	*Metric Tons (in thousands)*
Maize	73328
Banana/Plaintain	55199
Coconuts	32664
Sorghum	31173
Yams, Taro, etc.	28777
Potatoes	26909
(Pulses)	25997
Citrus	22040
Millet	21452
Barley	22775
Sweet Potatoes	17630
Soybeans	13842
Groundnuts	13502
Tomatoes	12755
Grapes	12720
Mangoes	12556
Watermelon	10436
Dry Beans	8537
Onions	6474
Percentage of total Developing Country Food Crop Production	94.39

Most of the substrates mentioned above are very rich in carbohydrates. However, for certain microbes, these substrates might have to be enriched with nitrogen and trace elements. The best source for supplementing the additional nutrients is the whole yeast.

This could be done either by using traditional yeast or active dry bakers yeast. The most important advantage of active dry bakers yeast over traditional compressed yeast is its greater stability at room temperature. A useful storage life of 6 to 21 months at 32°C to 21°C respectively has been reported. In some countries, canning residues could also be an excellent supplement.

Besides the cultivated plants, there are several weeds, or local plants which could prove to be a good raw material for microbial processes. For example, winged bean *(Psophocarpus tetragonolobus)* has recently become quite popular in southeast Asia and West Africa.

Table

Elemental Composition of 100 g of Compressed Yeast.,

Carbon	*Nitrogen*	*Phosphorus*	*Potassium*	*Magnesium*	*Calcium*
12.3 g	2.3 g	0.28 g	0.54 g	0.03 g	0.01 g

Some tropical plants including basella and amaranthus are also becoming important sources of raw material. Some plants such as salt brush *(Atriplex sp.)* and buffalo gourd *(Cucurbita foetidissima)* and tamarugo *(Prosopis tamarugo)* are being introduced in arid and salt tolerant conditions in several countries for the active production of biomass for microbial processing. The major advantage of these plants would be their ability to produce biomass. Their ability to grow in conditions not suitable for regular crop eliminates competition. Certain other aquatic plants that grow suitable in saline environments include reed *(Phragmites communis)*, papyrus reeds *(Cyperus sp.)*, Cattails *(Typha sp.)* and eelgrass *(Zostera marina)*.

Recently, water hyacinth *(Eichornia sp)* has been exploited as a raw material for microbial processes. It is grown in several countries and in India it is a weed blocking several ponds and large natural water reservoirs. The same is true in the Nile Basin and the United States where it is being utilized to treat domestic sewage. Because of its prolific growth, (about 900 kg/ha/day), water hyacinth is a good potential raw material.

Bioconversion

The biological conversion of ethanol into a dilute solution of acetic acid (vinegar) was done in Babylon by 5000 B.C. Other important bioconversions are the transformation of isopropanol into acetone, glucose into gluconic acid and sorbitol into sorbose. This conversion of sorbitol into sorbose is the only biological step in the otherwise nonbiological manufacture of vitamin C. In the pharmaceutical industry, a good example of bioconversion is the production of steroids.

The most dramatic results of bioconversion methods have been obtained in the manufacture of the steroid hormones. In the early 1930's, Edward C. Kendall of the Mayo Foundation and Tadeus Reichstein of the University of Basel isolated cortisone, a steroid secreted by the adrenal gland. About a decade later, Philip S. Hench of the Mayo Foundation showed that administration of cortisone could relieve the pain of patients with rheumatoid arthritis. There were immediate demands for the hormone; chemical methods for its synthesis were developed as the market has tremendous potential. The chemical synthesis, however, was elaborate, and required 37 steps, many of them under extreme conditions. Cortisone made this way cost $ 200 (Rs 2400) per gram.

One of the major complications in the chemical synthesis of cortisone is the need to introduce an oxygen atom at a position in the four-ring steroid structure designated position 11; this step is crucial in determining the physiological activity of the molecule. In 1952, D.H. Peterson and Herbert C. Murray of the Upjohn Company discovered that a strain of the bread mold *Rhizopus arrhizus* is able to hydroxylate progesterone, another steroid, thereby introducing an oxygen atom at position 11. Progesterone is an early intermediate in the synthesis of cortisone; by means of microbial hydroxylation (which is accomplished industrially with microorganisms closely related to R. *arrhizus*) the synthesis was shortened from 37 steps to 11. As a result, the price of cortisone was reduced to 56 dollars per gram.

The microbial hydroxylation of progesterone had economic consequences beyond shortening the chemical synthesis of cortisone. The fermentation could be achieved at 37 degrees Celsius, with water as the solvent and under normal atmospheric pressure. Reactions under these conditions are much cheaper than those carried out under extremes of temperature and pressure, with solvents other than water, which had been required in the chemical synthesis of cortisone.

Several other uses for microorganisms have since been found in the industrial synthesis of the steroids. The commercially important steroids include the corticosteroids cortisone, hydrocortisone, prednisone and dexamethasone, the androgen testosterone and the estrogen estradiol (the last two used in contraceptives) and spironolactone (a diuretic). The raw material for all of them are the complex alcohols called sterols. There are two common sources of sterols: the production of soyabean oil leaves, a waste rich in stigmasterol and sitosterol; the roots of the Mexican barbasco plant contain diosgenin.

The first step in steroid production from plant sterols is the degradation of the side chain of the sterol molecule. Several pharmaceutical companies have found it economical to rely on mycobacteria (a group of aerobic, Grampositive bacteria) to perform this step, which was formerly accomplished nonbiologically. The mycobacteria use sterols as a source of carbon and energy; mutant strains that lack certain enzymes cannot complete the degradation. These mutants are exploited in partial degradations that yield valuable intermediates. Other bacteria modify the steroid nucleus to yield a number of derivatives.

The introduction of these microbiological processes has been of great significance in the manufacture of the steroids, first in making steroid synthesis commercially feasible and later in progressively lowering the unit cost of production. In 1980, the price of cortisone in the U.S. was 46 cents per gram, a 400-fold reduction from the original price. The identification of new uses for steroids (in contraception and in treating hormonal insufficiencies, skin diseases, inflammation and allergies) and more efficient production has created a substantial demand for them. Bulk sales of the four major steroids in (cortisone, aldosterone, prednisone and prednisolone) in the word amounted to about $300 million in 1978. These are the most important commercial products of bioconversion processes but by no means the only ones.

Recently, in the production of semi synthetic penicillins, it became possible to replace a chemical reaction that creates pollutants with a nonpolluting bioconversion. Commercially, microbial products fall into four type:

1. The microbial cells themselves,
2. The microbial enzymes and other macromolecules synthesized by microbes.
3. Their primary metabolic products (compounds essential to their growth).
4. Their secondary metabolic products (compounds not required for their growth).

In general, both the primary and secondary metabolites of commercial interest have a fairly low molecular weight (About 1,500 dalton) compared with the molecular weight of an enzyme which may range from 10^4 to several million daltons.

Microbial cells are used to carry out biological conversions; processes in which a compound is changed into a structurally related compound by means of one or more enzymes supplied by the cell. These microbial transformation or bioconversions can be accomplished with growing or nongrowing cells, spores or even dried cells.

Microbial transformations or biotransformations have a selective advantage over nonbiological chemical reactions.

1. Many nonbiological chemical reactions call for a considerable input of energy to heat or cool the reaction vessels.
2. Several solvent and inorganic catalysts are required in chemical reactions.
3. In nonbiological chemical reactions, a few unwanted by-products which require additional purification are also formed.
4. Bioconversions, unlike nonbiological reactions proceed at ordinary cell temperatures with water as the solvent.
5. Bioconversions are high yielding and can reach 100 per cent.

Production of Wine and Beer

Alcoholic fermentation is known since the prehistoric era. In fact, alcoholic beverages have been a part of several civilizations. The credit for the discovery of the process goes to Louis Pasteur, almost 100 years ago, when he reported yeast mediated fermentation of sugar described in Chapter 2.

The manufacture of alcoholic beverages involves the fermentation of sugars by S. Cerevisiae, but here the main requirement is the alcohol rather than the carbon dioxide. Alcoholic beverages are grouped into three categories: (i) Wines and beers-which are made by fermenting fruit juice or a sugary extract of grain with yeast. (ii) Fortified wines in which brandy is added to the wines (iii) Spirits, made by distilling wines or beers.

Archaeological evidence shows that the fermentation of grain extracts was already an advanced art more than 6,000 years ago. In fact, beer is totally free from pathogenic organisms as its acidity and the presence of antimicrobial compounds derived from hops prevent their growth. At present, more than 700 million hectolitres of beer is produced in the world. The maximum consumption is in West Germany, United States and Australia.

Beer is primarily made from barley. Recently, however, fermentation of potato starch, corn and other cereals are also employed for beer production. The grains of barley are first malted (allowed to germinate for a short time). Malting activates enzymes in the grain which are helpful in breaking down starch. The malted barley is then crushed and simmered with hot water (around 67°C). During this process, enzymes break down the long chain of starch-into smaller carbohydrate molecules. The resulting extract (malt wort) is separated from the crushed grains and boiled (generally with hops) to provide flavour to the beer. The boiling of wort stops further enzyme action in the wort and precipitates proteins from it, besides imparting the flavour. Next, the hopped wort is inoculated with S. cerevisiae. This is the crucial biochemical step when sugar is converted to alcohol anti carbon dioxide. Yeast also grows during this conversion. Other minor products of yeast metabolism such as amyl alcohol, isoamyl alcohol and phenyl ethyl, alcohol also affect the flavour of the final product. In addition, traces of acetic-acid and butyric acid are also formed. When the fermentation is complete, the yeast cells are separated from beer which is then matured for a particular period of time. Maturation is followed by filtration and possibly pasteurization and then the beer is ready to be released in the market.

In the market, particularly in the West, beer is classified as lager and bitter. The lager beer is made by yeasts that settles to the bottom of fermentation tank during fermentation. This yeast is different from chat routinely used and was first isolated about 105 years ago by Danish botanist Emil Christian Hansen at Carlsberg Institute in Copenhagen and named as Saccharomyces carlsbergensis.

To make the other beer, the yeast used is S. cerevisae which is the major strain used in the alcoholic industry. Taxonomists do not generally distinguish between these two strains.

Normal fermentation requires approximately seven days in a fermentation tank after which the "young" beer is transferred to vats for primary and secondary ageing, which may take an additional six months. Some yeast is left in the beer that is to become keg beer and the product is refrigerated to preserve it. The thick wall of the keg traps the carbon dioxide produced during continual fermentation. If the beer is to be canned or bottled, it may be pasteurized at 140°F for 15 minutes to kill the yeast or filtered through a membrane filter to remove the yeast. Some beer is used to seed new wort, the remainder may be dried for animal feed or pressed into tablets for human consumption. The alcoholic content of beer is usually 4 per cent.

Another beer which is less commercialized is the oriental sake. It is produced by allowing Aspergillus oryzae or a species of Rhizopus to form a mycelium in steamed rice and to convert the rice starch to fermentable sugar. Saccharomyces species then ferment the sugar until the alcohol level is approximately 14 per cent at which point the sake is ready for consumption.

Wine technology is relatively simple. It may be made from ripe fruit, fruit juice or extracts of plants such as dandelions. Nearly all the world's wine is made from one species of grape, Vitis vinifera. Wine grape juice is a perfect medium for wine production because of its high nutrient concentration, a natural acidity which inhibits undesirable microbial growth, high sugar concentration and the production of pleasant aromas and flavours.

Juice is extracted from red or white fine grapes. Earlier, grape juice was fermented spontaneously by microbes present on the surface of the grapes. This practice was and still followed for preparation of many "asvas" (a medicated liquid) in Indian systems of medicine where alcohol is self generated by covering and ageing the kegs containing the fruit juice. The microbes present on the grape fruit include several variables of yeasts including Saccharomyces (Table 3.13). Several of the yeasts responsible for the first part of the fermentation are killed off by the alcohol released when strains of S. cerevisiae ferment sugars in the juice. Recently, however, induced fermentation is preferred over natural fermentation and many vintners now add specially selected cultures of S. cerevisiae to the grape juice and incubate it at 7-14°C. In recent years, the continuous fermentation method has also been employed when juice is fed steadily into a fermentation process and wine is continuously removed. Fermentation normally begins with crushing of the fruit to produce must. In controlled processes, even sulphur dioxide is added to inhibit the action of non-wine yeasts and to allow the wine yeasts (which are more tolerant to sulphur dioxide) to carry out the necessary fermentation. In spontaneous fermentation, however, sulphur dioxide is omitted. Oxygen may be bubbled through initially to promote aerobic growth of the yeast but anaerobic conditions are gradually established. The stems, seed and skin remain intact with the fermenting must for red wine production until the end of the alcohol fermentation whereas in the white wine production only the juice is used. The grape variety is responsible for some flavour compounds. For example, the flavours of the Muscat varieties are attributed to the terpene alcohols, Linalool and geraniol.

Generally alcohol production requires only a few days. However, ageing may continue for months or even years. It is during ageing that wine develops flavour, aroma and bouquet. Red wines (which are commonly drunk with lunch and dinner in the West) turns red because alcohol extracts the colour from the grape skin. Besides, there are several sparkling wines such as champagne that are naturally carbonated with additional carbon dioxide production. Sherry wines (apetizer wines) have a slightly different flavour due to its seeding by a particular strain of yeast.

In the open market, wines are generally referred as Dry or Sweet. Dry wines are not very sweet because almost all the sugar is metabolized.

The alcohol percentage in wines vary from 12-16. Some fortified wines may have an alcohol content upto 22 percent and this is achieved by adding brandy. Addition of brandy to wine was originally done to stop the yeast fermentation process and to make wine biologically active. Except for certain wines (such as fino and amontillado sherry), brandy is added and then wine is stored, followed by a final adjustment of the alcohol content again with brandy. In Spain, special sherries are prepared with a nutty flavour. After fortification, the wine is matured in contact with air to encourage the growth of surface flora (primarily yeast) which imparts this nutty flavour. Some wines are fortified in which some of the alcohol is derived from the yeast fermentation of grapes and some from the addition of distilled spirits.

Distilled spirits contain much more alcohol than beer or wine. Distillation is necessary to produce a beverage with an alcohol content of 40% by volume. Modifications of the still which was designed in 1830 by Coffey continue to be used today. Generally, spirit is designated with a proof number which is twice the percentage of alcohol. A bottle of whisky marked 80 proof means the product contains 40 percent alcohol. Spirit distillation differs from beer and wine production because there is no boiling stage as in latter. Consequently, the enzymes active in the mash continue to operate during fermentation, breaking down more sugar and producing more alcohol. Several strains of Saccharomyces are used for distillery fermentations and most distilleries maintain their own selected yeasts. Although the choice of raw materials may be varied, certain spirits are usually produced from particular materials (Table). Brandy, a distilled wine product, is prepared from grapes and malt whisky from barley. Maize is also used to produce grain whisky, gin and vodka. Rum is generally prepared from cane or beet molasses. In Rum production, Schizosaccharomyces strains are usually used for the heavy aroma rums and the quick-fermenting Saccharomyces are better for the lighter rums. In addition, the bacterium Clostridium saccharobutyricum accelerates the formation of alcohol during fermentation by yeast and the best rum yield is obtained when the ratio of bacteria to yeast is 1:5. The bacteria are added when the ethanol concentration is 3.5-4.5% and the sugar content is 6%(w/v). Also contributing to the flavour of the final distilled product are compounds extracted by the liquid from the wood barrels in which such spirits are aged for several years. Distilled spirits differ from one another in the nature of the distillation process. Maturing of spirit is followed by addition of various chemicals such as aldehydes, ethers and volatile acids for flavouring. Scotch malt whisky is distilled in small pot stills, where as most other whiskeys are distilled in plants that operate continuously.

Table

Types of Spirits

Brandy	Prepared from fruit juice generally grapes.
Neutral spirit (Gin, Vodka)	Generally prepared from potato starch Gin is flavoured with junior oil vodka is unflavoured.
Rum	Prepared from Molasses
Whiskey	Prepared from various malted cereals including barley, rye and corn.

Production of Cider and Vinegar

Cider is fermented apple juice and the production steps resemble those of wine.

Vinegar, on the other hand, is a fermentation product which contains more than 4% acetic acid (w/v).

In the preparation of Cider, apples are crushed to a fine pulp from which juice is extracted. This is treated with sulphur dioxide to inhibit the growth of Kloeckera apiculata which would otherwise affect the final flavour of cider. Fermentation is facilitated using either the natural yeast flora or by adding yeast starter cultures. As the natural growth of Saccharomyces is slow, the large cider manufacturers prefer to add pure yeast cultures to the sulphated apple juice. A particular characteristic necessary in a yeast is the production of polygalacturonase, to degrade the de-esterified pectin to galacturonic acid. After fermentation, cider is removed from the yeast and clarified.

In vinegar production, two fermentation steps are necessary: (i) Conversion of sugar-containing material to alcohol using S. cerevisiae. Generally, any substrate capable of being converted to alcohol can be utilized. The fermented liquid is allowed to settle down, the yeast and supernatant are removed, (ii) The second step is the conversion of ethanol to acetic acid via acetaldehyde. Sulphur dioxide, if added, is removed. All vinegar processes require mixed Acetobacter microflora including A. schnetzenbanchii, A. curvim, A. oreleanense and other related microbes. The reaction is exothermic and requires excess of oxygen. These days this submerged fermentation technique is very much practiced for vinegar production.

Microbial Production of Cheese

The cheese making begins with curdling of milk. Generally cow's milk is preferred but these days buffalo's milk is also used. A starter culture of bacteria (lactic acid bacteria) is added to the milk and the mixture is incubated for some time. This is followed by addition of a proteolytic enzyme to coagulate solids in the souring milk. The nature of the bacteria serving as the starter is one of many factors resulting in a vast variety of cheeses. Earlier calf rennet was the source of enzymes. However, in recent years, it is being replaced by microbial enzymes. The coagulated curd is then separated from the whey-the watery fluid which separates during the formation of the curd. The coagulated curd is then pressed to squeeze out the remaining water and wrapped in cloth to dry up. The moisture content varies from cheese to cheese and is adjusted according to the requirement. Cheese is salted by adding sodium chloride. It may be mixed with the drained curd or applied to the surface of the pressed cheese. Salting is useful for preventing the growth of undesirable microbes.

Cheese is generally ripened (with the exception of creamy cottage cheese and ricotta) with the help of bacteria or molds. The ripening time varies from 1-16 months depending upon the type of cheese and the ripening agent. Hard cheese (such as Cheddar or Swiss) is ripened with bacteria inoculated throughout the interior mass of cheese. Soft cheese (Camembert and Limburger or Brie) is ripened with the surface growth of inoculation. A semihard cheese, cooked briefly to lower the content of the curd to about 45 percent, which makes the curd finer. Hard cheese, including cheddar, with water content less than 40 percent bears the same bacterial flora. Other hard cheeses however, differ in that the curd is inoculated with spores of mold (usually Penicillum roquefortii). The growth of the mold in the cheese generates compounds that impart a flavour and aroma characteristic of individual cheese such as Danish blue, Stilton, Gorgonzola and Roquefort. (Tables). Gruyere is another hard cheese in which bacteria (Propionibacterium shermanii) produces propionic acid in the starter mix. These bacteria also generate carbon dioxide gas giving rise to the holes typical of such cheeses. Lipolysis takes place during ripening of Roquefort and Blue cheese. Roquefort cheese is made from whole milk (a term used in the West referring to unskimmed milk) and the fat remains during curdling. This is inoculated with spores of Penicillum roqueforti (either before or after curdling). During ripening, several substances including butyric acid, caproic acid, capric acid and caprylic acid are released contributing to the flavour.

Table

Various Categories of Soft (Ripened And Unripened) Cheese

Cheese	***Microorganism***		***Origin***
	SOFT, UNRIPENED		
COTTAGE	Streptococcus lactis	Leuconostoc citrovorum	CENTRAL EUROPE?
CREAM	Streptococcus cremoris	U. S.	
NEUFCHATEL	Streptococcus diacetilactis		FRANCE
	SOFT, RIPENED		
BRIE	Streptococcus lactis	Penicillum camemberti	FRANCE
	Streptococcus cremoris	Penicillum candidum	
CAMEMBERT	Streptococcus lactis	Penicillum camemberti	FRANCE
	Streptococcus cremoris	Penicillum candidum	
LIMBURGER	Streptococcus lactis	Brevibacterium linens	BELGIUM
	Streptococcus cremoris		

Table

Various Categories of Semisoft and Ripened Cheese

Cheese	***Microorganism***		***Origin***
	SEMISOFT, RIPENED		
ASIAGO	Streptococcus lactis Streptococcus cremoris Streptococcus thermo-philus	Lactobacillus bulgaricus	ITALY
BLUE	Streptococcus lactis Streptococcus cremoris	Penicillum requeforti or Penicillum glaucum	FRANCE
BRICK	Streptococcus lactis Streptococcus cremoris	Brevibacterium linens	U. K.
GORGONZOLA	Streptococcus lactis Streptococcus cremoris	Penicillum roqueforti or Penicillum glaucum	ITALY
MONTEREY	Streptococcus lactis Streptococcus cremoris	U. S.	
MUENSTER	Streptococcus lactis Streptococcus cremoris	Brevibacterium linens	GERMANY
REQUEFORT	Streptococcus lactis Streptococcus cremoris	Penicillum roqueforti or Penicillum glacucm	FRANCE

Table

Types of Hard, Ripened Cheese

Cheese	***Microorganism***		***ORIGIN***
	HARD, RIPENED		
CHEDDAR	Streptococcus lactis Streptococcus cremoris Streptococcus durans	Lactobacillus casei	BRITAIN
COLBY	Streptococcus lactis Streptococcus cremoris Streptococcus durans	Lactobacillus casei	U. S.

Cheese	*Microorganism*			*ORIGIN*
EDAM	Streptococcus lactis Streptococcus cremoris			NETHERLANDS
GOUDA	Streptococcus lactis Streptococcus cremoris			NETHERLANDS
GRUYERE	Streptococcus lactis Streptococcus thermophilus	Lactobacillus helveticus	Propionibacterium shermanii or Lac tobacillus Bulgaricus and Propionic bacterium freudenreichi	SWITZERLAND
STILTON	Streptococcus lactis Streptococcus cremoris	Penicillum roque-forti or Penicillum glaucum		BRITAIN
SWISS	Streplococcus lactis Streplococcus thermophilus	Lactobacillus helveticus	Propionibacterium shermanii or Lac tobacillus bulgaricus and Propionic bacterium freudenreichi	SWITZERLAND

Production of Ethanol

Ethanol production became significant after the discovery of fermentation. These days, alcohol production is primarily synthetic. However, the rising cost of petroleum has renewed the need for production of ethanol by fermentation for use as a fuel. Several countries including Brazil and India have undertaken large programmes

Table

Types of Hard Ripened and Pasta Filta Cheese

Cheese	*Microorganism*		*Origin*
	VERY HARD, RIPENED		
PARMESAN	Streptococcus lactic Streptococcus cremoris Streptococcus thermo philus	Lactobacillus bulgaricus	ITALY

Cheese	*Microorganism*		*Origin*
ROMANO	Lactobacillus bulgaricus	Streptococcus thermophilus	ITALY
	PASTA FILTA (PLASTIC CURD)		
MOZZARELLA	Streptococcus lactic Streptococcus thermophilus	Lactobacillus bulgaricus	ITALY
PROVOLONE	Lactobacillus bulgaricus		ITALY

Table

Selected Enzymes Used in Cheese Ripening

Cheese Type	*Enzyme added*	*Obtained* **From**
Chedder	Proteinaxs (Acid & Netural) Peptidases Lipases decarboxylases	Several commercial and animal enzymes
Edam Cheddar	Proteinases and Peptidases	Pstudomonas fluoresscxns
Parmesan Romano	Proteinase	Gastric tissues
Gouda	Proteinase	Aspergillus oryzae
Mozzarella	Lipases Esterases	Calf (Pregastric secretions)
Rossiisltii	Proteinase	Pancreatin
Blue	Lipases	Aspergilus sp.

for production of fuel alcohol on a large scale. Synthetic production of ethanol is carried out from ethylene (C2H4), which is derived from petroleum or natural gas. As a matter of fact, till recently, about 70 per cent of ethanol was produced by synthetic methods in the world.

In recent years, a large number of plants are being used or are in the process of being used for ethanol production. These include maize (corn), cassava, Jerusalem artichoke and potatoes. Except for Jerusalem artichoke, the storage carbohydrate in these plants is primarily starch. In Jerusalem artichoke, however, the storage product is insulin. Some sucrose containing plants such as sweet sorghum, pineapple, sugarbeet and sugar cane are also used. Several plants including corn, wheat, rice, cassava and sugarcane (molasses) is already being used on a mass scale in Africa, South America, Near East and Far East.

Ethanol has wide commercial applicability. It is an important organic solvent besides an extractant. It is used in the manufacture of detergents, adhesives, lubricants, pharmaceuticals, dyes, pesticides, explosives, fabrics, resins and as an antifreeze. It is also used as a fuel in internal combustion engines as anhydrous alcohol (99.8 per cent) or mixed with petrol (20 percent ratio). The economic value of alcohol as a petrol blend is 15 to 20 per cent higher than that of alcohol used as a direct petrol substitute.

When sugar cane is used as a source of alcohol, the bagasse is separated from the sweet juices. Bagasse is dried, then burnt to generate the energy required. The juice is concentrated and sterilized followed by fermentation. Ethanol is recovered from the fermentation solids and from 8-10 per cent alcohol solution by distillation. Methods have been developed to achieve continuous fermentation to yield a higher alcohol concentration (12 per cent) besides utilization of agricultural wastes and by-products for fuel or feed stock purposes. Since 1975, Brazil has been producing alcohol on a mass scale from sugarcane and cassava. Brazil could easily afford to do so because being the world's second largest sugarcane producer. In Brazil over 5 x 109 litres of ethanol were produced in 1982, and 11 x 109 litres in 1985. In France, research (in the Bouches-du-Rhone department) has been focussed on the production of alcohol from sugar-beet, Jerusalem artichoke or sugarcane. In India, ethanol production is primarily by fermentation of molasses.

The flow sheet for Ethanol production as carried out by fermentation of molasses or corn is shown. (Fig.). The fermentation is generally done at atmospheric pressure. It is facilitated by rapid fermentation using a vacuum followed by recycling of microbial cells. Ethanol can continuously be recovered by applying a vacuum of about 50 mm of mercury to a fermenter operating at about 30-35°C. Vacuum distillate from the fermentation liquid contains upto 20 per cent ethanol. The cost of

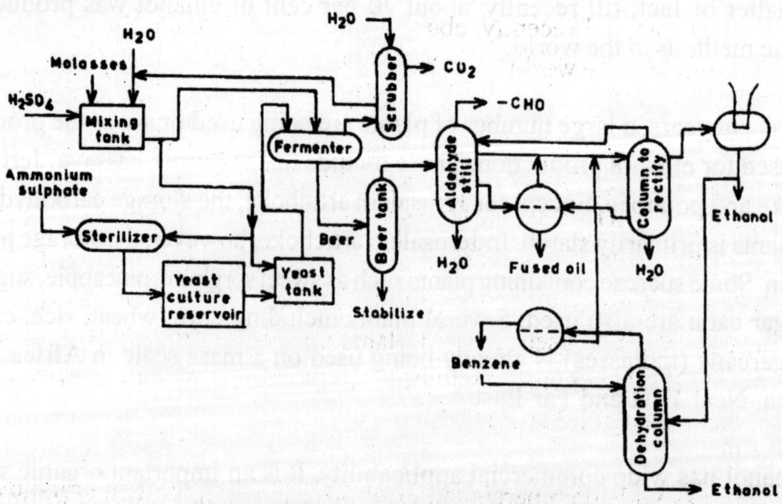

Fig. Flow sheet for production of Ethanol

distillation to achieve an ethanol concentration of 80-95 percent is then significantly less than that by conventional methods. Currently, commercial alcohol is produced using yeasts (Saccharomyces sp.) including S. cerevisiae, S. uvarum and S. diastaticus). Various substrates are used by individual yeast species.

Table

Specific Substrates Used by Yeast Sp.

Species	*Substrate*
S. cervisiae	Glucose, Fructose Maltose Maltotriose
˜. diastaticus	Dextrins as well
Kluyveromyces fragilis K. lactus	Lactose

Ethanol and methane production from cellulosic wastes by thermophilic microbes has gained importance recently. A host of lignocellulose rich materials including corn stover, bagasse, rice straw are becoming more suitable material for ethanol production.

New Steps in Antibiotic Production

Alexander Fleming, a student of Almroth Wright (the British investigator who described opsonins), first discovered the value of antibiotics. The antibiotic was, in principle, used therapeutically even by the traditional European classes. A doctor travelling through Europe in 1908 had observed that in the kitchens of several houses hung loaves of moldy bread. When he asked its significance, he was told that when a wound or abrasion was sustained, a sliver of the bread was cut and mixed with water to form a paste that was applied to the tissue. Infection was less likely to occur in a wound treated this way. By chance, however, the first scientific investigation was carried out in 1928 by Fleming who was awarded the Nobel Prize in 1945 with Howard Floret' and Ernst Boris Chain for this discovery. In 1928, Alexander Fleming noted that the mold *Penicillum notatum* killed his cultures of the bacterium *Staphylococcus aureus* when the mold accidentally contaminated the culture dishes. After growing the mold in a liquid medium and separating the fluid from the cells, he found that the cell-free liquid could inhibit many species of bacteria. He named the active ingredient in the liquid penicillin, but soon afterward discontinued his work on the substance. Eventually a study begun in 1938 at the University of Oxford by

Howard W. Floret', Ernst B. Chain and their colleagues, leading to the successful preparation of a stable form of penicillin and to the demonstration of its remarkable antibacterial activity. Floret' later on moved to the United States and U.S. Department of Agriculture and a few other American pharmaceutical companies helped in the commercialized production of penicillin from *Penicillum chrysogenum.*

A practical definition of an antibiotic is a microbial product of low molecular weight that specifically interferes with the growth of microorganisms when it is present in very small amount. Most of the substances that satisfy this definition are fungal or bacterial metabolities which have no obvious role in the growth and maintenance of the cell. These molecules are called secondary metabolites to distinguish them from the primary metabolites needed for the growth of organisms. Alkaloids, toxins and pigments are also secondary metabolites. Like other secondary metabolites, an antibiotic is the end product of a long series of enzymatically catalyzed reactions. All antibiotics satisfy the principle of selective toxicity as put forward by Paul Ehrlich. The principle holds that an effective chemotherapeutic agent should be safe for human tissues but toxic to the infecting organisms.

Commercially, antibiotics are the most important class of pharmaceuticals made by microbial biotechnological process. In 1978, the Worldwide bulk sales of the four most important groups of antibiotics—the penicillins, the cephalosporins, the tetracyclines and erythromycin—amounted to more than 60 billion rupees. Another commercially important group of antibiotics consists of aminoglycosides which include streptomycin. The microbes that carry out fermentation in the antibiotics industry are drawn from a rather narrow taxonomic range. More than 1000 antibiotics are derived from six genera of filamentous fungi, the important being the genus *Cephalosporium* which yield cephalosporins, and *Penicillium* the main source of penicillins. Since the 1940s, penicillin has remained the most important of the antibiotics because of its low cost and the broad use of thousands of penicillin derivative. Penicillin G or benzylpenicillin is currently the most popular of the penicillins, the one referred to when the physician prescribes "penicillin". Other types that may be specified are penicillin F or penicillin V. All have the same basic structure with a beta-lactam nucleus (described in the following pages) and several attached groups (Fig.). The largest number of antibiotics are derived from actinomycetes-a group of filamentous bacteria. Three genera of actinomycetes account for almost 3,000 'antibiotic agents. The most important antibiotic yielding actinomycete is *Streptomyces*. Among the nonfilamentous bacteria, two genera synthesize about 500 antibiotics. The penicillins and cephalosporins are the most important in the market.

Unlike a protein which is the immediate product of a single gene, an antibiotic is made by the joint action of 10-30 genes. For several commercial antibiotics, the entire pathway has not been worked out. In recent years automatic methods for examining the survivors of the process of inducing mutations and identifying new strains with higher yields have been developed. The automatic equipment screens tens of thousands of survivor types per round of mutation. Instead of isolating new antibiotics, for the last several years work has been directed towards modifying the structure of existing antibiotics to increase their potency. Most of the efforts are focussed on the penicillins and the closely related cephalosporins. In both groups, the central structure of each molecule is the four-member beta-lactam ring, composed

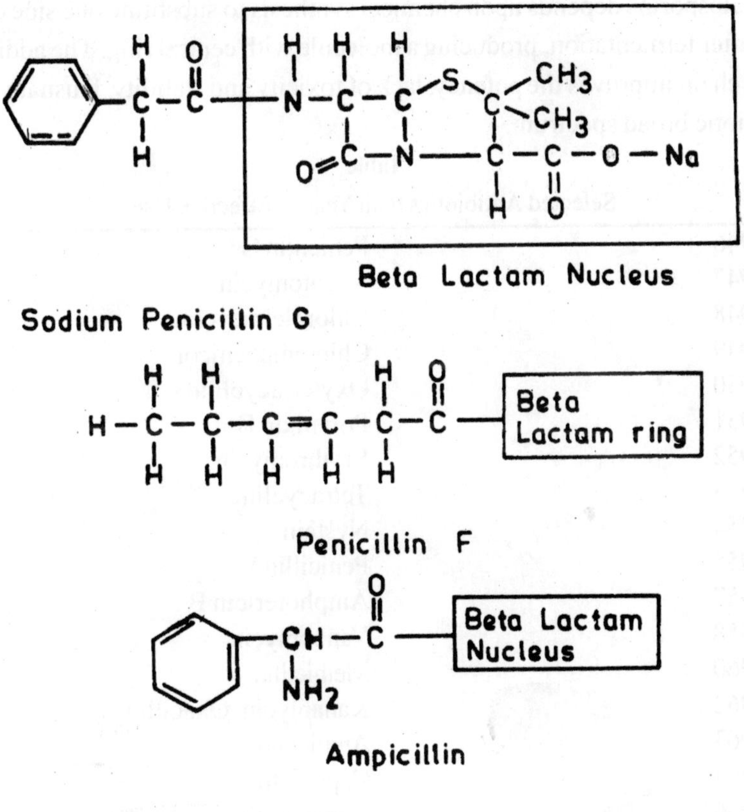

Beta Lactam Nucleus

Sodium Penicillin G

Beta Lactam ring

Penicillin F

Beta Lactam Nucleus

Ampicillin

Fig. Some members of penicillin group with beta lactam nucleus

of three carbon atoms and a nitrogen atom; the penicillins and the cephalosporins are collectively known as the beta-lactam antibiotics. In addition to having a broad spectrum of antibacterial activity, the beta-lactam antibiotics are probably the least toxic of all the major groups of antibiotics. In the late 1950s, the betalactam nucleus of the penicillin molecule was identified and synthesized and it soon became possible to attach various groups to this nucleus and thereby create new penicillins. In the following years several hundred new penicillins emerged from the semisynthetic process (Tables).

Besides Penicillium chrysogenum and cephalosporium acremonium, Eli Lilly and company and Merck, Sharp and Dohme have reported new betalactam antibiotics in the fermentation broths of streptomycetes. These compounds, the cephalomycins, have a structure similar to that of the cephalosporins with the addition of a methoxyl group (CH_3O-) on the beta-lactam ring. For the last several years, efforts are being made to improve the beta-lactam antibiotics. Semisynthetic production of penicillins and cephalosporins depends upon chemical synthesis to substitute one side chain for another after fermentation, producing a molecule with central ring. The addition of a new side chain improves the potency, lack of toxicity and stability. It usually reduces the antibiotic broad spectrum.

Table

Selected Antibiotics with Year of Effective Use

Year	Antibiotic
1946	Penicillin G
1947	Streptomycin
1948	Chloroletracycline
1949	Chloramphenicol
1950	Oxytetracycline
1951	Polymxin B
1952	Erythromycin
1953	Tetracycline
1954	Nystain
1955	Penicillin V
1957	Amphotericin B
1958	Vancomycin
1960	Methicillin
1962	Kanamycin, Oxacillin
1963	Ampicillin
1964	Cephalothin
1966	Gentamycin
1967	Doxycycline
1968	Cephaloridine
1970	Carbenicillin, Clindamycin
1971	Rifampicin
1972	Spectinomycin

Table

Selected Antibiotics and their Site of Action

Antibiotic	***Target Effect***
Penicillin	Cell wall synthesis
Cephalosporin	
Cycloserine	
Polymxin B	Cell Membrane
Tyrocidin	
Valinomycin	
Bleomycins	DNA replication
Anthracyclines	
Itifamycin	
Actinomycin D	
Acridines	
Erythromycin	Protein synthesis
Tetracyclines	
Streptomycin	
Chloramphenicol	

The important example of a semisynthetic antibiotic is penicillin, which is produced when Penicillium chrysogenum is grown in the presence of phenylacetic acid. Phenylacetic acid is the precursor of the benzyl side chain of penicillin G. Currently, production of penicillin G is being carried out on a massive scale. The time required for fermentation is generally 200 hours. During the fermentation, sugar is added continuously. Penicillin G is then injected into another fermentation broth enriched with bacteria secreting acylases. The enzyme acylases remove the benzyl group from penicillin G producing 6-aminopenicillanic acid (Fig.). In the production of semisynthetic penicillins, this 6-APA forms a core structure to which side chains can be attached, yielding new antibiotics. Cephalosporins is also manufactured semisynthetically. Cephalosporin C which acts both against Gram positive and Gram negative bacteria, is produced by Cephalosporium acremonium. It has a very weak clinical activity. When the amide side chain of the molecule is removed, however, the remaining chemical nucleus 7-alphaaminocephalosporanic acid (7-ACA) is as useful as 6-APA forms a core nucleus to which side chains can be attached, yielding new antibiotics. The cephalosporins like Beta-Lactam antibiotics can fail to attack bacteria because of destruction of antibiotic by enzymes called beta lactamases which are recorded in bacteria, cyanobacteria and actimomycetes as well as yeasts. The gene for this enzyme is plasmid borne. A few years ago workers at Merck,

Sharp and Dohme discovered a new class of beta-lactam antibiotics called Thienamycins (Fig.) in the fermentation broths of Streptomyces cattleya. These are capable of inhibiting the beta-lactamases. The three thienamycins include Cefoxitin, Clavulanic acid and thienamycin. Workers at Harvard Medical school have reported that clavulanic acid and Olivanic acid (another beta lactamasa inhibitor) inactivate beta lactamase by the "suicide" method. Clavulanic acid is now being marketed in combination with the beta-lactam drug amoxicillin. This pharmaceutical hybrid known as augmentin is also resistant to beta-lactamase inactivation.

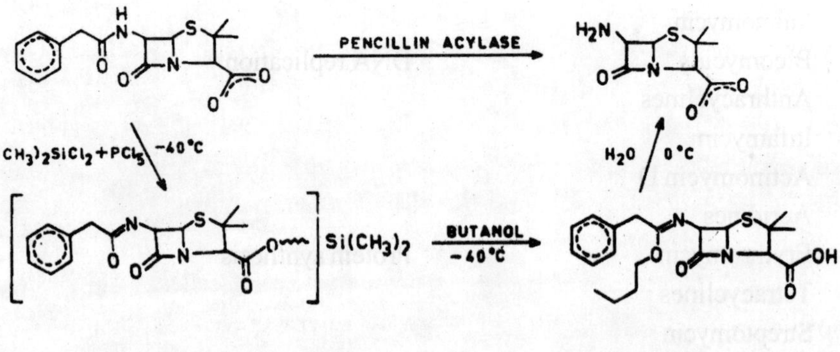

PENCILLIN ACYLASE

$(CH_3)_2SiCl_2 + PCl_5$ −40°C

H_2O 0°C

$Si(CH_3)_2$

BUTANOL −40°C

Fig. Conversion of penicillin G into b-aminopenicillanic acid

Production of Amino acids and Vitamins

Amino acids and vitamins are primary metabolities and low molecular weight compounds (up to 1500 daltons). Generally, microbes do not overproduce these primary metabolites. However, there are several strains which do over-produce and are thus

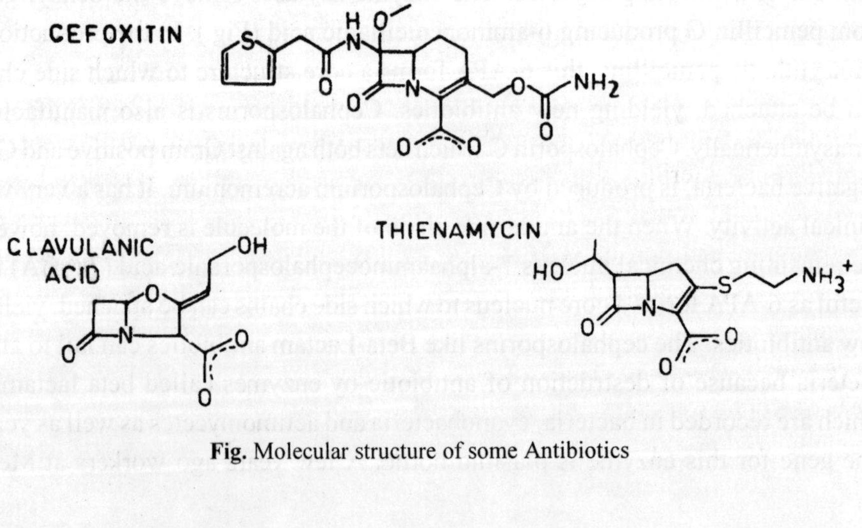

CEFOXITIN

CLAVULANIC ACID

THIENAMYCIN

Fig. Molecular structure of some Antibiotics

exploited for industrial purposes. Of the twenty amino acids that are constituents of proteins, eight cannot be synthesized in the human system and, therefore, these must be supplied in the diet. The two most important amino acids include methionine, which is primarily produced by synthetic methods and lysine, produced biosynthetically (about 80%). Most cereal grains are deficient in these two amino acids. Lysine is an essential amino acid in animal nutrition, but many plant proteins are deficient in it. Lysine is, therefore, produced by fermentation to serve as a supplement to animal feeds. Mutant strains of *Brevibacterium flavum* and *Corynebacterium glutamicum* have been selected that convert more than a third of the sugar in a fermentation medium into lysine. Upto 75 grams of lysine per liter of medium is produced. Microbial biosynthesis of amino acids is useful as it yields the biologically active isomer (L-amino acid), whereas the chemical synthesis yields equal amounts of both D and L isomers. It is difficult to separate both the isomers from each other. Lysine synthesis is accompanied by synthesis of methionine and threonine because of the branched metabolic pathway of which lysine is the end product, Asparatate kinase, the first enzyme of this branched pathway, is subject to feed back inhibition by threonine and lysine together. This organism has been manipulated by induced mutation by blocking the synthesis of homoserine dehydrogenase, thereby preventing the production of threonine and methionine; such mutants when grown in the presence of threonine and methionine synthesize and release large amounts of lysine. Amino acid producing bacteria have been used commercially since the 1950$ and the strains have been subsequently improved genetically by the generation of auxotrophs or regulatory mutants. The use of wild type bacteria for the production of amino acids, such as L-glutamate, L-valine, DL-alanine, L-glutamine and L-proline, depends on either inherent metabolic peculiarities within the organism, or stimulation of secretion by environmental parameters. Several bacteria including Corynebacterium, Brevibacterium, Aerobacter Bacillus, Microbacterium and Escherichia possess the necessary cellular metabolic activity to commercialize amino acid production.

Glutamic acid is another important amino acid produced by microbial activity. It is produced by certain species of Micrococcus, Arthrobacter, Brevibacterium flarum and Corynebacterium glutamicum. These bacteria aminate a ketoglutaric acid, a Krebs cycle intermediate, to yield glutamic acid. (Fig.) Glutamic acid is used as a food supplement and its sodium salt, monosodium glutamate is utilized in food preparations. Because of heavy demand in the flavouring industry, monosodium glutamate is imported from Japan and South Korea. Konbu (Kelp like seaweed) has been widely used as an important traditional seasoning service in Japan. In 1908, the taste of Konbu was identified as being due to Glutamic acid. Based on this discovery, the industrial production of monosodium L-Glutamate was initiated by Ajinomoto

Co. in 1909. The industrial production of amino acids has been stepped up because of their utility both in human nutrition and as a food supplement for domestic animals. The Eurolysine company in Amiens, France is the leading producer of lysine. Besides glucine, the production of alanine and cysteine is also being enhanced because glycine and alanine are used as flavouring agents and cysteine as bread texturizer. Some other amino acids including tryptophan, valine, are also currently produced by microorganisms. It is expected that due to the increased understanding of cellular metabolism, all commercially valuable amino acids will be produced solely by fermentation.

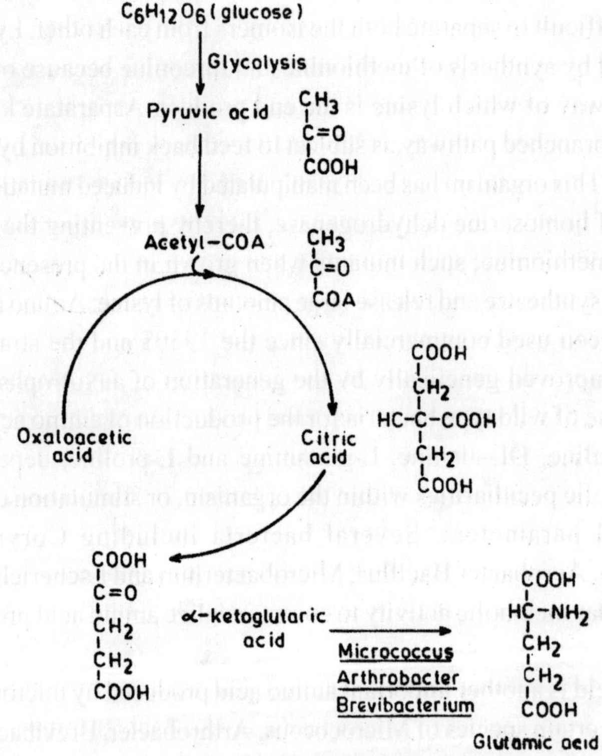

Fig Production of glutamic acid (chemical intermediates and product.)

Vitamin production by microbes is relatively less developed. Riboflavin (Vit B) and cyanocobalamin (Vitamin B,) are produced on a large scale by microbial activity. Riboflavin is a product of Ashbya gosspii a mold which produced 20,000 times the amount of the vitamin that it needs for its own metabolism. The mold is cultivated in corn steep liquor and excess vitamin is recovered from the medium. Cyanocobalamin is produced by selected species of Pseudomonas Propionibacterium and Streptomyces, grown in a cobalt supplemented medium of either malt extract or corn steep liquor. Pseudomonas denitrificans produces 50,000 times more cyanocobalamin than required for metabolic activity. The recovery of Vitamin B_2 and B12 from the sludge of sewage tanks is now being proposed.

The future of the amino acid industry is bright. From a clinical viewpoint demand has increased: Gastric ulcers are being treated with glutamic acid and histidine, and liver disorders with arginine. The production of amino acids by fermentation is developing fast. One can expect the commercial production of more amino acids and an improvement in yield to be brought about by the application of recombinant DNA techniques to the circumvention of intracellular regulatory mechanisms. The most significant contribution of genetic engineering to amino acid synthesis will be to make possible the fermentative production of methionine.

Production of Acetone and Butanol

Until 1914, acetone had been derived almost exclusively from the dry distillation or pyrolysis of wood in a retort, producing pryoligneous acid, which was neutralized with lime to give impure calcium acetate. This was then dry distilled to produce acetone (Fig.). Between 80 and 100 tonnes of birch, beech or maple wood were needed to yield 1 tonne of acetone. In 1901, the USA exported about 27,000 tonnes of calcium acetate and in 1914 produced about 74,000 tonnes as a by-product of methanol production.

Acetone was being used at that time, as a gelatinizing agent for nitrocellulose in the manufacture of cordite and also as a source of isoprene for the production of synthetic rubber. Substantial amounts were consumed in the dyestuff industry. When war broke out in 1914, the demand for acetone for the manufacture of smokeless powder (cordite) soared dramatically, outstripping the existing supply and creating an opportunity for the development of the fermentation process.

Pasteur first noted that the fermentation of starch by Bacillus species could produce butanol, but he did not follow up the observation. In 1910, Fernbach developed a process using potato starch as a feedstock and a bacillus which produced acetone and butanol, although yields were never achieved on an industrial scale. The breakthrough came from Chatim Weizmann who was working in Manchester, U.K., In a Basement room converted into a sterile laboratory.

Here he developed anaerobic fermentations of starch to yield acetone and butanol using a range of bacteria. The work was carried out for the firm of Strange and Graham Ltd. who needed butanol for the production of artificial rubber. At this stage, acetone was a secondary product, but as his research continued, he discovered

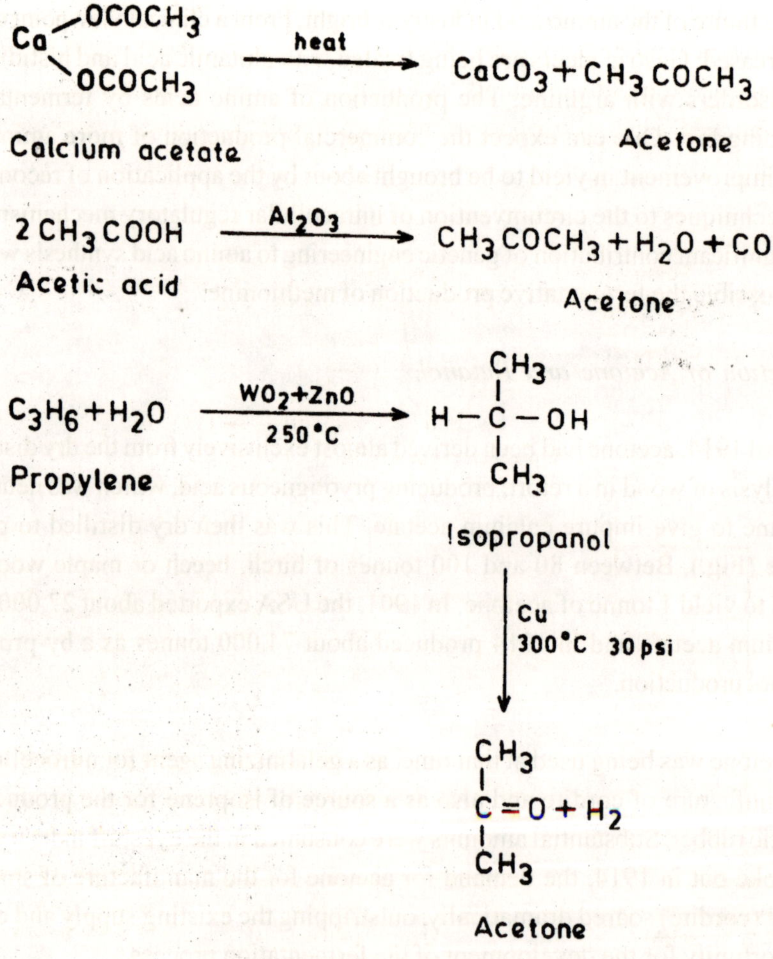

$$Ca(OCOCH_3)_2 \xrightarrow{\text{heat}} CaCO_3 + CH_3COCH_3$$

Calcium acetate → Acetone

$$2\,CH_3COOH \xrightarrow{Al_2O_3} CH_3COCH_3 + H_2O + CO_2$$

Acetic acid → Acetone

$$C_3H_6 + H_2O \xrightarrow[250^\circ C]{WO_2 + ZnO} H-C(CH_3)_2-OH$$

Propylene → Isopropanol

$$\xrightarrow[300^\circ C \quad 30\,psi]{Cu} (CH_3)_2C=O + H_2$$

Acetone

Fig. Production of Acetone

an organism which was highly vigorous and tolerant of the acid conditions, giving good yields from a variety of feedstocks, including cornmeal mash and starches from maize, rice and horse chestnuts. This bacillus was four times as active as those previously used and was originally named Bacillus granulobacter pectinovorum, but later became known as Clostridum acetobutylicum.

The process was initiated with a cornstarch mash diluted to a 5% suspension which was heat-sterilized, cooled to 30°C and pumped into a fermenter, together with nutrients and an inoculum of Clostridium acetobutylicum. The chemical equations for the process are shown in Fig. Later fermentations used molasses as feedstock, the price and availability of which became an important factor. The fermentation was allowed to proceed for upto 48 h and the solvents were recovered by distillation and fractionation. Solvents in the ratio of 6 : 3 : 1 for butanol, acetone and ethanol, were obtained. Thus, 100 tonnes of molasses would give approximately 12 tonnes of acetone.

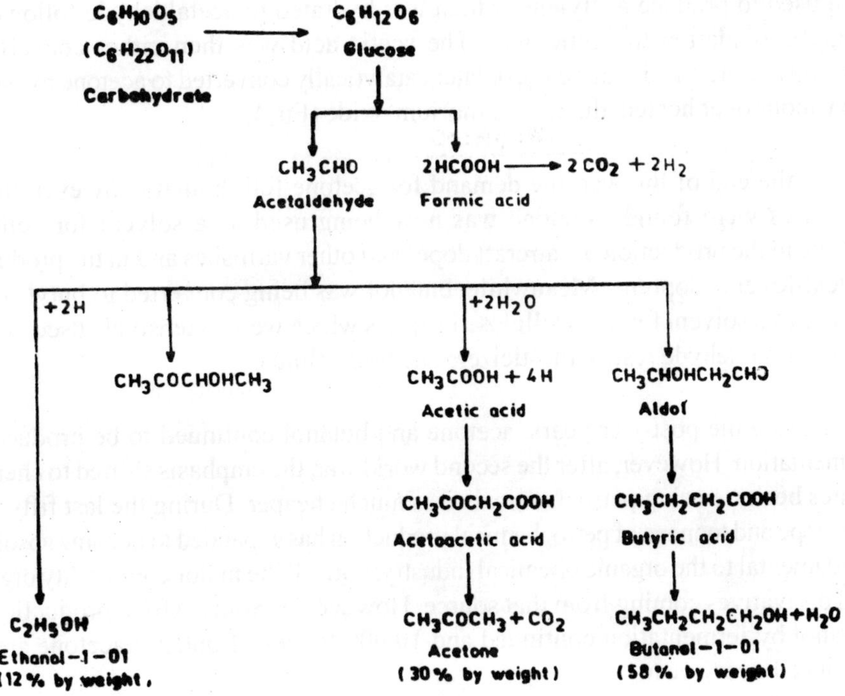

Fig. Chemical steps in fermentation of Acetone/Butanol

An important off-shoot of the Weizmann process was the recovery of riboflavin (Vitamin B) as a by-product. In the development of his process, Weizmann had improved the organism by 'heat shock'. Each subculture from the spore state was treated and this served to both eliminate weak strains and to deoxygenate the culture.

Weizmann patented his process, and offered into the British Government, but he was ignored until February 1915 when the war effort demanded tonnages of acetone which could not be met by traditional supply. Then, on behalf of the Admiralty, the Nobel Company visited Weizmann who showed that his process could convert 100 tonnes of grain to approximately 12 tonnes of acetone and 24 tonnes of butanol. On the recommendation of the Nobel Company, the process was adopted. The government promptly commandeered brewery facilities and rapidly built new plants and installed the process. The growth was dramatic; by 1916 the UK had an installed fermentation capacity for acetone production of about 700 m. However, the traditional route was still used and two other processes were being developed.

In the UK, Joseph Crosfield was pioneering the dehydrogenation of fermentation alcohol to acetaldehyde and then its oxidation to acetic acid from which acetone was produced, using the calcium salt. Another route, favoured by both Germany and Canada, was via calcium carbide produced in electric arc furnaces. This was then used to produce acetylene, which was hydrated to acetaldehyde followed by catalytic oxidation to acetic acid. The acetic acid was then either converted to calcium acetate and dry distilled, or later catalytically converted to acetone by passing the vapour over heated alumina or thorium oxide (Fig.).

At the end of the war, the demand for acetone fell dramatically even though new uses were found. Acetone was now being used as a solvent for cellulose acetate in the production of 'aircraft dope' and other varnishes and in the production of textiles and isoprene. Meanwhile, butanol was being converted to butyl acetate for use as a solvent for nitrocellulose lacquers which were extensively used on cars, urea-formaldehyde resins, plasticizers and brake fluids.

During the post-war years, acetone and butanol continued to be produced by fermentation. However, after the second world war, the emphasis shifted to chemical routes because using petrochemicals was much cheaper. During the last fifty years the scope and tonnage of petrochemical production has expanded to become absolutely fundamental to the organic chemical industry, with all the major commodity organics and derivatives coming from that source. However, in South Africa production of - butanol by fermentation continued and 10,000 tonnes of butanol/acetone are still produced every year.

It is important to note that there are special factors involved in this operation. For example, South Africa has had political problems in importing oil and thus any process which minimizes oil usage is of particular value. Secondly, South Africa has an available supply of feedstock and, equally importantly, it has cheap coal for raising steam.

The economic viability of any process for producing large tonnage, low added-value materials is particularly dependant on the cost of feedstocks, the capital cost of the plant and the expense of effluent treatment. It is difficult to predict future costs of oil and gas, which are currently subject to substantial change.

Classically, grain or molasses have been used as feedstock for fermentation; both commodities are expensive. If other feedstocks such as woodwaste and paper are considered, then a different picture emerges. Considerable tonnage of paper waste is generated, most of which is still incinerated or used as landfill. Likewise, substantial amounts of sawdust and wood chippings are produced by the forestry industry. Demand for these cellulosic materials is not high, thus the price is low. Sawdust and wood chippings can be obtained at a low price, depending on seasonal demand. Unfortunately, the costs of collection and transport are high because these materials have a low bulk density and relatively high moisture content.

Over the last few years, researchers have studied the breakdown of lignocelluloses with a view to its use as a fermentation feedstock. A particularly interesting approach has recently been promulgated by D.I.C. Wang of MIT. He has demonstrated the use of a mixed culture system using *Clostridium thermocellum* to degrade cellulose to hexose and pentose sugars, together with *Clostridium thermosac-charolyticum* to produce ethanol and traces of acetic and lactic acids, from the monoaaccharides. Both organisms are obligate anaerobes operating effectively at 60°C. It is possible that such an approach might be successfully modified to produce acetone and butanol. For such a process to have any real chance of success, it would have to be a continuous operation. An outline of the process is shown in Fig.

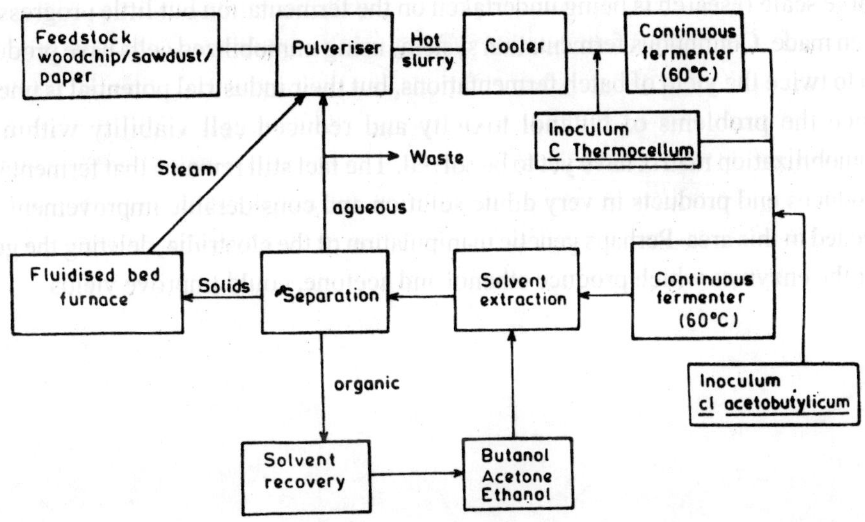

Fig. Flow sheet for the production of Acetone and Butanol from cellulose waste.

Some problems still remain. Firstly, butanol is very toxic in relatively low concentrations and thus markedly limits the fermentation. The first industrial strain of C. acetobutylieum fermented at the most 3.8 per cent of the starch substrate to yield 1.2 per cent n-butanol. At higher concentrations of butanol, the bacteria are killed. In the laboratory, a tolerance of 2.85 per cent has been achieved by adding activated charcoal to the fermentation medium. However, this has not been very successful at a large scale. Thus, technology for the continuous removal of solvents

Table

Some Selected Yeast Varieites found on Grapes/Wines

Variety of yeast	*Presence recorded*
Candida pulcherrima	Common in wines and musts; on grapes
Leucosperidium scottii	In Czechoslovakia; on green grapes from New Zealand
Endomycopsis lindneri	On grapes and in must in Brazil
Hansenula saturnus	Wines made from grapes
Saccharomyces bailiff	In spoiled wines and grapes
S. cerevisiae	The most common wine yeast and widely distributed
S. baillii (S. elegans var. baillii)	On grapes in Brazil
S. italicus	Grapes and grape Juices from tropical regions
Torulopsis stellata	On grapes and in wine.

must be developed, as must improved systems for dealing with stillage and effluent. Large scale research is being undertaken on the fermentation but little progress has been made. Continuous fermentation systems using immobilized cells have produced up to twice the yield of batch fermentations, but their industrial potential is unclear since the problems of butanol toxicity and reduced cell viability within the immobilization matrix have yet to be solved. The fact still remains that fermentation produces end products in very dilute solution and considerable improvements are needed in this area. Perhaps genetic manipulation of the clostridia, deleting the genes for the enzymes which produce ethanol and acetone, could improve yields.

SECTION - 2

HEALTH AND ENVIRONMENT

FOUR

Public Health Engineering

For a society, the two most important and immediate requirements are good food and health, besides other commercial requirements. Therefore, reorienting Public Health Engineering has been a crucial step towards health for all. According to World Health Organization, no less than four-fifths of the population still do not have access to health services. It is undesirable, for example, that while mortality in the first year of life is less than two per cent in advanced Western countries, it is 25 per cent in underdeveloped and just developing Asian countries. The reasons are multiple and may be as small as shortage of a particular drug, or as big as an inability to handle a particular health situation. However, DNA biotechnology is likely to make substantial contributions in solving both small and big health problems and it may not be overoptimistic to envisage a recommendation of WHO, and UNICEF to be that the main social target of governments, international organizations and the whole world community in the coming decades should be the attainment by all people of the world by the year 2000 of a level of health that will permit them to lead a socially and economically productive life. Brief and eventful discoveries relating to DNA genetic engineering and public health are outlined here.

Production of Human Insulin

In 1916, Sir Edard Sharpy-Schafer, while studying the Islets of Langherans of the pancreas, discovered and named the substance secreted, insulin. The antidiabetic role of insulin was reported in 1921 by Banting and Best in Canada. Insulin was for the first time applied to a diabetic nine year old boy in 1922 with success. It was immediately set on for production by Eli Lilly in 1923 and animal insulin (from Pancreas) was marketed. About 800-1000 Kilograms of Pancreas produce 100 gm of insulin. Since then, it has been produced commercially by several pharmaceutical concerns.

In the history of protein chemistry, the year 1954 will go down as a landmark, for that year Frederick Sanger and his coworkers at Cambridge University finally succeeded in achieving the first complete description of the structure of a protein molecule insulin.

Insulin is one of the smallest proteins. However, its formula is sufficiently formidable (Fig.). The molecule of the animal (Cattle) insulin is made up of 777 atoms, in the proportions 254 carbon; 377 hydrogen, 65 nitrogen, 75 oxygen and 6 sulphur (C254H377N65O75S6). The total number of amino acid units in the molecule is 51 in two polypeptide chains, 21 and 30 amino acids units long (Containing 17 different types) Table Sanger's work was not only to discover the overall chain configuration of the insulin molecule but also to learn the sequence of all the amino acids in the chain. It is made up of two polypetide chains A and B. It has already been synthesized.

Table

Composition of Insulin Molecule (Amino Acids)

Formula	***Name***	***Abbreviation***
$CH_2(NH_3^+)$. COO^-	Glycine	Gly
CH_3—$CH(NH_3)$. COO^-	Alanine	Ala
CH2OH—$CH(NH_3^+)$. COO^-	Serine	Ser
CH_3 . CHOH—$CH(NH_3^+)$. COO^-	Threonine	Thr
CH_3 CH—$CH(NH_3^+)$. CCO^- CH_3	Valine	Val
CH_3 CH. CH_2—$CH(NH_3+)$. CCO^- CH_3	Leucine	Leu
CH_3 . CH_2 CH—$CH(NH_3^+)$. CCO^- CH_3	Isoleucine	Ileu
CH_2—CH_2 CH_2 CH — COO^- NH^+	Proline	Pro
CH CH CH C. CH_2—$CH(NH_3^+)$. CCO^- CH CH	Phenylalanine	Phe
CH CH HO—C C. CH_2—$CH(NH_3^+)$. COO^- CH CH	Tyrosine	Tyr
NH_2CO. CH_2—$CH(NH_3^+)$. COO^-	Asparagine	Asp (NH_2)
COOH. CH_2 . CH_2—$CH(NH_3^+)$. COO^-	Glutamic Acid	Glu
NH_2 . CO . CH_2 . CH_2—$CH(NH_3^+)$. COO^-	Glutamine	Glu (NH_2)
NH_2—C—NH . CH_2 . CH_2—$CH(NH_3^+)$.COO^- NH	Arginine	Arg
CH C . CH_2 — $CH(NH_3^+)$. COO^- NH N CH	Histidine	His

Formula	***Name***	***Abbreviation***
$CH_2NH_2.CH_2.CH_2.CH_2—CH(NH_3^+)\ .\ COO^-$	Lysine	Lys
COO- $\quad\quad\quad\quad\quad\quad\quad\quad$ COO^- CH - CH Z - S - S - CH2 - CH NH 3+ $\quad\quad\quad\quad\quad\quad\quad\quad$ NH_3^+	CyS Cystine CyS	 I

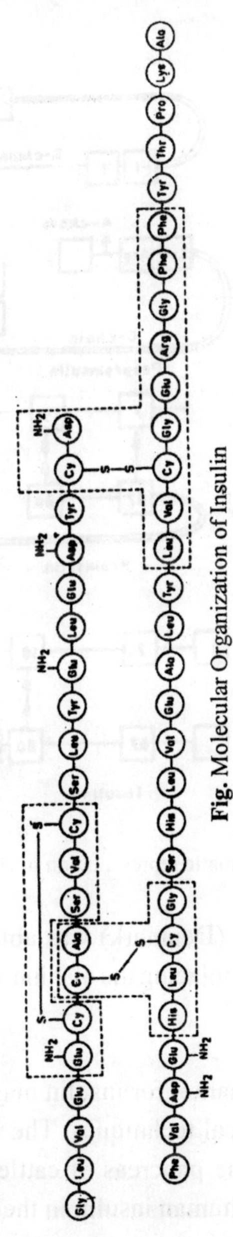

Fig. Molecular Organization of Insulin

The two polypeptide chains of insulin were originally derived from preproinsulin (with 109 amino acids). As preproinsulin is synthesized in the beta cells of the pancreas, the first 23 amino acids serve as a signal to allow the passage of the molecule through the cell membrane. These amino acids are cleared off, leaving the chain of proinsulin (86 amino acids). The proinsulin molecule folds up to bring the first and last segments of the chain together and the central portion of the molecule is cut out by enzymes to leave insulin. The role of this central portion is to align the two chains of insulin correctly (Fig).

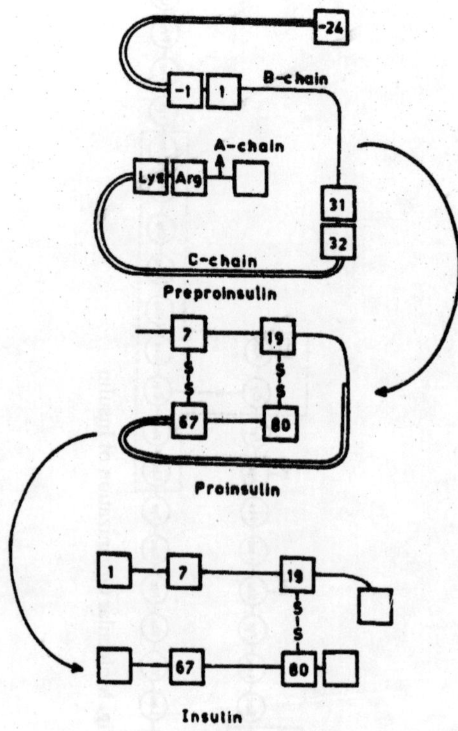

Fig. Schematic representation of Biosynthesis of insulin

In 1980, Novo Industry (Denmark) was able to transform swine insulin into 99% pure human insulin by replacing the alanine residue which is the thirtieth amino acid of the B chain.

The economics of the market for insulin may be fundamentally altered by the application of microbiological techniques. The insulin currently used in diabetes therapy is extracted from the pancreas of cattle and swine. The insulin of these species differ slightly from human insulin in their amino acid sequence. Although animal insulins are effective in controlling the major symptoms of diabetes, they do not prevent some of the ancillary effects including deterioration of the kidneys and the retina. More than 20% of the diabetics all over the world are dependent on bovine and porcine insulin injection. Porcine insulin is similar to human insulin in that the C-terminal threonine on the cx chain is replaced by alanine, while bovine is different at three positions. Besides bovine, insulin shows a greater immunogeneticity.

Harvesting DNA Biotechnology for Public Health Engineering is not easy as compared to porcine. All patients treated with injections of bovine insulin develop circulating anti-insulin antibodies which are responsible for subcutaneous fat atrophy and also neutralize insulin in the circulation. Consequently, the patient needs a higher dose in due course. Moreover, some diabetics are allergic to animal hormones.

A few years ago Roberto Crea, Adam Kraszewski, Tadaaki Hirose and Keiichi Itakura of the city of Hope National Medical Center in California synthesized genes coding for insulin polypetide. Eighteen fragments of a few nucleotides each were assembled to make the gene for the longer chain and 11 fragments were joined into a gene for the shorter chain. Each synthetic gene was linked to a plasmid near the end of a betagalactosidase gene of E. coli. After gene expression and the translation of mRNA into protein, the two polypeptides were cleaved from the enzyme and linked to form the complete insulin molecule. Approximately 1 x 105 insulin molecules were synthesized per E. coli cell.

Other workers (Gilbert and Vitta-Komaroff) also separated mRNA from the beta cells of the pancreas of rat; complementary DNA was prepared with the help of reverse transcriptase, and a copy of this DNA was inserted into pBR322 plasmid in the middle of plasmid in the middle of a gene for penicillinase. The plasmid also contained the structural genes for proinsulin. The hybrid protein synthesized in the bacterial cell was penicillinase + proinsulin from which insulin could be separated by trypsin.

If human insulin manufactured by bacteria proves effective in controlling these pathologies, it will almost certainly gain a substantial share of the world market for insulin. Eli Lilly of the United States is already in the forefront for human insulin production by gene transfer mechanisms. (Fig. 9.3). A conservative estimate is that if the yields can be increased to the level of those of other industrial processes that employ E. coli, more than 2000 litres of fermentation broth could yield 100 grams of purified insulin. This is the amount extracted from some 1,600 pounds of animal pancreatic glands. Recently, at the centre for Applied Microbiology and Research,

Porton Down (in U.K.), high yields of insulin were obtained; upto 200 grams of protein from 1,000 litres of fermentation broth. Several other biotech industries such as Genetech, Novo Industry and Biogen have also launched organized programmes for bacterial production of human insulin. Eli Lilly has started selling Humulin the commercially produced human insulin since 1982, and it has been approved by United States Food and Drug Administration department.

Production of Interferon

Interferon was discovered by Alec Issacs and Jean Lindenmann in 1957. When interferon was discovered, it was realized that it was possible to interfere with viral replication without necessarily endangering cellular metabolism and many thought that interferon might prove to be a useful antiviral agent. However, a major problem was in material supply and it remained so until the cloning of interferon genes in E. coli in the late 1970's.

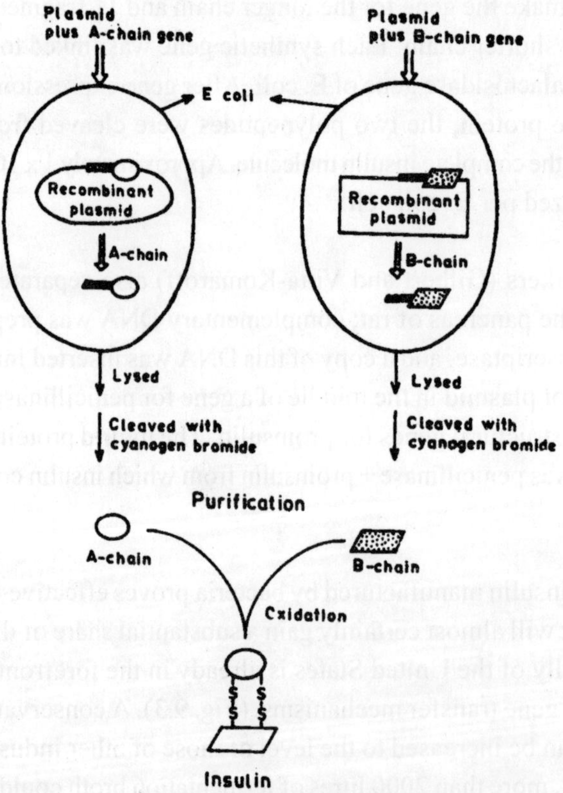

Fig. Production of human insulin

The term interferon originated in connection with viral interference and it generated a lot of interest because of the enormous potential of its action against viral invasion. It consists of a group of small proteins secreted by cells in response to viral infection. It can also be produced by sensitised lymphocytes when exposed to a particular antigen under laboratory conditions. Interferon can be produced by treating the cells with double stranded RNA. It is produced by most body cells on exposure to viruses and triggers a reaction that neutralizes not only the viruses that induced its formation but also many other viruses.

Interferons are a group of small proteins with molecular weights between 20,00030,000 daltons, and are broadly classified on the basis of physicochemical and antigenic properties into three types:

Interferon — (IFN-α)
Interferon — (IFN-β)
Interferon — (IFN-γ)

subtypes of any interferon are classified as IFN 1 or IFN 2.

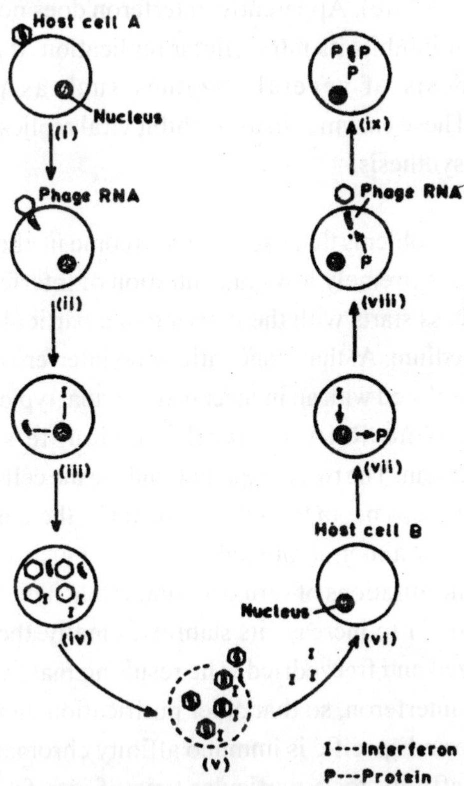

Fig. Steps in production of interferon

Interferon (originally referred to as leucocyte interferon) is best understood and was the first interferon to be purified. It is produced by separating WBCs from human blood and exposing them to Sendai Virus. The viral capsid stimulates a human cell chromosome (presumably number 5) to produce interferon. However, these interferons are unable to affect viral replication in the cell in which they were produced. These interferons are absorbed by adjacent cells where they induce chromosome number 21 to produce an unknown protein that binds to mRNA molecules specified by the virus (Fig.). Interferon is finally released in the culture medium and is purified using monoclonal antibodies as described in the previous chapter. It seems that IFN-a consists of a family of at least five proteins with different but homologous amino acid sequences (IFNa1-IFNa5). The complete amino acid sequence of IFNa1 (166 amino acids) and IFNa2 (165 amino acids) was determined in 1980. Partial sequence analysis of IFN-B has also been made. Interferon-y is produced by sensitised lymphocytes in response to the sensitising antigen or to non-specific mitogens. It is a mediator of cellular immunity and is one of the major lymphokines. Interferon is a very powerful antiviral agent (effective at concentrations 10-'2 to 10-" moles per litre). Apparently, interferon does not prevent a virus from infecting the cell, but inhibits its intracellular replication. It acts as a hormone and triggers the synthesis of several enzymes such as protein kinase, and phosphodiesterase. These enzymes in turn inhibit viral replication by degradation of mRNA and protein synthesis.

One of the major problems that had to be overcome in obtaining interferon from cultured cells was the extremely low concentration of interferon made by the cells. The production process starts with the growing of a particular cell line for about a week in a nutrient medium. At that stage little or no interferon, is being synthesized. The nutrient is then replaced with an inducer medium that typically contains a mixture of polyinosinecytosine RNA (a synthetic double-stranded RNA) and diethylaminoethyl dextran. The two compounds induce the cells to start manufacturing interferon. Before the ceps begin to excrete interferon the medium is replaced once again with a medium containing additional substances (such as insulin and guanosine phosphate or low concentrations of serum albumin) that have been found to increase the yield of interferon or to increase its stability. Finally, the medium is collected, concentrated; dialyzed and freezedried. The resulting material contains only about 0.1 per cent of pure interferon, so that other purification methods are needed. One of the most effective and specific is immuno affinity chromatography. Monoclonal antibodies witn an affinity for a particular type of interferon can be attached to polysaccharide beads, which are placed in a glass column. When the crude interferon solution is passed through the column, the interferon molecules are absorbed on the beads while the impurities pass through the column. The interferon is released from the column by altering the pH of the column with a suitable washing solution. By this method, it can be purified by 5,000 fold.

Human interferon was successfully produced in 1980 by two American scientists Gilbert and Weissmann by cloning the genes in colon bacilli. Similar results were also obtained by Japanese workers a few months later, with a comparatively higher yield. The work triggered a wave of experimentation in Israel (Weizmann Institute) and France (Pasteur Institute). At the same time, the nucleotide sequence of the genes of IFN-a and IFN-B was determined by Taniguchi from Japan, Houghton from England, and Derynck from Belgium: Both IFN-a (Leucocyte interferon) and IFN-B (Fibroplastic interferon) had quite a similar structure and 14 different genes coding for human IFN-a are known as reported by Weissmann from Biogen.

The molecular biology of synthesis of interferon took a vital position when human leucocyte interferon was engineered by yeast cells for the first time. A DNA sequence coding for human leucocyte interferon (LeIF-D) was attached to the yeast alcohol dehydrogenase gene in a plasmid and introduced into cells of Saccharomyces cerevisiae (The plasmid could also replicate independently in E. coli.). These yeast cells could synthesize about a million molecules of interferon (on a per cell basis) and the yield was several fold higher than that produced earlier by Japanese workers (About 10,000 per cell). The production was relatively slow in E. coli (1 x 105 per cell) because in yeast, it is easy to grow and replicate glycoproteins derived from mammalian cells. In yeast the mechanism of glycosylation (addition of carbohydrate group) is similar to that in animal cells. The DNA sequence for 7-interferon has also been determined and is different from a-interferon. However, in contrast to β-interferon, 'y interferon gene is split with three introns and four axons located on human chromosome (12), y interferon gene has also been cloned in E. coli.

In 1982, a host of scientists from Imperial Chemical Industries (ICI) and University of Leicester in England reported the complete synthesis of a gene coding for human leucocyte interferon (IFNa1) comprising 514 nucleotides being the largest gene synthesized. The synthetic gene has been demonstrated to function normally when integrated into a plasmid and introduced into E. coli and Methylophilus methylotrophtis.

Interferon is perhaps the body's first line of defense against viral attacks. It is likely to boost clinical trials in several ways. Workers from the Central laboratory of Public Health in Helsinki (Finland) procured 0.1 gm of pure interferon from approximately 50,000 litres of human blood. It has been calculated that processing of 2 litres of human blood, can produce about 4 million units of interferon (A unit is the amount of interferon that will protect 50 per cent of the cells in culture against virus infection). A large number of firms now are manufacturing leucocyte interferon in the United States including Genetech, Interferon Sciences, Schering Plough as well as Roche Institute of Molecular Biology. In Japan also, interferon has been manufactured by Yamanouchi Pharmaceutical Company, Tokyo. In Europe, there are a large number of public and private concerns which are attempting/have produced interferon such as Searle Laboratories and Wellcome Laboratories in U.K., University of Louvain in Belgium, Sanofi Company in France, Hoechst in Federal Republic of Germany and Kabi Vitrum AB in Sweden. Israel is also on the world market for successful production of Interferon (Table) However, the major difficulty is how to reduce the cost of production to produce interferon to be utilized by an average customer.

Table

Various International Biotech Companies producing Interferon

Cetus & Shell	U.S.A.	α–2
Biogen	U.S.A.	β
Collaborative Genetics	U.S.A.	β
Genetech	U. S. A.	α, β, γ
Dauchi Seiyaku	Japan	γ
Kyowa Hakho Koygo	Japan	β
Toray Industries	Japan	β, γ
Pasteur Institute	France	β
Transgene	France	γ
Searle	U. K.	β
Kabigen AB	Sweden	γ, β
Weizmann Institute	Israel	β

Potential Importance of Interferon

1. Interferon has a direct role to play in regulating the immune system and suppresses the multiplication of both B and T lymphocytes in vitro. It is also known to inhibit antibodies (formation) vitro.

2. Interferon has been known to inhibit further multiplication of cancerous cells. It is assumed that interferon kills the cancerous cells by increasing host immunity. There are a special type of cells known as natural killer cells (NK) which are specialized lymphocytes which offer major resistance against the formation and

multiplication of cancerous cells. Basically, NK cells differ from lymphocytes and macrophages. Interferon is now considered to stimulate the activity of-NK cells both by increasing the potential ability of preexisting cells and by the induction of new cells.

3. Interferon is very effective against viral diseases. As a matter of fact, it acts directly by stopping the virus. It has been used for quite a while in Russia, for the treatment of common colds and influenza. It is also possibly tried against hepatitis and herpes zoster. Interferon can be inoculated by in travenous, intramuscular injections or even as nasal spray (as done in U.S.S.R.)

Production of Somatostatin

Hormones are chemical substances made in particular cells that function metabolically by interaction with target cells or organ elsewhere in the system. Many hormones are proteins. Recent studies have provided important new insights regarding protein hormone biosynthesis and function. Some important hormones and their triggering functions are listed in Table

Table

Several Hormons with Regulatory Functions.

Metatonin	Regulates circadian rthythims
Somatoliberin	Stimulates somatotropin secretion
Somatostatin	Intutrats gastrin seretion Instute Somatotripic secretion
Lipotropin	Fatty acid release from adipocytes
Somatotropin	General anabolic effects, stimulates release of growth factors
Thyroxine and Triodothyromine	General stimulation of many cellular reactions
Pastrin	Stimulates and secretion from stomach and pancreatic secretion
Secretin	Regulates pancreas secretion
Insulin	Glucose uptake, Lipogenesis, General anabolic effects
Glucagen	Glycogenolysis, release of lipid
Estrogen	Maintenance of pregnancy
Androgen	Maturation and Function of secondary six organs
Relaxin	Muscle tone
Epinephrine	Smooth muscle contraction, heart function, glycogenolysis and lipid release.

Initially, it was thought that hormones bind directly to specific rate limiting enzymes, thereby either activating or inactivating them. However, this idea has now been abandoned and the concept of a receptor introduced.

Receptors are proteins that bind a hormone with high specificity and affinity. All hormones act by binding to macromoleculer receptors that are located either on the cell membrane or on the inside of responsive cells. The receptors for all polypeptide hormones and most amino acids derivatives are located on the cell membrane. The first human peptide hormone to be synthesized in a bacterial cell was somatostatin, which has been studied in detailed by workers from Hope National Medical Centre, at the University of California at San Francisco. Somatostatin is one of a group of hormones made in the hypothalamus at the base of the brain. The somatostatin gene is inserted into the bacteria by means of an expression vector constructed partly from a plasmid. Somatostatin is a hypothalamic hormone, 14 amino acid units long, that controls the release of several hormones from the pituitary. The gene was synthesized from eight 'blocks of single-stranded DNA fragments made up of a few nucleotides each, which are indicated in this simplified diagram as A-H. The fragments had overlapping complementary sequences to allow for correct assembly of the gene. Of the 52 base pairs in the gene, 42 make up the code for somatostatin; the remainder provide the two "sticky ends" that allow the gene to be inserted into the plasmid and include the information needed for proper expression of the gene and recovery of the hormones. The expression vector was constructed from the plasmid pBR322, to which was added the control region and most of the betagalactosidase gene from the bacterial lac operon. Beta-galactosidase is an enzyme involved in lactose metabolism; the control region contains the regulatory elements needed for expression of the beta-galactosidase gene. The somatostatin gene was inserted into the plasmid next to the beta-galactosidase gene. After the plasmid was introduced into cells of the bacterium Escherichia coli, the human hormone was synthesized as a short peptide tail at the end of the enzyme. Cleavage with cyanogen bromide freed the hormone, which has been shown to be identical with the molecule in human beings (Figs.). The synthesis of many hormones is ultimately regulated by the brain through neurohormones. Somatostatin is then transported in the blood to the pituitary gland, where it acts to inhibit the release of insulin and human growth hormone.

Production of human growth hormone (Somatotropin)

A deficiency of the pituitary growth hormone (Somatotropin) results in a form of dwarfism that can be cured by administering the hormone. The hormone is species

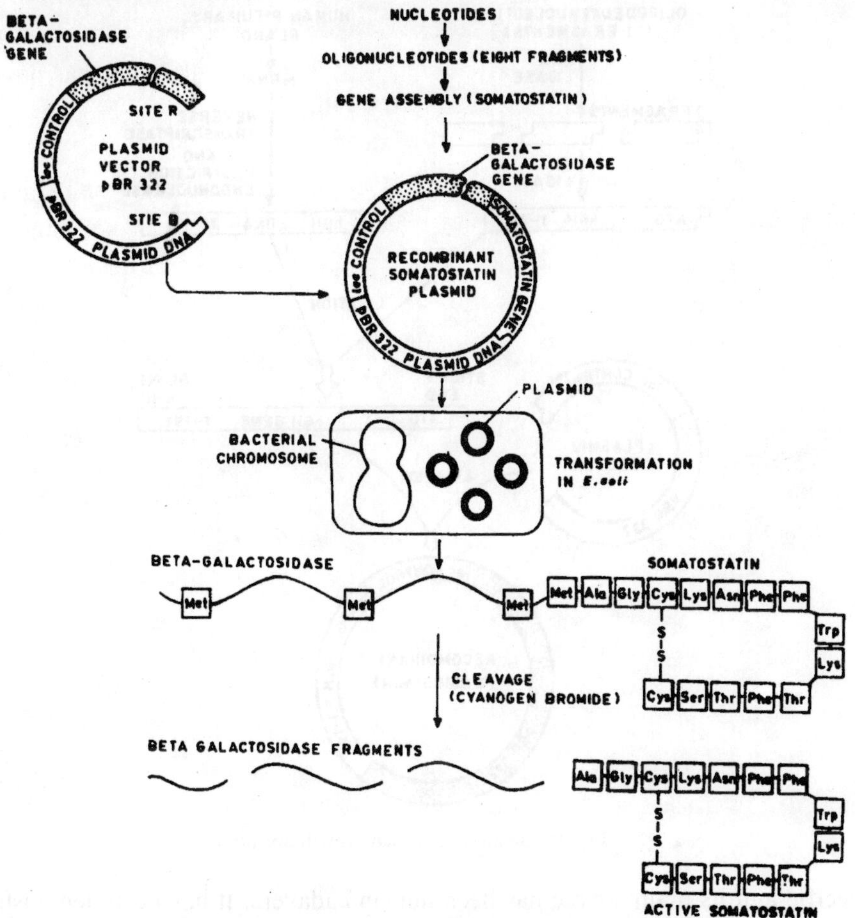

Fig. Production of Somatostatin by Recombinant DNA Technology

β-galactosidase

NH_2 —~~~~~~— Met — Ala — Gly — Cys — Lys — Asn — Phe — Phe — Trp — Lys — Thr — Phe — Thr — Ser — Cys — COOH (Cys–S–S–Cys)

Cleavage with Cyanogen bromide

NH_2 — Ala — Gly — Cys — Lys — Asn — Phe — Phe — Trp — Lys — Thr — Phe — Thr — Ser — Cys — COOH (Cys–S–S–Cys)

β-galactosidase fragments + Somatostatin (14 amino acids)

Fig. Cleavage with CNBr and production of Somatostatin

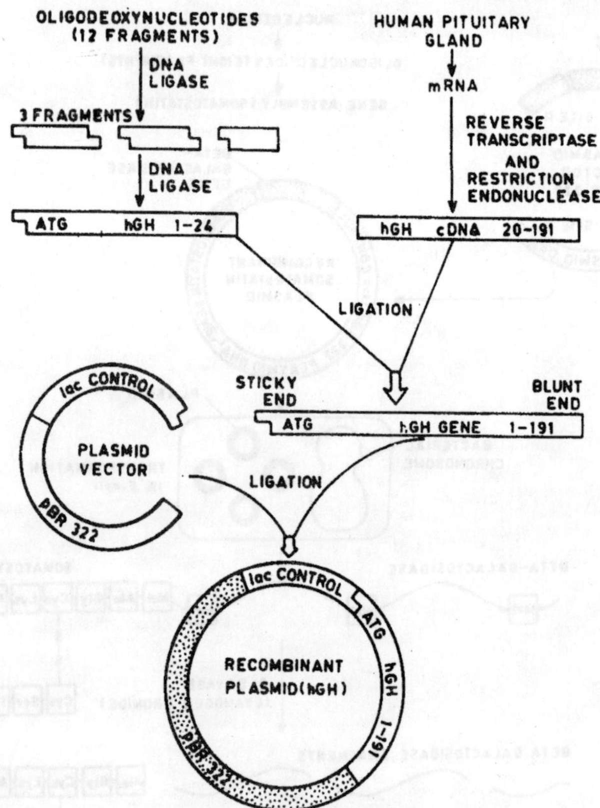

Fig. Production of human growth hormone

specific and its main source has been human cadavers. It has been demonstrated that an intramuscular dose of 6-10 mg per week can increase the stature by more than 6 cms during the first year of treatment. Recently, Stomatotropin gene was synthetically produced by a combination of chemical synthesis and isolation of natural molecule. This is a polypeptide of 191 amino acids (worked out in detail by Goeddel and workers at Genetech). The segment of gene that codes for the first 24 amino acids of the peptide was constructed chemically from blocks of nucleotide. As shown in Fig. recombinant hGH plasmid was constructed by usual genetic engineering techniques. Reverse transcriptase was employed to copy the gene for the hormone from mRNA obtained from human pituitary tissues. Restriction endonucleases cut out the needed fragment and DNA ligase was then used to join the natural and synthetic fragments. The complete gene was inserted into a modified version of plasmid pBR 322 incorporating the !ac operon. The synthetic part of the growth-hormone gene had been constructed with its own initiation Codon (ATG). The hormone could, therefore, be produced independently in bacterial cells, without the need for attachment to a bacterial protein. This growth hormone may have many clinical uses, but the extremely limited availability of the substance is a serious limitation. Microbiological production may not only increase the commercial availability of the drug but also make it possible to investigate in detail its potential applications. Currently, several industries including Genentech, Pasteur Institute, Kabi Vitrum AB Biosynthesis of somatotropin might reduce the cost of treatment.

Production of other hormones

With the advent of techniques of recombinant DNA and gene cloning, several other human hormones are being produced on a commercial scale by isolating specific DNA sequences coding for those proteins/hormones (Table). This is likely to enable clinical applications and better economic provision for their utility for several deficiencies. Therefore, production of Beta-endorphin (a 31-amino acids brain opiod) by gentically engineered colon bacilli has been widely acclaimed (see Chapter). Workers from Australian National University Canberra; University of California, San Francisco, and University of Columbia, New York were able to clone the DNA of beta-endorphin in E. coli and transcribed it alongwith Bgalactosidase. The beta endorphin which was released from the hybrid protein and purified had the biological activities of endorphins and reacted specifically to antiserum against beta-endorphin.

Table

Some Other Hormones produced by Inserting and Gene Cloning in Bacteria

Somatomedia A	Stimulate the fixation of sulphur in cartilage
Somatomedia B	Stimulating the synthesis of DNA in cyclical cells
Somatomedia C	Stimulates the incorporation of proline into collagen and of uridine into DNA
(Hypothesic Growth hormone-releasing Factor).	Could make up the deficiency of Somatotropin
Throotropin (THRF)	Regulates Thyroid gland
Pancreas growth Hormone-releasing factor (hpGRF)	Possible treatment of Pituitary dwarfism to accelerate the reconstitution of tissues of people with severe burns
Thymo poietin	Thymic hormone

Production of Vaccines

Vaccines are preparations of either dead organisms or living attenuated microorganisms that may be administered to man or animals to stimulate their immunity to infection by the same or closely related organisms (Table). The word immune is derived from the Latin stem immuno, meaning for safe or free. It suggests that once an individual's immune system has responded to a particular antigen, a state of resistance will remain for a long period of time. A vaccine is an important immunizing agent. For example, viral vaccines consist of either inactivated viruses that are incapable of multiplying in the body (such as Salk poliovaccine) or of attenuated viruses that multiply at low rates (Sabin oral polio vaccine) in the body, but do not cause the symptoms of disease. Bacterial vaccines fall into similar categories; the whooping cough vaccine consists of dead cells and the tuberculosis vaccine is composed of live, attenuated organisms. Except for the whooping cough vaccine that gives long term immunity, bacterial vaccines provide temporary protection. Small pox vaccine, commonly prepared from the vaccinia lesions on the skin of inoculated calves or sheep or from the allanoic membranes of inoculated chick embroys, induces a high degree of immunity in vaccinated individuals and has been highly effective in controlling the disease. Measles vaccine, also prepared from an attenuated live virus, is likewise highly effective and produces a degree of immunity that is demonstrable by measurements of antibody titers usually for a period of 8-10 years or more. Both these vaccines produce immunity by infection with live virus and thus are long acting. Influenza, on the other hand may be prevented by the

Table

Current and Future Vaccines or Protective Agents

Diseases	*Current Vaccine or Protective Agent*
Cholera	Killed whole organism
Diphtheria	Toxoid-absorbed
Tetanus	Toxoid-absorbed
Pertussis	Killed bacteria or bacterial fraction
Viral hepatitis	
Type A	Passive immune serum globulin (ISG)
Type B	Passive immune serum globulin (ISG)
Influenza	Egg-grown virus, formalin inactivated, highly purified by zonal ultracentrifugation
Mumps	Vaccine
Measles (rubeola)	Live attenuated virus, chick embryo or canine tissue-culture grown
Plague	Formaldehyde-inactivated Yersinia pesos
Poliomyelitis	Inactivated virus Attenuated virus monovalent or trivalent
Rabies	Active: (I) B-propiolactone inactivated virus in embryonated duck eggs (2) Phenon-inactivated virus grown in rabbit brain Passive: equine hyper-immune serum
Rubella	Vaccine
Typhoid	Whole organism killed by several different techniques
Typhus	Formaldehyde-inactivated Rickettsia prowazekii grown in embryonated eggs.
Yellow fever	Live attenuated virus prepared in chick embryo: Dakar strain or 17D strain

parenteral injection of vaccines of influenza viruses that have been grown in chick embryos and rendered noninfectious by formalin or ultraviolet irradiation. With such a dead vaccine, the duration of immunity rarely exceeds a year and since the causative agent (virus) of influenza is capable of mutating frequently, different vaccines usually must be prepared to combat each mutant strain. While antibiotics and vaccines are both used in modern medicine to counter infectious diseases, vaccines are used as prophylactic agents to increase immunity and to prevent subsequently infection with more virulent strains. Antibiotics are employed most often as therapeutic agents to control disease after its onset, although they may also be used prophylactically against some organisms.

Transgene, a French enterprise, has been aiming at producing a vaccine against rabies; in 1982 the workers strived to synthesized surface protein of the rabies virus. They isolated a clone of E. coli that produced a 58,000-dalton viral protein of which the encoding DNA had been introduced into the bacteria. Rabies is a zoonois that occurs endemically in animals and humans in several countries of Africa, South America and Asia and is increasing in European countries. Preventive vaccination is impeded by the high cost of the current vaccines which is primarily due to difficulties countered in virus propagation in human cells besides difficulties in viral purification. Therefore, the production of a cheap vaccine from a viral protein with the required immunogenic properties would be useful from the industrial and public health view points.

Rickettsial vaccines for Rocky Mountain, spotted fever, and typhus fever are also available but allergic reactions often arise from residual egg protein in which the rickettsiae have been cultivated. This has not been achieved successfully for the causative microbe of syphilis, Treponema pallidum, nor has it been possible to make vaccines with conventional techniques relating on the extraction and purification of antigens. The DNA of this spirochete was isolated from infected rabbit testicles and cloned in E. coli cells with the aid of a coliphage used as a vector. This has enabled the development of more specific diagnostic screening tests for syphilis and the manufacture of an effective vaccine. The need for such a vaccine is particularly important as the strains of Treponema tend to become resistant to the antibiotics currently in use for the treatment of syphilis. In veterinary medicine, efforts are also being made to produce vaccine in genetically modified microbes which could be effective against animal diseases such as food and mouth disease and swine fever. Foot and mouth disease is highly contagious and afflicts about thirty species of cloven-hoofed animals worldwide and is especially threatening for cattle, sheep and swine. The virus causing this disease belongs to the Aphthovirus (single stranded

RNA) genus of the family Picornaviridae. Recently, German scientists have been successful in cloning double-stranded DNA copies of the viral RNA into plasmid pBR 322 and grown in E. coli. It was found that more than 1,000 molecules of viral protein were synthesized per bacterial cell with a positive serological test.

The virus is surrounded by a capsid made of sixty copies of our polypeptides, VP1, VP21 VP39 VP4. Treatment of the virus with trypsin results in the cleavage of VP, and a considerable decrease in infectivity and a loss of immunizing activity. It has, therefore, been advantageous to synthesize the polypeptide procursor of VP, and to use it as the protein conferring long lasting immunity. A lot of work is currently going on to improve the immunizing ability of VP, protein. A vaccine made up of VP, has also been synthesized by genetically engineered bacteria by workers from Plum island Animal disease Center, New York. Intervet International, a pharmaceutical company from The Netherlands has produced two vaccines, cloned in bacteria, to prevent infections diarrhoea of Swine and calves caused by pathogenic strains of E. coli. A similar vaccine has also been produced by Cetus in the United States.

The development of synthetic vaccines, produced from genetic recombination and gene cloning, is likely to increase their medical utility and availability. The development of synthetic vaccines has long term prospects as far as replacement of current bacterial and viral vaccines. Producing vaccines using peptides with specific immunogenic properties is also being investigated.

Recently, the new approach for producing vaccines has been the cell culture technique. The new vaccines based on cell culture are purer and safer than some of the existing vaccines. In the existing vaccines, all the antigens obtained after treatment of animals with particular bacteria or virus causing the disease are used. This causes "overloading." In the new method, cells are cultured with the specific bacteria or virus and only those antigens or antibodies, that are specifically required against that disease are injected into the human body.

FIVE

Environment

Pollution Control

Environment related biotechnology is associates with both the implications and applications of biotechnology in the wider context of environment. Due to rapid industrialization, urbanization and other developments, there is a constant threat to the clean environment and to the depleting natural resources. In this connection, a reference may be made to the following two conferences: (i) The First Conference on the Human Environment was held in 1972 at Stockholm, where Late Smt. Indira Gandhi, the - then Prime Minister of India called poverty to be the biggest pollutant; (ii) After 20 years, in June 1992, United Nations Conference on Environment and Development (UNCED) was held in Brazil, where heads of the states from 166 countries participated to examine the issues involved and the solutions possible. While on the one hand, there is an increasing problem of control of environmental pollution, there is also a problem of conservation of nature and natural resources. Both these problems are receiving constant attention of environmentalists. Among implications, there is also an alarm due to release of genetically engineered organisms in the atmosphere and also due to the release of effluents from biotechnological companies, so that the environmentalists are having a debate on the effects of developments in biotechnology on the environment. There is also a debate on the safety of the use of the products of biotechnology, an area described as biosafety. Among applications, on the other hand, efforts are also being made to use biotechnology to protect the environment from pollution and to conserve natural resources. At a time, when the gap between those who have plenty and those who do not have even the minimum is widening, both ends of this spectrum i.e. plenty and poverty are contributing to environmental degradation. It is, therefore, necessary that the developing and developed countries jointly find a path of development which "meets the needs of the present without compromising the ability of future generations to meet their need" (World Commission on Environment and Development). Efforts are being made to achieve this objective through a variety of approaches, and biotechnology is certainly one of them. In this and the next three chapters, environmental implications and applications of biotechnology for environment will be discussed.

In recent years, we have witnessed a debate on the environmental implications of biotechnology. In this debate, risks involved in the use of biotechnological approaches have often been emphasized (or even overemphasized) and the adequate guidelines for safety have been suggested and enforced by law. These implications have partly been discussed. However, there have also been rapid developments in the applications of biotechnology, which may help in controlling environment pollution, thus giving a cleaner and sustainable environment in future. According to one estimate in USA, the US market for environmental clean up applications was expected to grow at an average rate of 17%, while that for microbes and enzymes was expected to grow by only 7% every year. Besides others, these applications for environment clean up include biotreatment methods for effluents and toxic wastes. However, these treatments, it is feared, could be problematic, where they involve deliberate or accidental release of genetically modified microbes to the environment. These applications of biotechnology in environment management (pollution control) and the risks involved in their use will be briefly discussed in this chapter.

Cleaner Technologies

Fermentation technology

In several fermentation technologies used currently, there are serious environmental implications, so that cleaner technologies for fermentation have been devised using biotechnological approaches, thus reducing the risk of environmental damage. This will be illustrated using the example of Provesteem, a single seed protein (SCP). This was initially produced by yeast fermentation and patented in 1972 by Provesta Corporation, a subsidiary of Phillips Petroleum in USA. Provesteem is 60 per cent protein and also contains essential vitamins and minerals. It can be fed directly to farm animals or used as a dietary supplement for human consumption by mixing it with flour, meal or locally packaged food. This SCP is derived from feedstocks like methanol or ethanol, which in their turn are obtained from petroleum or natural gas. However, in countries with little or no hydrocarbon reserves, biomass in the form of agricultural or foresty wastes can be converted into alcohol feedstock to be used for production of SCP. Sugars like glucose or sucrose can also be used. Biotechnological processes have been devised, in which all nutrients introduced for fermentation are retained in the final product, ensuring high conversion efficiency and low environmental impact.

Paper industry

In paper industry, the current pulp bleaching technologies are being replaced by a better technology. The objective of all pulp-processing operations is to efficiently remove lignin without damaging valuable cellulosic fibres. Various chemical and mechanical processes are currently used to release cellulose from its encasing lignin matrix (this process is called pulping), but they suffer from serious disadvantages, including damage to cellulosic fibres, high costs, high energy use and corrosion. In this connection, a legnin degrading and modifying enzyme (LDM) isolated from Phanerochaete chrysosporum was found useful, since it may reduce energy costs and corrosion thus increasing the life of the system and also reduce environmental hazards associated with bleach plant effluents. Bioreactors have been designed, where environmentally problematic processes have been replaced by those which are environment friendly. Arrangements have also been made to treat effluents from this system.

Plastic industry

Conventional process for plastic industry uses oil-based raw materials. Alkenes like ethylene and propylene, produced from these raw materials, are first converted into alkene oxides, which are then polymerized to form plastics, such as polypropylene (used for making containers) and polyethylene (commonly called polythene). The use of these raw materials has inherent danger of escaping into the atmosphere, thus causing pollution. Therefore, it is suggested that in plastic industry raw materials like sugars (glucose) should be used, which can be enzymatically (or through direct use of microbes) converted into alkene oxide to be used in this industry. Methylococcus capsulatus has also been successfully used for converting alkene into alkene oxide.

Reducing Impact—Industrial Effluents

In order to control pollution, industrial effluents, which contain toxic substances, must be treated to render them harmless. Alternatively, sources of the toxicity of effluents may be attacked. The undesirable waste characteristics of industrial effluents include the following: (i) Suspended solids and soluble organic compounds, which undergo progressive decomposition and thus result in oxygen depletion and production of noxious gases; (ii) Heavy metals, cyanides, and other toxic organics (e.g. chlorinated compounds in effluents from paper industry), which are deleterious to aquatic life; (iii) Undesirable levels of nitrogen (N) and phosphorus (P), which enhance eutrophication (excessive plant growth, which kills animals due to deprivation of O_2) and stimulate undesirable algal growth; (iv) Nonbiodegradable chemical and volatile materials like H_2S and SO_2. The subject of industrial effluent treatment, in view of the above four classes of pollutants, is vast and can not be treated at length here. Therefore, a few examples will be used to illustrate the subject.

Treatment of distillery effluent

The distillery effluent does not contain toxic substances but has high oxygen demand and is, therefore, harmful. It consists of two components : (i) spent wash, containing 9-10% solids, 80,000-90,000 ppm C.O.D. and 40,000-60,000 ppm B.O.D.; (ii) fermentation sludge (0.3 kg per kilo litre of rectified spirit produced) with 25-30% solids and 100,000-125,000 ppm B.O.D. (B.O.D. = biological oxygen demand: C.O.D. = chemical oxygen demand). The disposal of these two wastes into land or water is a great environmental hazard. The standard laid down by the Government of India for disposal are 100 ppm B.O.D. for land disposal and 30 ppm for disposal in natural water streams. The treatments used for distillery effluents are summarised below

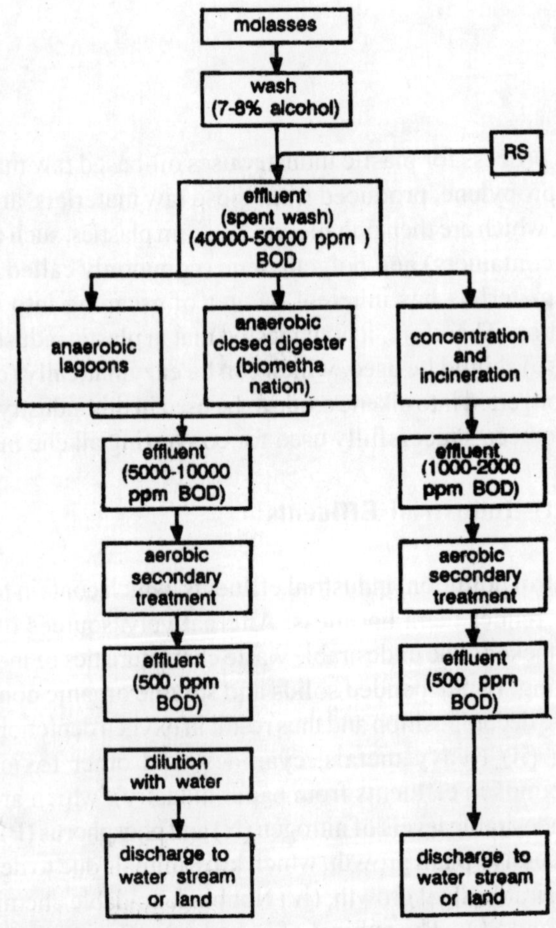

Fig. Different methods of distillery effluent treatment.

Table

Primary and secondary treatments used for disposal of distillery effluents.

	Primary treatments		*Secondary treatments*
1.	Anaerobic digestion (in lagoons or closed digestors)	1.	Anaerobic filtration
		2.	Primary and extended aeration
2.	Concentration and incineration (burning to ashes in a furnace, etc.)	3.	Sunbed drying

Primary treatment— Anaerobic digestion of effluents in lagoons reduces B.O.D to an extent of 80-90%. This process has advantages of relatively low capital investment and ease of adoption and operation, but requires large area of land for lagoons. Therefore, anaerobic treatment in closed digestor involving generation of biomethane gas is more suitable (methane can also be used as a fuel for domestic and industrial uses). However, concentration and incineration requires high capital investment, so that this may not be teehno-economically viable under Indian conditions.

By primary treatment, BOD is brought down to 5,000-10,000 ppm, which is still too high for disposal, keeping the standards fixed (100 ppm for land and 30 ppm for water).

Secondary treatment— Secondary treatment involves aerobic treatment and may be used to bring down further the BOD. However, this is energy-intensive and capital-intensive. The standard fixed as above, can not be achieved even after secondary treatment.

Under the Biotechnology Programme, anaerobic digestion (biomethanation) and secondary treatment have been recommended by DGTD as a national programme in India. This will be techno- economically viable and will help in pollution control and generation of non-conventional energy in the form of methane. A number of Indian distilleries, therefore, have already entered in approved collaboration with foreign companies/ institutes for generation of methane gas from distillery effluent.

Reducing impact of paper industry

Bleaching of chemical pulp is an important step in paper industry. Pulp bleaching is normally performed as a sequence of treatments in order to achieve brightness. Chlorination, alkaline extraction, chlorine-dioxide treatment are some of the most common bleaching stages. In recent years, oxygen bleaching under alkaline conditions has also been used. During bleaching of pulp, therefore, huge amounts of chlorine are ultimately converted to chlorinated organic compounds, which, to a substantial degree , are discharged to receiving waters, imposing a great threat to the environment.

Toxic chlorinated compounds are measured as total organic chlorines (TOCI) or absorbable organic xalogens (AOX). Normally 5 kg of TOCI/AOX is discharged per tonne of bleached pulp, in addition to 300 different organic compounds and dioxins. In view of this, more stringent restrictions are being imposed on release of waste bleach waters, so that there is an urgent need to reduce the impact of pulp bleaching on the environment. Following two approaches are suggested to deal with this problem.

Purification of waste bleach waters— ***(a) Biological methods:*** In paper industry, waste bleach waters are currently treated using aerated lagoons (having bacteria) and activated sludge plants. These methods, however, can not remove high-molecular mass chlorinated materials. In view of this, the use of white-rot fungi is recommended, since they can degrade polymeric lignins. However, white-rot fungi have not been used on a commercial scale, due to their complicated physiological demands (energy source), which are difficult to satisfy on a large scale.

(b) Ultrafiltration combined with biological treatment: Ultrafiltration of the effluents of paper industry reduces TOCI from 50% to 35-45% (3 kg/tonne pulp), which can be further reduced to 2 kg/tonne pulp, by combining ultrafiltration with oxygen bleaching. The effluent, however, is still acutely toxic, so that further biological treatment is needed. Following biological methods are available : (i) use of aerated lagoons, which are expensive; (ii) fungal treatment, particularly filamentous fungi (e.g. white-rot fungus = Phanerochaete chrysosporum), which can be filtered off and recirculated to the process.

New techniques for pulp bleaching— ***(a) Substitutes for chlorine:*** Chlorine is being replaced by oxygen or by chlorine dioxide for bleaching pulp for paper industry. This replacement may decrease the colour of the effluent, the amount of total chlorinated organic compounds (TOCI/AOX), and the amount of dioxins. Chlorine dioxide may still produce chlorate, which is harmful. Oxygen and chlorine dioxide may also be replaced by lignolytic fungi.

***(b) Enzymes facilitating pulp bleaching*:** Lignolytic enzymes and xylanases are two sets of important enzymes which reduce lignin content in pulp and thus, if used for pulping, help in pulp bleaching. While lignolytic enzymes degrade lignin, xylanases attack lignin-xylan complexes commonly found in hardwoods. Xylanases attack xylans and remove it thus making it easier to remove lignin. It is believed that xylanases will prove extremely useful in pulp bleaching in future.

(c) Trees easier to pulp and bleach: In future, with better understanding of lignin biosynthesis and deposition, one may be able to obtain forest trees with (i) lower lignin content; (ii) lignins of modified methoxyl content or (iii) reduced association between lignin and the hemicelluloses. These superior trees will be obtained using genetic engineering and tissue culture techniques and should prove useful for paper industry.

Reducing heavy metals' pollution

The waste effluents of several modern industries also contain heavy metals like mercury, lead and cadmium, which cause poisoning. For instance, mercury causes metal-poisoning, which attacks the nervous system of patients. Lead causes mental retardation of children. The use of lead free petrol is one measure to check this pollution, but these heavy metals are also present in factory-effluents, and need to be removed, before these effluents are discharged to natural land or waters.

The property of some species of bacteria and algae, to extract metals from their surroundings, has been utilized to purify industrial effluents. Biotechnological approaches are recommended, where metal extracting forms (mainly algae) can be grown in ponds, where factory effluents (rich in heavy metals) are discharged. The microbes will extract the heavy metals and sequester them inside their cell-membranes. The metal can be subsequently recovered from these microbes.

Biodegradation of pollutants

Biological treatment of effluents is a long-established practice in many countries, but some constituents of these effluents are calcitrants and are thus not amenable to conventional treatments. Biotechnology helps in overcoming this problem. Following examples will illustrate the use of biotechnology in biodegradation of pollutants : (i) In USA, a company BioTechica is using lignin degradation, for treatment of substances like polychlorinated biphenyls (PCBs) and dioxin; (ii) In Europe, companies like ICI and Ciba-Geigy are working on enzymatic detoxification to breakdown substances such as cyanides and also the byproducts from the synthesis of S-triazine herbicides; (iii) Microbial transformation of biarylethers, cyclic biarylethers, biarylketones, halogenated dibenzodioxins and dibenzofurans are used to avoid the problem of release of effluents with pollutants. For instance, Pseudomonas strains have been isolated, which selectively deoxygenate the 1,2 positions of substituted biaryethers anqwqd biarylketones (e.g. biphenylether, xanthene, dibenzofuran, dibenzodioxin). (iv) Microbial degradation of chloro-, dichloro- and trichloromethanes and carbontetra-chloride is also used to deal with the problem.

Toxic Site Reclamation

Since the toxic constituents of industrial effluents are known in most cases, their treatment is easy. However, the waste disposal sites are more difficult to deal with, since their chemistry is often complex. Although incineration (drying and then burning to ashes in a furnace) or chemical treatments are suggested for treatment of these sites, biotechnological approaches involving biodegradation is considered more attractive. Companies like BioTechnica are working on treating polluted sites in situ. In this connection, it is interesting to note that the first patent ever awarded was given to Dr. Ananda Chakrabarty of General Electric (USA) for developing a microbe designed to biodegrade spilled oil (see later in this chapter for details).

There has been a debate on whether or not to release genetically engineered microbes for treatment of toxic sites. It is argued that biotechnology products parallel toxic chemicals in many respects and unlike chemicals, released engineered organisms have the capacity to reproduce and spread beyond the initial release sites and may even transfer novel genes into other naturally occurring organisms, thus causing the risk of developing what are described superbugs.

In view of the above, 'Occidental Chemical Corporation' company of USA is conducting their work, on effluents from toxic waste sites (Love Canal and other polluted sites in USA) in closed reactors. One such experiment was conducted at Hyde Park landfill site at the northwest corner of Niagara, New York. This site occupies just over 6 hectares of land and contains an estimated 73,000 tonne of chemical waste. The site was used from 1953 to 1975 to dispose off a wide range of waste materials including chlorinated organics. For treatment of this site, a clay cover was placed on the landfill in 1978, and a leachate collection system was installed around the perimeter in 1979. From this system, 230 m^3 of leachate per week was shipped to a nearly treatment facility in Niagara, where the treatment involved carbon absorption, thus making the effluent suitable for discharge. This solution was not found suitable on a long term basis. Therefore, the Hyde Park wastewater was subjected to biotreatment in sequencing batch reactors (SBRs), where biodegradation rate of some more persistent chemicals was accelerated by adding bacterial strains isolated from the landfill site. Further, genetic manipulation of these organisms produced more than 100 recombinant bacteria, which have been tested for their usefulness in reactors. It was demonstrated that significant cost savings will be possible, if SBR treatment is used before the carbon absorption stage.

Several other companies that are reported to be active in toxic site clean up in USA include the following : Detox Industries, Kerr-McGee, Polybac and Sybron Chemical. The developments in this field have regularly been highlighted in several international symposia held during the last decade.

Removal of Spilled Oil and Grease Deposits

Oil spills from oil tankers on land (e.g. beaches) and from distant oil spills have been recognized as "environmental hazard". This spilled oil is believed to not only destroy the habitats of aquatic animals and fish, but also create health problems for local residents and causes long term damage to the environment. Therefore, remediation efforts in this connection have been made, chemical dispersant are believed to cause major pollution problems in shallow waters due to their toxic nature and persistence in the environment. Efforts have been made in the past to wash the oil off the gravel and rocks in the area of oil spill, but this has been found to be expensive and time consuming. In view of this, if possible, environmentalists would prefer the use of microbiological degradation of oil, which will allow slow removal of oil from the environment, particularly from the beaches or from the shallow waters. In order to achieve this objective, following methods have been used:

Use of Oleophilic Fertilizers

In early years, application of oleophilic (oil loving) fertilizers as food for oil utilizing microbes was considered useful, since this would allow rapid growth and multiplication of idigenous microbes, thus speeding up the biodegradation process for removal of oil. When, 6000 gallons of petrol was leaked from a company in Pennsylvania in USA, it posed a serious health problem due to contamination of underground water supplies. Addition of oleophilic fertilizers on the sites expedited the biodegradation of petrol, thus cleaning the toxic site in a year, which would have otherwise taken several years. In recent years, however, using genetic engineering oil utilizing microorganisms have been successfully produced, which would grow rapidly on oil. These genetically modified strains could not be tested in the field for removing the oil due to the debate concerning the release of genetically engineered microorganisms in an open environment.

Many oil utilizing microorganisms may also produce surface active compounds that can emulsify oil in water and thus facilitate removal of oil. A strain of Pseudomonus aeruginosa has been developed by Ananda Chakrabarty at University of Illinois, Chicago (USA), which produces a glycolipid emulsifier, that reduces the surface tension of an oil-water interface and thus helps in removal of oil from water. This microbial emulsifier is non-toxic and biodegradable, thus being ecologically sound. It was demonstrated by laboratory experiments that this microbial surfactant releases oil to an extent of 2 to 3 times that by water alone. Thus, it has the potential for future use, whenever its release is allowed.

Use of a mixture of bacterial strains

A mixture of bacterial strains has also been used to clean oil contaminated water reservoirs (due to oil spills from ships) and water supplies. This could be illustrated using the following example. When 800,000 gallons of oil water was once accumulated in the bilges (bottom of ship) of a ship 'Queen Mary', there was a danger that this oily water will be discharged at the harbour causing death to marine life and spoiling beaches, rendering them unsuitable for human use. When a mixture of bacterial strains was introduced in the bilges of the ship, oil was decomposed within six weeks, leaving water with innocuous chemicals, safe to be discharged at the harbour.

A mixture of bacterial strains is also being used for removal of grease deposited on the inner side of tanks in a ship. The technique may also prove useful for cleaning deposits of grease in pipes and vessels of a variety of industries.

Use of genetically engineered microbes

The use of genetically engineered microbes for cleaning spilled oils will be facilitated by mixing the microbe with straw, which can he stored. When needed, the straw with mixed microbes can be scattered over the spilled oil, so that the straw will first soak oily water and then the microbe will break up the oil into harmless, non-polluting materials, rendering the water and the site harmless.

Reducing Environmental Impact of Chemical Herbicides, Pesticides and Fertilizers ('Birational' or 'Environmentally Safe' Weed and Pest Control)

Majority of chemical herbicides, pesticides and fertilizers that are used commercially cause environmental hazards, because the toxic chemicals used as herbicides, pesticides or fungicides, and fertilizers undergo degradation by microorganisms or ultraviolet light releasing pollutants in the environment. Following solutions to this problem of pollution are being tried under biotechnology programmes.

Weed control and herbicides

Herbicides are being developed which will be environmentally safe. To allow the use of these herbicides for crop protection programme, genetically engineered herbicide resistant plants have been produced in a number of crops, which should eventually allow the use of environmentally safer herbicides. These plants are being field tested. Similarly, genetically engineered insect resistant plants have also been successfully produced in some crops thus raising the hope for reduced use of insecticides in future.

Pest control and biopesticides

Bacterial pesticides are also being developed (even though they are slow in action and are expensive) and a company in U.S.A. with the name Ecogen Inc. established in 1983, made some progress in this direction. Heliothis complex, which lives in close association with plant roots consists of two major crop pests, namely budworm and ballworm. Biological insecticides against both these insects are being prepared by transfer of a gene from Bacillus thuringiensis (Bt), into either a naturally occuring soil bacterium, or into a strain of Pseudomonus. Bt insecticides are already in the market for 30 years and in future these will be modified through techniques of genetic engineering and will be utilised against a variety of insects. Several companies in USA including Ecogen, Monsanto, Mycogen, Repligen and Zoecon are involved in development and marketing of biological pesticides. Among these, while Monsanto was using genetically engineered live soil bacteria for coating seeds before planting, Mycogen kills recombinant bacteria and applies them to the leaves of crop plants. Both these approaches protect the toxin from degradation by microorganisms or ultraviolet light, once applied to crop plants.

Viral pesticides have also been successfully developed and used for the control of pests. These pesticides are environmentally safe and avoid risk of toxicity. These pesticides can also be used against the pest strains which have become resistant to chemical pesticides. A number of entomopathogenic viruses (viruses which can

Table

Some viral pesticides, with the names of pests they control

Class of virus	*Pest*	*Trade name*	*Crops*
1. Nuclear polyhedrosis viruses (NPVs) (subgroup A of Baculovirus)			
(a) Baculovirus heliothis	Heliothis	Elcartm, Biotrol VHZ. Viron/H	Cotton & other crops
(b) Gypsy moth baculovirus	Lymantria dispar	Gypcheck	Hardwood trees
(c) Sawfly baculovirus	European sawfly (Neodiprion sertiferis), N. lecontei and other sawflies	Polyvirocide	Forest or tree plantations
(d) Brassica NPV	Trichoplusia ni	Biotrol-VTN	Brassica
(c) Cotton NPV	Spodopter litura	Biotrol-VSE	Cotton
2. Granulosis viruses (GVS) (subgroup B of baculoviridae)	Cydia pomonella	-	Apple, vegetables
3. Cytoplasmic polyhedrosis viruses (CPVs)			
CPV of pine	Thaumetopoea pityocampa Dendrolimus spectabilis (and others)	-	Pine, spruce, etc.

cause infection to insects) belonging to the family Baculovirdidae (baculoviruses or nuclear polyhedrosis viruses = NPV) and family Reoviridae (cytoplasmic polyhedrosis viruses or CPVs) have been used as safe and effective pesticides. These viruses kill specific pest species and have no adverse effects on useful insect pollinators, insects yielding useful products, parasites or predators, warm blooded mammals or even man. They are safe even in large scale spray operations. However, large scale production of these viruses demands either a living host or its cells for multiplication. Virus infects and kills host (pest) by gaining entry through mouth into the digestive tract. Some of these entomopathogenic viruses, which are used in pesticides are listed in the table on the previous page.

Biofertilizers

Biofertilizers are also being used to decrease the cost of fertilizer application and to reduce the environmental hazards caused by chemical fertilizers.

Biosensors to Detect Environmental Pollutants (Environmental Monitoring and Biomonitoring)

Biosensors represent biophysical devices, which will detect the presence and measure the quantities of specific substances in a variety of environments. These specific substances may include sugars, proteins or hormones in human body, pollutants in abiotic component of the environment including air, soil and water and its variety of toxins in the industrial effluents. Since living organisms are capable of sensing the presence of a variety of substances, they have been profitably used for developing devices, which will help in environment management, in health care and also in developing more efficient industries. In designing a biosensor, an enzyme or any antibody or even microbial cells are associated with microchip devices, which are used for quantitative estimation of a substance. The subject is so important that an international congress, described as 'Biosensors 90' was held in Singapore during May 2-4, 1990. Second World Congress on Biosensors, i.e'Biosensors 92' was held in 1992. A journal called 'Biosensors and Bioelectronics' is also being published to give the desired emphasis to this subject.

A biosensor equipment consists of (i) its biological sensing agent; (ii) a device for collecting the product obtained from the interaction between the substrate and the biosensor and (iii) a device for measuring the quantity of this product, thus indirectly giving an estimate of the substrate. While biosensors have found their use in medicine as well as in industry, their use in environmental monitoring is particularly attractive. Based on biosensors, kits have been designed to identify specific pollutants. For instance, enzymatic tests have been designed to detect pesticide contamination in water, and cost only 3.00 U.S. dollars per test as against 150300 U.S. dollars for a laboratory test. Thus using biosensors, pollution problem can be detected at an early stage, so that the environmental monitoring is possible.

In India, the Central Electrochemical Research Institute, Karaikudi has developed a glucose biosensor based on glucose oxidise enzyme. The enzyme is immobilized on an electrode surface, acting as an electrocatalyst for oxidation of glucose. The biosensor gives a reproducible electrical signal for glucose concentrations as low as 0.15 mM and works for several weeks with no appreciable degradation of enzyme activity.

Another related aspect is 'biomonitoring' or 'biological monitoring (BM)', which is defined as the "measurement and assessment of work place agents or their metabolites either in tissues, secreta, excreta or any combination of these systems" in occupationally exposed subjects (human beings). While biomonitoring refers to uptake, distribution, biotransformation, accumulation and excretion of toxic agents, their biological effects in workers exposed to these agents are often described as `Biological Effect Monitoring (BEM)'. Fully evaluated biological monitoring methods exist today to assess the risk to which employees are exposed due to exposure to the toxic agents present at the working place. They can provide an important complement to environment monitoring in research as well as policy recommendations.

Environment and Energy

The economic prosperity of any region is reffected by the level of its energy consumption. With the Industrial Revolution in the Western Hemisphere, there was a quantum jump in the consumption of fossil fuels (non-renewable energy sources) like coal, petroleum and gas. Although during 1920s, in the total energy supply of the world, coal was at the peak, but at present, it accounts for only 26% while oil shares 40% of the world energy needs. These fossil fuels, according to one estimate, are depleting at a rate 100,000 times faster than they are being formed and are therefore, classified as non-renewable energy sources. The depleting energy resources coupled with increasing demand for energy are posing a threat to the global ecosystem in terms of climatic changes and health hazards. In view of this, there is a conflict between economic development and environmental issues, so that in the interest of protecting the environment, various policies for energy supply, use, and conservation are being framed, both in the industrialized and developing countries.

In the year 1973, when the first oil crisis was witnessed, the oil price was doubled and quotas were introduced by OPEC countries. At this time oil constituted nearly half of the world's annual share of primary energy consumption. The growth rate of oil consumption during the period 1950-1973 was 7.5% per annum, this rate being higher (8.8%) in developing countries than that (6.9%) in the industrialized nations. In India, the demand for petroleum products is growing at a compound rate of 7% per annum. With two consecutive oil shocks in 1973 and 1979, oil consumption pattern changed in developed countries, where more efficient use of oil and its substitution by other fuels led to reduced consumption. However, in countries like India, due to non-availability of viable alternatives, the growth rate of demand for oil consumption has been inelastic despite the oil prices being the highest in international market.

Natural gas is another source of energy, and is emerging as an attractive substitute for oil in India. Thus, the future discoveries are likely to be more in the form of natural gas. At present, natural gas accounts for 17 million tonnes of oil equivalent and geological estimates indicate that the gas availability will increase to 30 million tonnes of oil equivalent by the year 2000. The natural gas is also preferred over oil, because it burns cleaner and offers great potential for conservation of energy.

India's per capita consumption of energy is 206kg of oil equivalent (oe) compared to world average of 1540kg. Oil continues to be the most important source of energy in India, which continues to import oil (in 1990-91, India imported Rs 10,000 crores worth of oil) and meets only 52% of its requirements through indigenous production. Projections on the energy demand by the end of the present century are alarming and are estimated to be 400 million tonnes per year for coal, 100 million tonnes per year for petroleum and 1,00,000 MW per year for power. This energy scenario poses a great challenge, not only to our technical and managerial abilities, but also to our environment, which is under great pressure.

In view of the above, although strategies are being discussed and implemented to meet our basic energy needs on short and medium term basis, we need to recognize that our hydrocarbon reserves can not last indefinitely. At the present rate of production and consumption, our oil reserves may hardly last for another 20-30 years. In view of this, we need to participate in developing non-conventional sources of energy, such as solar energy, wind geothermal energy, etc. The solar energy can be harvested in the form of biomass through the use of waste material and by growing energy crops (petrocrops). It can also be harvested through the use of microorganisms, which can convert solar energy into chemical energy in the form of molecules like glycerol (from salt, H_2O, and CO_2) and hydrogen (from H_2O). These energy sources are renewable and offer less polluting alternatives, since petroleum consumption and also production of varied organic substances by fermentation (this leads to production of CH_4 and CO_2) lead to environment pollution. These aspects make an important component of biotechnology programmes and will be discussed in this chapter. (There are several other aspects of energy including 'conservation of energy' and 'energy technology policy', which are important for a complete discussion on energy. These aspects are not relevant for a discussion on biotechnology and, therefore, will not be discussed in this chapter.).

Renewable Sources of Energy

Biomass production and its utilization for energy

In recent years, a great deal of research effort has been witnessed for increasing the production and utilization of biomass as a source of renewable energy. The biomass is defined as living matter or its residues, which is a perpetual or a renewable source. The common examples of biomass are wood, grass, gerbage, grains or bagasse. The main sources of biomass can be classified in two groups: (i) waste materials including those derived from agriculture, forestry and municipal wastes; and (ii) growing energy crops involving short rotation forestry plantations.

Waste materials for energy— The waste material includes a variety of wastes originating from plants, animals and domestic or municipal wastes. The plant wastes include wood, green plant matter, straw, paddy husk, rice bran, saw dust, etc. Waste wood can be converted into methanol liquid fuel. Similarly, green plant matter and straw are used as animal feed or converted into an efficient energy source through anaerobic digestion. In 1989, in India a rice straw bred thermal plant at a cost of Rs 35 crore was inaugurated at Jalkheri, Punjab. This will generate 62 million units of electricity per year and is the only plant of its kind in the world. Paddy husk (obtained during milling of paddy) can be converted into smokeless solid fuel briquettes suitable for use in domestic cooking, hotels, kilns and boilers (in India, this is done at Briquetted Fuel Plant installed at Alternate Hydro Energy Centre, Roorkee). This process is based on principle of pyrolysis and involves heating the raw material at 250-300°C in absence of air. The fuel briquettes are sold at Rs 0.60 per kg, which is cheaper than firewood and coal. Similarly, rice bran (obtained during polishing of rice) is used as a source of oil, which is converted into a methyl-ester for use as fuel. Saw dust is also converted into low calorific value producer gas for thermal power generation, based on the principle of partial combustion. Animal waste is also a potential energy source, although the high water content in the animal waste makes it unsuitable for many conversion processes except anaerobic digestion used for biogas production (see later in this section). Domestic and municipal wastes are also used as alternative energy sources. This may not prove to make any significant contribution in western countries, but will certainly prove significant in Asian countries where it is available in huge quantity due to population explosion. Since 1988, Central Mechanical Engineering and Research Institute, Durgapur has successfully run India's first pilot plant using solid garbage (leafy waste) to produce electricity. The plant used about 500 kg garbage and can product 5 kwh of electricity per hour. The garbage undergoes anaerobic digestion to produce biogas which in turn is used to produce electricity.

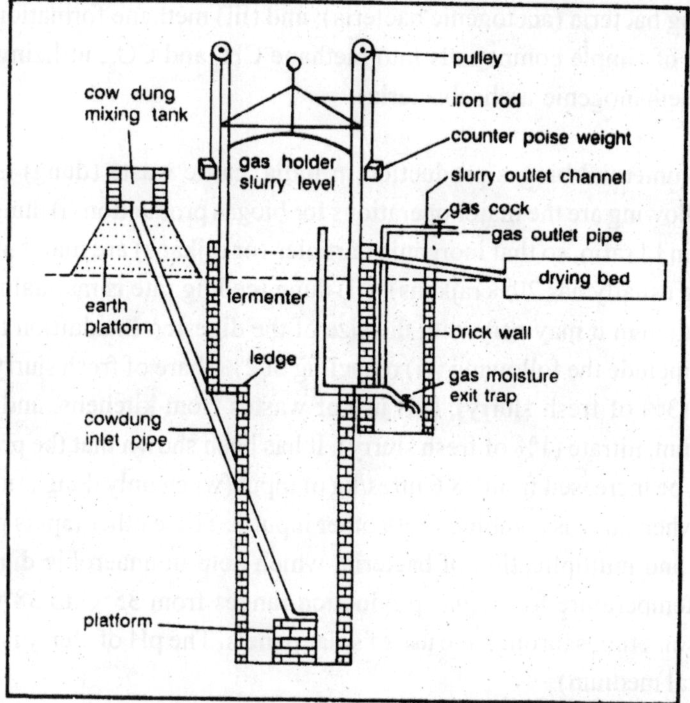

Fig. Design of a gobar gas plant developed at Indian Agricultural Research Institute (IARI), New Delhi.

Biogas production— Biogas is a term used to represent a mixture of different gases (in varied composition) produced as a result of action of anaerobic microorganisms on domestic and agricultural wastes. It contains methane in bulk (50-68%) and other gases in relatively low proportions viz., CO_2 (25-35%), H_2 (1-5%), N_2 (2-7%) and O_2 (0-0.1%). It is believed that in China there are five million simple installations of biogas in rural areas. Similarly, in India a large number of 'Gobar Gas Plants' are being used by villagers not only as cheap source of energy, but also as a measure for pollution control, and for improvement of health and sanitation conditions. Biogas plants also provide highly enriched organic fertilizer. In 1961, a Gobar Gas Research Station was also established at Ajitmal (district Etawah in U.P.), which has designed a variety of 'gas plants' suited to Indian conditions and supplying electricity at a cheap rate. The digesters in 'gas plants' are almost entirely buried underground, with a fixed dome that serves as a gas holder.

Biogas is produced by anaerobic digestion of animal waste, as discussed above. It involves three steps: (i) hydrolysis, which converts organic polymers into monomers (with the help of hydrolytic bacteria); (ii) acid formation, which involves conversion of monomers into simple compounds such as CO_2, NH_3, and H_2, using a group of acid forming bacteria (acetogenic bacteria); and (iii) methane formation involving conversion of simple compounds into methane CH_4 and CO_2, utilizing anaerobic bacteria (methanogenic archaebacteria).

For commercial biogas production in India, cattle waste (dung) is the prime source. Following are the major operations for biogas production (i) dung and water are mixed in l:l ratio, so that inorganic particles (or rajkans) are maintained at 10% level (dung usually has 20% rajkans); (ii) dung feeding rate is maintained at 3,500 kg/ day, although it may vary with the size of the digester. In addition to dung, the input may include the following : (i) recycling of 2% share of fresh slurry, (ii) use of night soil (3% of fresh slurry), (iii) use of wastes from kitchens, and (iv) use of calcium amm. nitrate (1% of fresh slurry). It has been shown that the production of biogas can be increased from 38.6 litres/kg of input (when only dung is used) to 59.6 litres/kg (when dung is combined with other inputs). These other inputs prove useful in growth and multiplication of bacteria, which help in anaerobic digestion. The optimum temperature for biogas production ranges from 35°C to 38°C, which is achieved sometimes through the use of solar panels. The pH of slurry is maintained at 7 (neutral medium).

Large scale growing of energy crops—Hydrocarbon (energy) producing plants include a large number of Euphorbia and Asclepias species. These species bear latex and 35% of their dry weight can be extracted as organic extracts. After the water is removed, the latex gives rise to a liquid oil. Chemical analyses of extracts from E. lathyris show that 5% of its dry weight is a mixture of terpenoids and 20% is a simple sugar (hexose). The terpenoids can be converted into a gasoline like product and hexoses may be fermented to ethanol. The conversion of certain biomass derived gasoline-like materials into high quality transportation fuels has also been demonstrated. Consequently, 8-12 % of its dry weight can be obtained as oil, giving 20 barrels of oil/acre/year. Similarly, E. tirucalli can give upto 5-10 barrels of oil/ acre/year.

Sugar crops (like sugarcane and sugar beet) and starch crops (like corn) are also valuable as solar energy converters. An effective use of these renewable resources leads to conversion in several products that can be used as fuels, chemicals and other products. These crops not only give sugar which may be used directly or converted into ethanol (liquid fuel), but also give valuable by-products such as bagasse, which can be used as solid fuel.

The technology of conversion of sugar and starch into ethanol has been developed to utilize this ethanol as fuel alcohol in motor cars. In Brazil,.motorists fill their tanks with alcohol (95% pure) instead of petrol (motor engine can be modified at a small price, so that it can use alcohol). In other cases, a mixture of petrol and alcohol called gasohol can be used, without altering the engines. The 'gasohol programme' in Brazil, which started in 1975, showed tremendous success and is the world's largest single biotechnology programme. Many countries of the world are now trying to substitute alcohol for petrol (partially or fully). These countries include USA, East Europian Countries and Several countries in Africa.

'Short Rotation Forestry' is another approach used for biomass production. It involves growth of forest species (e.g. Eucalyptus, Populus, Pinus), which give rapid juvenile growth and can be regrown from harvested slumps. This type of biomass production is suited to a variety of climates and gives a very high yield. In recent years micropropagation using tissue culture techniques has proved very useful in propagation of these trees.

Cellulose as a source of energy— Cellulose is the most abundant organic material on the earth and is a major component of all plants. If fuel alcohol can be derived from cellulose, this will be a very flourishing industry. Each cellulose molecule consists of a few hundred to several thousand glucose units linked together. If cellulose can be degraded into glucose, the latter can be easily used for the production of fuel alcohol. Several fungi are known which possess the enzyme cellulase and, therefore, can degrade cellulose into glucose. However. cellulose in woody plants is intimately associated with lignin in the form of lignocellulose. An enzyme 'ligninase' has also been discovered, which can degrade lignin. Biotechnologists, therefore, are busy in developing the following two steps-process for converting cellulose (from trees, weeds, shrubs, straw, etc.) into fuel alcohol: (i) lignocellulose may be degraded with the help of certain fungi into sugars (action of ligninase and cellulase); and (ii) sugars may be fed to yeasts for fermentation to alcohol. Studies in several countries including USA and Sweden have shown great promise in this approach.

Energy and fuel using microorganisms

Hydrogen production— Hydrogen is also a renewable fuel, because the raw material in the form of water is abundant and solar energy used for the decomposition of water to produce hydrogen, will be available for millions of years. Furthermore, when used as a fuel, hydrogen gas causes no pollution and forms water thus renewing the raw material. In view of this, hydrogen (H_2) production is emerging as an important area of biotechnological research with respect to environment friendly non-conventional and renewable source of energy. In achieving this objective, the principle used in photosynthetic machinery by green plants is used. The chlorophyll helps in trapping solar energy in the form of ATP, which is utilized in photosynthesis. In this process, water is subjected to photolysis, which leads to splitting of water molecules into (i) oxygen, (ii) electrons and (iii) hydrogen ions (H^+). The hyrdogen ions do not get a chance to form hydrogen gas, but is used to form energy rich compounds like glucose. However, if these hydrogen ions can be converted into hydrogen gas (H_2), the latter can be collected and used as a fuel. Enzymes hydrogenase and nitrogenase have been tried to convert hydrogen ions (H^+) into hydrogen gas (H_2).

(a) Hydrogen gas using hydrogenase— Some microbes possess the enzyme hydrogenase, which helps the two electrons join two hydrogen ions to produce one molecule of hydrogen gas (H_2). These microbes include a number of algae (e.g. Scenedesmus, Chlamydomonas, Dunaliella. Porphylidium and Oscillatoria) and some photosynthetic bacteria (e.g. Rhodospirillium, Chromatium, Clostridium and Thiocapsa), which can be gainfully employed for hydrogen production, although their commercial exploitation is still at a very early stage of development.

The above microbes may be used for the isolation of hydrogenase enzyme, which can be used with isolated chloroplasts, so that the hydrogen ion (H^+) and electrons liberated due to the use of solar energy by the chloroplasts are made to combine into hydrogen molecules, with the help of hydrogenase enzyme. The hydrogen gas thus formed, will bubble out of the solution and can be easily collected. This system though presently works only for a few hours, but can be substantially improved to become commercially successful.

(b) Hydrogen gas using nitrogenase— In cyanobacteria (blue green algae), 'nitrogenase' enzyme is the chief hydrogen producing enzyme. This is considered better, because hydrogenase enzyme is sensitive to oxygen which is produced during the photolysis of water, while nitrogenase enzyme is relatively less sensitive to oxygen. Further, in the heterocysts of cyanobacteria, oxygen concentration is low or completely absent, so that hydrogen production becomes more efficient. The enzyme nitrogenase reduces nitrogen to NH_3 with associated release of hydrogen gas. Hydrogen evolution depends on light, which provides electron donors and also activates nitrogenase and ATP formation in heterocysts. It has been shown that light dependent hydrogen evolution can continue for several days in some strains of Anabaena (nitrogen fixing cyanobacteria).

Hydrocarbon production— Hydrocarbons produced by some microorganisms may also serve as liquid or gaseous fuels. For instance, microbial production of methane gas from polymers like carbohydrates, proteins, lipids, etc. involves following three steps: (i) breakdown of polymers into organic acids and alcohols by hydrolytic and fermentation bacteria; (ii) conversion of acids and alcohols into acetates, H_2 and CO_2 by the hydrogen producing acetogenic bacteria and (iii) conversion of acetate and $H_2 + CO_2$ to methane by methanogenic bacteria.

Hydrocarbons are also produced by unicellular alga Botryococcus braunii, which possesses a hydrocarbon content as high as 75% of its dry weight, the highest reported in any kind of biomass. Therefore, cultivation of this alga affords a direct and renewable source of 'solar fuel'. Several other algae and fungi are also involved in the production of hydrocarbons and can be used to produce hydrocarbons as a source of energy.

Conservation of Energy

A variety of approaches are also being used for conservation of energy to meet the energy demands more effectively. Even though most of these approaches do not fall within the area of biotechnology, a brief account is included here to develop awareness among readers that the development of new renewable sources of energy are not the only measures to deal with the problem of energy demand in this developing world. In order to understand the meaning of conservation of energy, we should clearly differentiate between the conservation of petroleum from curbs on petroleum consumption. The conservation measures should not lead to a situation where the economy and the developmental programmes start getting adversely affected. At times, the curbs on consumption imposed by the government could bring hardship to the consumers, for whom fuel is the most vital input/raw material.

Oil industry in India, in conjunction with other interested agencies, organized an 'Oil Conservation Week' during Feb 10-16, 1992. Similar other activities, from time to time, develop awareness about the problem of conserving energy. The use of a variety of lubricants and additives which increase the efficiency of fuels, employing efficient processes and equipment permitting low energy consumption and avoiding wastage of energy through leakage and proper insulation are some of the measures that can be adopted to conserve energy.

SIX

Restoration of Arid Lands

Increased human activity has affected the different ecosystems in a variety of different ways. One of these effects has led to degradation of habitats, including cultivated land. Land areas accounting for about 50 per cent in the world are arid with problems of salinity, acidity or aluminium toxicity. An area of 400 million hectares suffer with salinity alone. Restoration of these degraded lands, therefore, is a major concern both in the developing and developed countries. This is also essential due to rapid urbanization leading to decreasing area of available land for cultivation in rural areas.

However, the available conventional methods of reconstruction are relatively inefficient. With the advent of biotechnology, however, unprecedented opportunities for recovery of degraded ecosystem have become available through the manipulation of biological systems. In this chapter, various methods involving plant biotechnology for restoration of degraded land will be discussed. No attempt in this chapter will be made to discuss other conventional methods available for restoration of these degraded lands. The plant biotechnological methods will include the following: (i) reforestation through micropropagation and use of mycorrhizae; (ii) improvement of soil fertility through the use of nitrogen fixing bacteria Rhizobium in association with leguminous trees and Frankia in association with nonleguminous species; (iii) development of plants tolerant to abiotic stress, so that they may be grown on degraded lands; and (iv) use of selected and engineered microbes for removal and recovery of strategic and precious metals from contaminated degraded lands.

Reforestation through Micropropagation

The forests account for 29-34% of the land area on earth and for 19.52% (62.20 million ha) of the land area in India, where only 10.88% (35.79 million ha) is covered by closed forest. While on the one hand, forests are a source of goods and services (including firewood, fodder, pulp, and fibre industries), they also maintain climatic stability and conserve water and soil. Today, there is a widening gap between the supply and demand of forest based materials, and the forests are being denuded leading to degeneration of environment in a variety of ways. It has also been shown that the degraded lands can be effectively used and restored by planting forests using trees of wide adaptability and high productivity. In this manner, high yield of biomass and other plant products may be obtained from the degraded lands. For this purpose, it is recommended that clonal propagation of superior genetic stocks rather than seeds of uncertain genetic quality should be used. Even on the degraded lands, few trees grow well despite the adverse conditions, and these trees can be used for clonal micropropagation through the technique of tissue culture. It has been estimated that a multiplication rate of 10,000 to 100,000 per year is technically possible for many species. Towards this objective, superior genotypes of many species capable of growing well on degraded lands have been selected for mass multiplication. This clonal multiplication involves the following four steps: (i) initiation of aseptic culture; (ii) shoot multiplication using shoot apical meristem or buds; (iii) rooting of in vitro formed shoots and (iv) acclimatization and transfer of micropropagated plantlets in the field. It is estimated that in the year 1990, Std) minion plants of diverse nature have been produced in 60 countries. In New Zealand, large number of micropropagated plantlets of radiata pine (Pious radiata) are being used for reforestation and it is estimated that as many as two million plantlets were in the field in 1992.

Casuarina for Tropical Reforestation

In tropical and subtropical regions, where desert encroachment and soil degradation are major concerns, different species of Casuarina, which undergo nitrogen fixing nodulation, provide means to improve the fertility of nitrogen-deficient soils. Casuarina grows fast and yields upto 15 $m^3ha^{-1}yr^{-1}$ of wood even under poor conditions of soils. It provides excellent firewood with high calorific value and can be converted into excellent charcoal. Coastal sand dunes in tropical and sub-tropical regions can also be stabilized by some species of Casuarina. In Senegal and China, one million hectares of plantations with Casuarina equisetifolia have been established effectively. Similar efforts are also underway to transform some species of Casuarinaecae (e.g. Allocasuarina verticillata) to develop genotypes with wide adaptability.

Development of Stress Tolerant Plants

Efforts are also underway for the development of plants resistant to abiotic stresses through the use of techniques of tissue culture and genetic engineering. Salinity, acidity and aluminium toxicity are the most important among these abiotic stresses. Cell lines exhibiting resistance or relative tolerance to salt stress have been selected in a large number of plant species and can be used for plantation on degraded lands suffering with salinity. These plant species include the following: Brassica spp., Capsicum annum, Cicer arietinum, Citrus aurantium, C. sinensis, Daucus carota, Hordeum jubatum, H. vulgare, Nicotiana tabacum, N. sylvestris, Oryza .sativa. Prunus hybrids, Solanum tuberosum, Triticum aestivum, Vitis rupestris and Sapindus trifoliata.

There are also tree species, that exhibit moderate resistance to salinity and, therefore, can be used for plantation on saline soils. These include the following : Prosopis spicigern, Butea monosperma, Terminalia bellerica, etc. Efforts are also underway to understand the molecular basis of salt tolerance and to isolate genes responsible for this attribute, so that salt tolerant plants can be developed through the use of genetic engineering.

In vitro selection for tolerance to aluminium toxicity has also been successful in tomato, soybean, rice, barley, and wheat. Regenerates of selected lines can be used for growing plants in acidic soils with aluminium toxicity. Man made synthetic crop 'triticale' has also be found suitable for growing it (i) on acid soils in countries like Poland, Kenya, Ethiopia, India, Ecuador, Brazil, Mexico, etc.; (ii) on dry and sandy soils (e.g. Savanna region of Brazil and some marginal lands in Asia); (iii) on alkaline and calcareous soils (e.g. in Mexico, Spain, Portugal, USA); and (iv) on mineral (e.g. Cu, Mn, Zn) deficient and high boron soils (e.g, Australia). In China, a number of new varieties of rice, wheat and tobacco have been developed through anther culture, some of them showing stress resistance with broad adaptability. For instance, rice cultivars Zhong Hua 10 and Zhong Hua 11 have a high degree of drought resistance and have been planted on more than 1, 30, 000 hectares of land in China.

Use of Mycorrhizae in Reforestation

On degraded lands, that suffer with drought, poor nutrient supply and other abiotic stresses, there is usually only a brief period which is favorable for growth. If tree seedlings do not get established during the window period, they are unlikely to survive. Mycorrhizae can improve seedling survival and growth by enhancing uptake of nutrients (particulary phosphorus) and water, by lengthening root life and by providing protection against pathogens. Mycorrhizae are symbiotic non-pathogenic associations between plant roots and fungi. Trees in temperate regions form Ectomycorrhizae (eM) and 70% plant species including most of the commercial crops, growing in the tropical region, also always have mycorrhizae. Ectomycorrhizae are generally crucial for acceptable growth and of particular interest are vesicular-arbuscular mycorrhizae (VAM), which are very useful for desired growth of many species including those predominant in tropical forests.

Even though there are reports that fewer mycorrhizae are formed on highly disturbed than on undisturbed sites, abundant mycorrhizae (both ectomycorrhizae and VAM) have been reported from highly disturbed sites such as coal mine spoils. In view of this, a list of such fungi forming mycorrhizae has been prepared, which are efficient in forming mycorrhizae and are useful. These fungi can be used as inocula to be applied to roots of seedlings to allow formation of mycorrhizae. Inoculants like VAM fungi for mycorrhizae may be supplemented with Rhizobium / Azotobacter (Rhizobium for legumes and Azotobacter for not-legumes) and phosphorus solubilizing microbes (PSM) to allow better establishment and growth of the seedlings. The mycorrhizae formation depends on several factors including environmental factors, host physiology, soil, microflora and microfauna and finally on the inoculum, Therefore, these factors can be monitored effectively to enhance the formation of mycorrhizae on plantations made on any degraded land region.

Inoculation of hardwood and conifer seedlings with specific fungi in nurseries and glasshouse has shown dramatic improvement in growth and survival following transplantation in routine and adverse afforestation sites in Canada. Similarly, experimental infection of micropropagated plants during rooting should increase their survival chances in the field, which is very important in case of plantations on degraded lands. Growth regulators and nutrients supplied by microorganisms like VAM may promote root and shoot growth in tissue cultured plants. In this connection, Pisolithus tinetorius mycorrhizae programme has been tested. Large batches of viable P. tinctorius inocula have been prepared and used, demonstrating the success of the use of fungal inocula for the formation of mycorrhizae leading to the better establishment of plantations on newer and degraded sites.

Use of Microbes for Improving Soil Fertility

In the previous section, we discussed the use of mycorrhizae to enhance the uptake of nutrients and water for establishment of seedlings on degraded lands. This will, however, not lead to improvement of soil fertility by mycorrhizae as such, although the success of plant growth will eventually lead to reclamation of degraded lands. However, nitrogen fixing bacteria like Rhizobium or actionomycete genus Frankia can be used to induce nodule formation in a variety of plant species, so that these can be used for improving soil fertility of degraded lands. This nodule formation can be induced, both in leguminous and non-leguminous plant species comprising annuals (cereal and legume crops) and perennials (trees). Efforts are underway to manipulate the genes of both host and rhizobia to obtain maximum efficiency of nodule formation. Strains are also being tailored for unusual soil environments representing degraded lands.

Nitrogen fixing bacteria for nodulation in legumes

It is well known that Rhizobium forms symbiotic association in the roots of leguminous crops leading to the formation of nodules, which help in improving the fertility of soil through fixation of nitrogen from the atmosphere. Extensive studies on the genes of host as well as the symbiont, responsible for nitrogen fixation have been undertaken during the past few decades giving valuable information for future genetic engineering exercises. This may eventually lead to better 'legume - Ririzohium' symbiotic nitrogen fixation. The use of 'legume - Rhizohium' system for restoration of degraded lands may involve the following steps: (i) produce inoculum of selected rhizobial strains on a commercial scale; (ii) evolve effective and efficient carriers of inoculum for field inoculation; and (iii) evolve suitable technique for rapid seedling establishment and nodulation. Different Rhizobium strains may be suitable for different soils. For saline soils, R8 and for acidic soils R36 and R73 were found useful.

Trees associated with rhizobia can also be used for reforestation on degraded lands. Micropropagation has been used wish several leguminous tree species to be used for plantations on degraded lands. These tree species include the following: Acacia nilotica, Albizia lebbeck, Dalbergia latifolia, D. sissoo, Leucaena, leococephala, Prosopis cineraria, Sesbania grandiflora and S. sesban.

Nitrogen fixing bacteria for nodulation in non-legumes

For the last few decades, efforts have been underway to induce nitrogenfixing bacteria to form root nodules in non -legumes like cereals including rice, wheat and barley or in oil seed crops like rape, etc. These efforts in the past often met with failures. However, during the last few years, groups led by E. Cocking in UK, by Y. Jing in China and by B. Rolfe in Australia were successful in inducing the bacteria to form nodules in the roots of important crop plants like rice, wheat and oilseed rape. Oilseed rape produces high quality industrial oil as well as canola oil, a current favorite for health due to its cholesterol lowering property. It is possible that in future, these crops may be grown on degraded lands in association with rhizobia, forming nodules.

Nitrogen fixing actinomycetes - Frankia for nodulation in non-legumes

Biotechnological methods are also being developed, which will help nonleguminous plants (other than conventional crops) to survive under conditions of adverse nutrient supply. There are about 160 species of angiosperms, which are known to form nitrogen fixing root nodules with the actinomycetes bacteria belonging to the genus Frankia which has become a subject of intensive studies in recent years. Frankia helps in nitrogen fixation in non- leguminous plant species and therefore can be used for land reclamation through reforestation due to high biomass production without the need of expensive nitrogen fertilizers. Of these non- legume nitrogen fixing systems, alder trees (Alms spp.) have been found to supply 12-300kg N ha^{-1} yr^{-1}, through leaf fall, excretion from roots, and nodule decays. As a result, they improve soil conditions and also suppress pathogens by excreting phenolics or other organic compounds. Several species of Alnus (A. crispa, A. glutinosa, A. incana, A. japonica, A. rubra A. sinuata, A. viridis) have been used, where rooted plantlets, after their transfer to nitrogen free substrate, were inoculated with pure cultures of Frankia. In all cases, effective actinorrhizae were formed. These symbiotic associations of actinorrhizae not only fulfilled completely the nitrogen requirements of the trees, but also helped in fixing nitrogen in the virgin soil thus improving the fertility of the degraded lands.

Restoration of Soils Contaminated with Heavy Metals

Industrial and domestic effluents, containing variable amounts of heavy metals, are being increasingly used for irrigation and lead to soil contamination. Biotechnology has been used to deal with this problem, both for prevention of contamination and restoration of contaminated soils. We have earlier discussed the problem of prevention of contamination, while dealing with pollution control. Government of India is also developing a pollution control policy to deal with this problem. Methods are also available for removal and recovery of heavy metals from contaminated soils. It has been demonstrated that selected and engineered microbes can be used for this purpose.

Plasmids have also been constructed which may enhance the recovery of gold from arsenopyritic - pyritic ores, by Thiobacillus fen-oxidans. A wood rotting macrofungus, Ganodernla lucidum has also been identified for heavy metal pollution control, as a potential biosorbent material.

Biodiversity Conservation

Extinction of wild species and destruction of ecosystems has been a major cause of worry of policy makers and biotechnologists alike. One of the major efforts has been to conduct a survey and conserve country's biodiveristy, so as to save wild plants and animals from extinction. National parks and sanctuaries have been established in many countries to meet this objective. Under the auspices of the United Nations also, funds are being established and other efforts being made for conservation of germplasm at the global level. Biodiversity studies thus include the following : (i) a systematic examination of the full array of organisms on this globe; and (ii) a study of the methods by which diversity can be maintained and used for the benefit of mankind. A discussion on biodiversity in a book on biotechnology is relevant, because biodiversity is being utilized to provide genes from wild species for biotechnology exercises. In recent years, a discussion on biodiversity has become important also because countries in the North of the hemisphere (developed countries) have been utilizing biodiversity available in the South (developing countries) without paying any compensation. Several Biodiversity Conventions were held in 1992 for discussions on measures required to be taken by developing and developed countries to preserve the biodiversity at the global level. In this connection, the latest Biodiversity Convention was held in May, 1992 at Nairobi to formulate a treaty, that was desired to be signed at the UN Conference on Environment and Development (UNCED) later held in Brazil in June 1992. In this treaty, an agreement was sought by the developed countries to allow, as a matter of right, access to every country on the germplasm or biodiversity available anywhere in the world. Since tropical countries are far richer than temperate developed countries, such a treaty would have benefitted only the developed nations. In view of this, the South (particularly India) had rejected such a treaty desired to be signed in this convention. Such a rejection was also based on the argument that while the convention was meant to globalize the natural resources, it did not want to globalize the benefit derived from biotechnology inventions. Instead, the developed countries wanted to privatize biotechnology through patents and other intellectual property rights (IPR; consult next two chapters for details). Similarly, to the disadvantage of developing poor countries, the concept of Plant Breeder's Rights (PBRs) was recognized in the North ignoring the Farmers' Rights for compensation desired to be given to poor farmers in the developing countries. Although, most countries agree to the need of preserving biodiversity, there is disagreement on the issue of who will pay for it. While the developing countries feel that the conservation efforts will depend on flow of money from rich nations, the latter reject such a treaty to avoid payment of compensation for the losses, which the poor nations suffer due to inadequacy of their efforts towards biodiversity conservation. In view of serious implications of the biodiversity studies on developments in biotechnology, in this chapter, various aspects of biodiversity will be discussed.

Current Levels of Biodiversity

The life on the planet Earth still remains largely unexplored at the species and intraspecies levels. Therefore, a complete knowledge of the world fauna and flora is not available even to the systematists, who are responsible for the taxonomic description and nomenclature of all biological species. It is estimated that the total number of species available on this planet may be close to 100 million (108), although earlier estimates give a figure of 10 million or 30 million. From such a large number of species, only about 1.4 million species of plants, animals and microorganisms have so far been given scientific names.

It has been shown that the diversity of fauna and flora varies in different ecosystems, habitats, geographical regions and also among different taxonomic groups. Following are some examples : (i) Diversity is greater in terrestrial and fresh water species than in marine species; (ii) In marine water, all the 33 living animal phyla are available, while only 17 are found on land and fresh waters; (iii) Arthropoda among phyla and Insecta among arthropods have more species diversity than other groups; similarly nematodes, mites and fungi are highly diverse, the number of species approaching hundreds of thousands or millions; (iv) Among mammals, rodents have more species than other groups; (v) Among monocotyledons, orchids have more diversity; (vi) Among squirrels (Sciuridae), the genus Sciurus has more species than other genera; (vii) More diversity occurs in tropics (tropical rain forests and coral reefs) than elsewhere. For instance, 300 tree species occur in single hectare plots in Peru, while only 700 tree species are found in whole of North America. In the same area, a single tree yielded 43 species of ants belonging to 26 genera, a level of diversity found in whole of British Isles.

Hyperdiversity in certain habitats or geographical regions or among certain taxonomic groups depends on special adaptations which allow penetration of multiple niches by them. Following are some examples of these adaptations. (i) Ants appear to expand due to fungistatic secretions, series-parallel work operations and a highly altruistic worker caste; (ii) In insects, size, metamorphism and wings are adaptations that allowed development of hyperdiversity; (iii) Some other adaptations common to many groups include small size, photophagy, parasitism with specialization on hosts, specialized life stages, greater dispersal ability, etc.

Alpha (α) and Beta (β) Biodiversity

Biodiversity is described by two parameters: (i) Alpha diversity represented by the number of species in a specified area and (ii) β diversity represented by the turnover of species across space. The a-diversity increases with the total number of individuals encompassed and thus with the increase in the area sampled and the productivity per unit area. We know that α-diversity is less on remote islands and increases as we move towards the equator. However, there are exceptions to this general trend as illustrated by the following : (i) α-diversity peaks along gradient of productivity and declines in the most productive systems; (ii) some taxa are more diverse in the north than towards the equator; (iii) some deserts have unusually diverse plant communities. We know relatively less about β-diversity to predict its current patterns or its future when the natural areas will be surrounded by highly modified habitats. β diversity depends on how large are species' ranges and following two scenarios may be found: (i) If the range is large, α-diversity is independent of the area sampled, so that a national park to protect diversity can be placed anywhere; (ii) If a diversity is low with species' ranges being small (despite high total diversity) and non-overlapping, many parks will be needed to protect diversity.

Extinction and Endangered Species

In many groups of plants and animals, there are groups of species, which are prone to extinction due to environmental change so that the group is reduced to one or few threatened or endangered species (two grades threatened or endangered are recognized by conservationists). These threatened or endangered species have a high probability of extinction during the next few years or decades. Red Data Books are being prepared which will contain the lists of these endangered species of plants and animals. The extinction of these rare species may be natural or man made and both will be discussed separately.

Natural extinction

The species which are prone to extinction are characterized by small size of breeding population (thus promoting inbreeding) associated with high coefficient of variation. When the size of breeding population drops to a hundred or less, the likelihood of extinction is enhanced still further due to inbreeding depression. The overall extinction rate estimated from fossil data is one species per 10^7 species per year. This rate refers to true extinction, which means extinction of a species and its descendents and differs from 'pseudoextinetion", which is caused due to evolution of one species into another. The longevity of a species defined as above -(species and its descendents) varies from 1-2 million years (e.g. mesozoic ammonoid and Silurian graptolite) to as much as 10 million years (e.g. paleozoic and mesozoic invertebrates). It has also been shown that in general, planktonic and sessile marine animals, including corals and brachiopods, have had higher extinction rates than mobile benthic animals, such as gastropods and bivalves.

Human-caused extinction

Due to destruction of natural habitats by human interference, biodiversity is being lost at a fast rate, particularly in the tropical regions (e.g. tropical rain forests and coral reefs), where not only biodiversity is high, but also the human interference is high due to population pressure. The rain forest cower has been reduced to 55% of the original area due to human activities, and the rate of loss of these forests is increasing at an alarming rate. For instance, the rate of loss of these forests was doubled during the period 1979-1989, and about 18% of the remaining forests are disappearing every year. This leads every year to loss of 0.2 to 0.3 per cent of all species occurring in the forests at a particular time. In other words, if two million species are confined to the forests alone, 4000 species will be lost every year due to deforestation. This means, if there are 20 million species available in forests, as the current estimates indicate, about 40,000 species will be lost every year. Among these, if there are endemic species restricted to a particular area, the clearing of a single habitat or mountain ridge may lead to immediate extinction of the species. Therefore, it is estimated that if the current rate of clearing the forests and other habitats continued, 25% or more of the total species on the Earth could be eliminated within 50 years. It is also estimated that almost 40% of net primary productivity (NPP, which makes food supply of all animals) on land is now directly used or lost due to the activities of a single species Homo sapiens. Further the human population is projected to become 10 billion in the next 50 years, and this is supposed to be accompanied with 5-10 fold increase in economic activity. If anything close to this level of increase in population and economic activity is witnessed in future, World's biodiversity is destined to be completely wiped out from the surface of the Earth. This demonstrates the crisis of the loss of biodiversity and the need to conserve it.

Reasons of Concern for Loss of Biodiversity

Ethical and aesthetic rewards

Since other living organisms on Earth are our living companions in the universe, an ethical or aesthetic issue concerns the moral responsibility of human beings to protect the fauna and flora, which is facing an acute danger. Hobbies like gardening, pet- keeping, bird watching, etc., suggest that human beings do derive aesthetic reward from these companions.

Economic benefits

Human beings have already obtained enormous direct economic benefits from biodiversity in the form of food, cloth, medicines, industrial products, etc. But this has been only a very small fraction of the total biodiversity which can be exploited. Most of our crops were originally wild plants and could be developed into productive crops through selection and breeding. About 7000 plant species have been used as food, but several times this number has been reported to have edible parts and, therefore, can be used in future. Similarly, a large number of medicines have their origin in plants or microorganisms. Quinine and penicillin are two such examples which are used extensively against some of the most deadly diseases. With the advent of biotechnology, in future genes from many species will be utilized for a variety of purposes. But if the biodiversity is lost, these future possibilities will disappear.

Essential services provided by natural ecosystems

The natural ecosystems provide essential services to human beings, and diverse species occurring in these ecosystems contribute to these services. These services include the following : (i) Maintenance of gaseous composition of the atmosphere and thus preventing rapid changes in the mix of gases, which may destroy fauna and flora; (ii) Maintenance of soil ecosystems having rich hiota, where the existing biodiversity contributes to fertility of soil thus supporting crops and forests. The fungal associations in mycorrhiza, particularly VAM transfer essential nutrients into the roots of plants. Other microorganisms also fix nitrogen and transfer other nutrients to the soil. Many bacteria decompose organic matter and help in producing humus; (iii) Production of soil by weathering of rocks to which plants contribute in a significant manner; (iv) Disposal of wastes, through decomposition by microorganisms and making nutrients available for the growth of green plants; (v) Through vast biogeochemical cycles, cycling of nutrients resulting from decomposition; before they are reincorporated into living systems; (vi) Control of a large number of insects and other pests, which attack crops or domestic animals. This is made possible by the presence of numerous species of predacious and parasitic insects, who feed upon these pests; (vii) Help in effective pollination in some plant species - e.g. by honey bee, wild bees or other insects; (viii) Supply of food from fishes and other marine animals, in addition to many food crops.

The ecosystem services including the above are provided by biodiversity at such a grand scale and in such an intricate manner that there is no possible substitute for these services even with the advances in knowledge made in recent decades. Recent attempts for finding inorganic substitutes for these services have met with little success. These attempts include the following : (i) synthetic pesticides; (ii) inorganic fertilizers; (iii) chlorination for purification of natural waters; (iv) dams for flood and drought control; (v) air-conditioning for high temperatures, etc. All these substitutes destroy environment, and often require large energy subsidy, thus adding to environmental problems discussed in earlier chapters. Therefore, these are not satisfactory, thus forcing us to think now of biopesticides, biofertilizers, afforestation, etc., to deal with these problems. In other words, we need to restore biodiversity, which is being destroyed due to use of synthetic devices as substitutes for ecosystem services.

Steps to Preserve Biodiversity

Many ways are being suggested for preserving biodiversity. Some of them include the following : (i) No undisturbed land be used for development or urbanisation, because development of townships and shopping centres, use of forest trees for fuel and the use of swamps for paddy fields lead to loss of biodiversity; (ii) Catalogues of genetic resources and national biological inventories be prepared, so that the threatened and endangered species may be protected against extinction; (iii) In poor nations, birth rules should be lowered and sustainable, high yielding agricultural systems be developed, so that preservation and sustainable exploitation of biodiversity go hand in hand. The rich countries should provide massive assistance to meet this objective without any further delay, since we are already late by 50 years in starting this process; (iv) Measures should be taken to reduce emission of greenhouse gases and ozone-destroying compounds; (v) Effective measures for conservation of biodiversity be developed and strengthened in all countries. The details of these conservation methods make a separate subsdiscipline of biodiversity studies-called 'conservation biology'. These techniques of conservation of biodiversity include in situ and ex situ conservation and are being briefly discussed in the following section:

In Situ and Ex Situ Conservation-Gene Banks

Conservation of biodiversity can be achieved in a number of complementary ways. These methods, all of them falling within the broader concept of gene bunks can be broadly classified as follows : (i) In situ conservation, which includes conservation of plants and animals in their native ecosystems or even in man made ecosystems, where they naturally occur; (ii) Ex situ conservation, which includes conservation of samples of genetic diversity (particularly representing endangered species) away from their field habitats.

In situ conservation

This type of conservation applies only to wild fauna and flora and not to the domesticated animals and plants, because conservation is achieved by protection of populations in nature. It includes a system of protected areas of different categories, e.g. National Parks, Sanctuaries, Nature Reserves, Natural Monuments, Cultural Landscapes, Biosphere Reserves, etc.

In situ conservation of forest trees—Efforts are being planned by CGIAR's latest international agricultural research centre, CIFOR (Centre for International Forestry Research) to conduct research for in situ conservation of forest trees. It is argued that although the tropical foresters reserve forest areas against other land uses, these have been used for multiple purposes (for water catchment, protection of mammals or birds, etc.). Therefore, CIFOR will conduct research and find more rational location of in situ reserves for the conservation of woody planes germplasm. This work will be done in close collaboration with IUCN (International Union for Conservation of Nature and Natural Resources), and the national institutes in developing countries.

Ex situ conservation

Ex situ conservation, using sample populations, is done through establishment of 'gene banks', which include genetic resource centres, zoos, botanical gardens, culture collections, etc. Although the phrase 'gene banks' often refers to only ex-situ conservation facilities, they do include in-situ conservation methods, which include national parks and sanctuaries. This has become particularly important for conservation of crop varieties and wild genetic resources, because of their utility in future crop improvement and afforestation programmes. However, there has been a competition for relative allocation of efforts directed towards in situ and ex situ conservation. For instance in 1987 for ex situ programmes, USA allowed only 1% of a total of 37.5 million dollars meant for biodiversity conservation, excluding contributions to international systems of gene hanks. The United Nations Environment Program (UNEP) however, advocated equality for in situ and ex situ conservation efforts. In view of this, efforts and funding for ex situ conservation were enhanced in recent years.

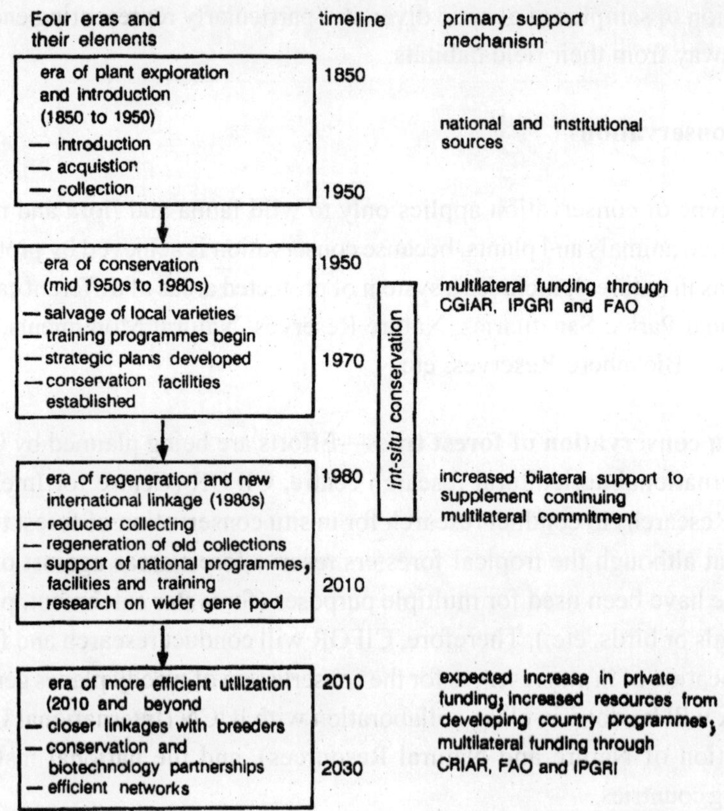

Fig. four eras of ex situ genetic resource conservation and use with timeline of conservation events. (After Cohen *et al.*, 1991).

Practical action on ex situ genetic resource conservation and its use can be divided into following four major eras (i) In the first era (1850-18611), utility of genetic resources was tested and genetic resources introduced; (ii) In the second phase (1950-1970), a wide spectrum of genetic resources was conserved due to their utility; (iii) In the third phase (1980-2011)), long-term viability of the investment in collection is ensured; (iv) In the fourth phase (2010-2031), there will be enhanced exploitation, through breeding programmes.

Ex situ conservation efforts at international level—Major efforts in the ex-situ conservation of crop genetic resources also became possible due to support provided to the crop-based research centres of CGIAR (Consultative Group on International Agriculture Research), which is a broadly based consortium supporting a world wide network of 17 (seventeen) international agricultural research centres

(IARCs). Most of these research centres are shown below. Since 1975, these IARCs have built up the world's largest ex-situ collection of crop gene pools,

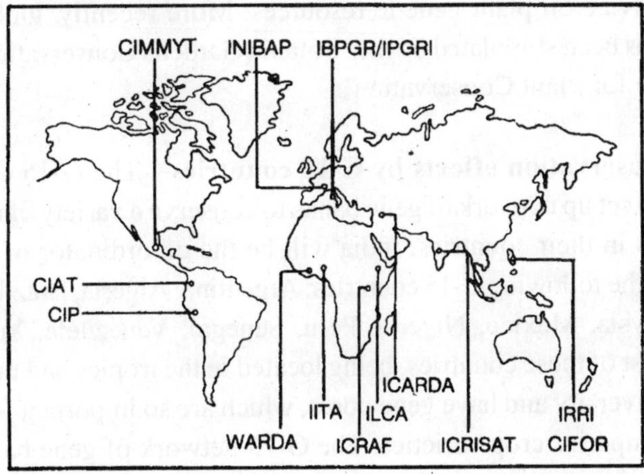

Fig. Different International Agricultural Research Centres (IARCs) supported by Consultative Group on International Agricultural Research (CGIAR).

approaching as many as 600,000 individual accessions. These germplasm collections are held in trust for the use of present and future generations of research workers throughout the world. CGIAR system has also helped to ensure the conservation of more than 140 species started in the gene banks of some 450 non-CGIAR institutions in over 90 countries. The CGIAR is also committed to strengthening national agricultural research systems in genetic resource programmes, which have now been established in more then IM countries,

In 1974 with the support of CGIAR, an international research centre exclusively devoted to plant genetic resources (PGRs) was also established. It was named IBPGR (International Board for Plant Genetic Resources). The mandate of this institute was "to advance the conservation and use of plant genetic resources for the benefit of present and future generations." IBPGR has contributed to (i) the establishment of ex-situ conservation facilities in over 100 countries; (ii) training of more than 1700 scientists and technicians; (iii) collection of over 200,000 samples of crops in 120 countries. However, the scope of activities of IBPGR has increased in recent years, so that in 1992-93, it was renamed as IPGRI (International Plant Genetic Resources Institute) with headquarters located at Rome. This institute will have following four major objectives: (i) to assist counries, particularly developing nations, to assess and meet their needs for plant genetic resources ; (ii) to strengthen and contribute to international collaboration in the conservation and use of plant genetic resources; (iii) to develop and promote improved strategic and technologies for plant genetic resources conservation; and (iv) to provide an international information service on plant genetic resources. More recently, global interest in conservation has been stimulated by the 'Botanic Gardens Conservation Secretariat' and the 'Centre for Plant Conservation'.

Ex situ conservation efforts by G-15 countries—The G-15 countries have also resolved to set up network of gene banks to conserve a variety of medicinal and aromatic plants in their countries. India will be the co-ordinator of this network, which includes the following G-15 countries: Argentina, Algeria, Brazil, Egypt, India, Jamaica, Malaysia, Mexico, Nigeria, Peru, Senegal, Venezuela, Yugoslavia and Zimbabwe. Most of these countries, being located in the tropics and the sub-tropics, are rich in biodiversity and have gene pools, which are so important for developing economically-superior crop varieties. The G-15 network of gene banks will offer conservation of important germplasm in the form of seeds, plant embryos, pollen and in vitro cultured tissues. The activity will also cover the conservation of plant species in the national parks and sanctuaries.

Ex situ conservation efforts in USA—In U.S.A, medium term storage facilities for seeds were established in 1940's by regional plant introduction stations. However, in 1960's due to widespread adoption of new high-yielding varieties, leading to green revolution, the local varieties were threatened. This was, therefore, the start of increased efforts to collect these local varieties and other germplasm, which could be valuable for future plant breeding programmes. In USA, therefore, in 1978 long-term seed storage (at -18°C) facilities were created at National Seed Storage Laboratory (NSSL) in Fort Collins, Colorado.

Ex situ conservation efforts in Europe—A European Community (EC) Programme on the conservation of plant genetic resources was initiated in 1992, with a meagre budget of 2-5 million US dollars. In January 1992, EC also convened a meeting of national coordinators to consider a proposal for conservation of plant genetic resources. The proposal was jointly prepared by (i) Centre for Genetic Resources (CGN), the Netherlands; (ii) Genetic Resources Action International (GRAIN), Spain; and (iii) International Board of Plant Genetic Resources or IBPGR (now called IPGRI = International Plant Genetic Resource Institute). This proposal recommends linking of European gene bank programmes and NGOs (non-government organizations) through crop networks. This will continue, build on and expand the work of the European Cooperative Programme for Crop Genetic Resources Networks, which has been coordinated by IBPGR since 1980's.

Ex situ conservation efforts in India—In India also an Indo-US project on plant Genetic Resources was signed in 1988, with an outlay of 23.95 million US dollars (approx. Rs 72 crores) for a period of seven years (upto 1995). This project involves construction of a National Gene Bank consisting of (i) Seed Repository (storage at - 20° C); (ii) Cryo Bank (storage at -196° C); and (iii) Tissue Culture Repository (10-25°C). It is also recognised that facilities in India are already available for the medium and long term storage of seeds of those crop plants that can withstand high drying and low temperature. However, in vitro conservation techniques including tissue culture and cryopreservation offer distinct advantages. In view of this, in 1986 with the financial support from the Department of Biotechnology (Govt. of India), NFPTCR (Notional Facility for Plant Tissue Culture Repository) was established at NBPGR (National Bureau of Plant Genetic Resources) located in New Delhi. For the development of this facility, initially a budget of Rs 2.96 crores was provided for a period of five years (1985-1990). An additional sum of Rs 2.80 crores has been sanctioned for the next five years (1991-1995) for continuation and expansion of tissue culture facility. This national facility aims at developing suitable in vitro conservation technologies for medium and long-term preservation of clonally propagated agri- biohorticultural and plantation crops including their wild relatives. The programme also aims at developing techniques for cryo-preservation of seeds, pollen and in vitro cultures; for this purpose liquid nitrogen (-196° C) container and a liquid nitrogen plant are already in operation for cryopreservation. A range of crops like millets (pearl millet, minor millet), oil seeds (Brassica spp., sunflower, sesame), vegetables (onion, carrot, chilli, amaranth, turnip, radish, tomato), pulses (Vigna spp.) and narcotics (tobacco, poppy) have been maintained for several years in liquid nitrogen without any decline in viability or vigour. The strategy of desiccating excised embryonic axes has led to the successful cryopreservation of even recalcitrant seed species (e.g. tea, orange, neem) and is being tried for cocoa, jackfruit and almond etc.

Short-term or medium-term conservation at NFPTCR are also being tried through long-term root cultures, which are capable of regenerating in shoots (e.g. Citrus = lime). Similarly, shoot cultures of banana, ginger, greater yam (Dioscoren alata), and sweet potato could be maintained for 8-24 months at 25° C . The sub-culture interval for this purpose can be extended by using low temperatures ranging from 4°C to 15°C, although chilling sensitivity varies between species. This technique also allows to keep the tissue cultures free from viruses.

Conservation efforts by private sector (community seed banks or gene banks etc.)—The preservation of diversity (particularly of endangered plants, plantation crops and industrial crops) is not solely the domain of large national and international programmes. In recent years, several organizations collectively called Non-Governmental Organizations (NGOs), 'Seed Savers Exchange', and industries (e.g. community and regional seed banks or gene banks) have emerged to support and expand local efforts on global biodiversity conservation. In many cases, they link traditional farmer with ex situ conservation programmes. 'Seed Savers Exchange' conserves and exchanges over 5000 native varieties. Ex situ conservation of crops like sugarcane, rubber, oil palm, pineapple etc. is also done by industries. Farmers also help in ex-situ conservation of forest trees and fruit trees through practice of agroforestry, which is facilitated and encouraged by ICRAF (International Centre for Reasearch on Agroforestry).

Ex situ conservation of gene pool of woody plants—CGIAR's new Centre for International Forestry Research (CIFOR) became operational in 1992-93. In addition to the approach of in situ conservation of woody plants/forest trees in forests, by finding more rational location of in situ reserves, CIFOR will also make efforts for long-term ex situ conservation of the gene pool of woody plants. Ex situ conservation will also be facilitated by growing trees in locations, where they do not naturally grow. This objective is also being achieved through agroforestry, where farmers grow trees with their crops on agricultural fields. The International Centre for Research on Agroforestry (ICRAF) is doing research in this area, particularly helping African farmers to use agroforesty as a means of their livelihood.

Conservation of germplasm of woody species having recalcitrant seeds (they germinate soon after they fall) has been found to be difficult. These seeds are difficult to dry and store. Naturally regenerating seedlings are often used in these cases for conservation and vegetative propagation. Since seed storage will be difficult in these cases (e.g. bud wood), in vitro collection and cryopreservation may be necessary, and CIFOR will take appropriate measures to achieve this objective.

Management of germplasm collections—With the availability of facilities for conservation of valuable germplasm, it is necessary that resources and facilities are also available for multiplication, regeneration, evaluation, characterization, documentation and distribution of this germplasm. It is also necessary to ensure that during conservation and management of the germplasm, its identity and genetic stability is maintained. Further, for safety, samples must be duplicated at different sites, particularly when older collections are transferred into modern conservation storage. If these precautions are not taken and facilities for management of collections are not extended, some of the germplasm repositories may become germplasm morgues or mortuaries.

A summary of the available accessions, regeneration of the available accessions, their regeneration capacities and facilities available for distribution of these accessions in the U.S. National Plant Germplasm System (NPGS) is presented in on the next page. Similar information for CGIAR centres is presented later on. The facilities for ex-situ conservation programmes, however, differ in developing countries, U.S.A. and in IARCs (or CGIAR centres). These differences are summarized on the coming pages and suggest a need for bilateral development and funding, so that the developing countries may join in the global network of base collections, without the need of excessive duplications of germplasm accessions.

Table

Accessions from NSSL (National Seed Storage laboratory) and regional plant Introduction and conservation facilities available for distribution after regeneration. Values shown are numbers of accessions or locations. (After Cohen et.al, 1991 ; Science 253: P. 870)

Crop : Facility		*Accessions*	*Accessions able to regenerate per year*	*Accessions sufficient for distribution*	*Locations available for regeneration*
Barley:	NSSL	721		436	
	NSGC	26,168	5,000	22,175	2
Maize :	NSSL	21,671		1,865	
	NCRPS	8,783	450	7,000	1-7
Peanut:	NSSL	121		29	
	SRPIS	8,165	1,100	2.000	1_4
Bean :	NSSL	6.427		1.065	
	WRPIS	11,030	600	10.200	2_4
Potato :	NSSL	-1.262		86	
	IRI	4,272	300	4,000	1
Rice :	NSSL	942		205	
	NSGC	16,010	5,000	13,899	1
Sorghum:	NSSL	15,043		8,797	
	SRPIS	17,604	2,000	15.600	4
Tomato :	NSSL	1.984		1,202	
	NCRPS	5,615	300	5,500	1
Cowpea :	NSSL	279		255	
	SRPIS	8,133	1,500	5,500	2
Wheat :	NSSL	1,597		619	
	NSGC	42,478	5,000	36.932	2
Totals					
	NSSL	52.047		14.559 (28%)	
	Regional	148,258	21,250	122,826 (83%)	

* Acronyms for regional plant introduction and collection facilities are as follows: NSGC, National Small Grains Collection ; NERPIS, North East Regional Plant Introduction Station; WRPIS, Western Regional Plant Introduction Station; NCRPIS, North Central Regional Plant Introduction Station; IRI, Inter- Regional Potato Introduction Project; and SRPIS, Southern Regional Plant Introduction Station. * Values for NSSI indicate number of unique accessions not yet in the regional plant introduction facilities.

Table

Accessions from crop germplasm collections, available of CGIAR Centres, for distribution following regeneration. Values shown are numbers of accessions or locations); N/A, not available.

(After Cohen el al., 1991 Science 253; p. 869)

*Centre**	*Crop*	*Accessions*	*Accessions able to regenerate per year*	*Accessions sufficient for distribution*	*Locations available for regeneration*
AVRDC	Mungbeait	5,373	1,000	3,200	2—5
	Pepper	3,471	250	1,000	2—5
	Soybean	12,303	1,000	8,000	2—5
	Tomato	3,814	350	3,000	2—5
	Total	24,861		15,200 (61%)	
CIAT	Phaseolus (bean)	25,000	1,800	20,000	3
	Tropical pastures	21,000	1,800	14,000	3
	Cassava	4,500	In vitro	In vitro	
	Total	50,500		34,000 (74%)	
CIMMYT	Barley	7,200	2,000	6,480	3
	Bread wheat	48,600	8,000	43,740	3—7
	Durum wheat	15,300	3,000	13,770	3—7
	Primitive wheat	4,320	4,000	3,888	3
	Triticale	11,700	3,000	10,530	3
	Wild relatives	2,700	500	2,430	3
	Total	89,820		80,838 (90%)	
	Maize	13,346	250	10,910	4—8
	Teosinte	93	In situ	N.A.	
	Tripsacum	90	Clonal	N.A.	1
	Total	13,529		10,910 (81%)	

CIP	Potato (clonal, in vitro, seed)	4,500	3,500 clones 300 seeds	3,500 seeds 650 clones	4—9
	Sweet potato (clonal, in vitro, seed)	5,507	2,000 clones 300 seeds	50 clones	1
	Total	10,007		4,200 (40%)	
ICARDA	Food legumes	17,900	2,000	10,740	1
	Total cereals	43,700	3,000	26,220	1
	Forage legumes	22,000	2,000	13,200	1
	Total	83,600		50,160 (60%)	
ICRISAT	Groundnut	12,841	1,500	12,500	10—15
	Pearl millet	21,919	2,000	21,800	10—15
	Pigeonpea	11,482	1,300	11,300	10—15
	Other millets	7,082	1,000	7,000	10—15
	Sorghum	32,890	2,600	32,000	10—15
	Chickoea	15,995	1,400	15,600	
	Total	102,209		100,800 (99%)	
IITA	Musa	412	412	412 (100%)	2—7
	Oryza species	12,500	4,500	9,000	3
	Vigna unguiculata	15,200	4,000	9,439	3—16
	Wild Vigna	1,450	1,000	1,000	
	Total	29,562		19,851 (67%)	
IRRI	Rice	86,000	10,000	77,400 (90%)^	2
WARDA	Rice	5,430	1,000	3,000 (55%)	3—16
CIFOR	Woody plants (forest trees), etc.	Started only in 1992	–	–	–

* AVRDC, Asian Vegetable Research and Development Center; CIAT; Centro International de Agriculture Tropical; CIP, International Potato Center; ICARDA, International Centre for Agricultural Research in the Dry Areas; ICRISAT, International (Tops Research Institute for the Semi-Arid Tropics; IITA, International Institute of Tropical Agriculture, and WARDA, West African Rice Development Association. Of its total 86,000 accessions, IRRI has successfully canned 43,500 accessions for medium and long term storage.

Table

Comparison of the current status of ex situ conservation programmes of crop plants in developing countries, the United States, and the IARCs (After Cohen et al., 1991; Science 253: p. 868)

	Characteristics	*Developing countries*	*U.S.*	*IARCs*
1.	Major regions of diversity represented within boundaries	Many	Limited	Some
2.	Participation in international germplasm exchange	Limited*	Yes	Yes
3.	*Ex situ* conservation policies developed	Very few	Yes	Usually
4.	Fully functioning germplasm system	Limited	Yes	Usually
5.	Long-term storage facilities	Yes	Yes	Yes
6.	Exploration interests	Yes	Yes	Yes
7.	Long-term storage facilities	Limited	Yes	Yes
8.	Regeneration capabilities	Limited	Some§	Some++
9.	Regeneration sites	Some	Some++	Some§
10.	Data management expertise	Limited	Yes	Yes
11.	Distribution capabilities	Limited	Yes	Yes
12.	Able to train scientists	Limited	Limited§	Yes
13.	Able to provide technical assistance	No	Limited§	Yes
14.	Hold global base collections	Some	Yes	Yes
15.	Operative germplasm quarantine facilities	Limited	Yes	Limited arrangement

* Often conditioned by national policies and by number of accessions with adequate material for distribution. Limitations imposed by lack of resources coordinating mechanisms, and availability of appropriate sites. ++ No one genetic resources programme has focal access to a complete range of sites for regeneration. § Limited by availiability of personnel and conflicting responsibilities. || Except in very few casts.

Species Conservation

In the depleting natural resources, there are many animal and plant species which have been classified as endangered and listed in Red Data Books by oragnizations like Department of Environment (Govt. of India), IUCN (International Union for Conservation of Nature and Natural Resources) and National Museum of Natural History (India). In 1980, IUCN in association with FAO (Food and Agriculture Organization), and UNESCO (Untied Nations Educational, Scientific and Cultural Organization) prepared a document entitled 'World Conservation Strategy'. For the preparation of this document, financial assistance was obtained from UNEP (United

Nations Environment Programme) and WWF (World Wildlife Fund). This strategy in the document has 20 sections, each section with several paras, containing priorities for conservation, which need action both at the national and international levels.

While a variety of measures are being used and suggested for conservation of both animal and plant genetic resources (PGR), biotechnology offers special opportunities for the conservation of a few remaining plants of an endangered plant species. For example, tissue culture methods may help in the multiplication of a threatened plant species. For conservation of animals also, techniques like artificial insemination and embryo transfer, can be utilized for multiplication of endangered species. For instance, scientists from the Institute of Food and Agricultural Sciences at the University of Florida have been investigating ways in which artificial insemination (which is normally very expensive and achieves poor results) can help in the captive breeding of alligators.

It is satisfying to note that a Keystone International Dialogue Series on Plant Genetic Resources (PGR) was initiated to promote an international commitment to PGR. The dialogue series had its first meeting in August 1988 at Keystone (Colorado, USA), and third and final meeting in May/June, 1991 at Oslo (Norway). The series involved participants from international and intergovernmental organizations, corporations and research institutions from developed and developing countries. The chairman of the series was India's Dr. M.S. Swaminathan.

SEVEN

Nitrogen Fixation

A distinctive property possessed by a select group of organisms by virtue of which the molecular nitrogen from the atmosphere is Biological nitrogen fixation which is converted to a usable form such as ammonia. The process is confined to the intracellular environment and is an extremely significant one because nitrogen play an important role in the synthesis of several molecules within plants and microbes. The topic of nitrogen fixation is expanding rapidly and developments occurring constantly.

Besides, interest in the fundamental aspects of nitrogen fixation from the point of view of understanding chemistry and biology of this process, biotechnological operations have also been gaining momentum during the last few years. The major thrust areas include (i) transfer of nitrogen fixing gene to ether plants (ii) improving upon existing nitrogen fixing plants and microbes (iii) Production, release and utilisation of ammonia by nitrogen fixing microbes. The present chapter outlines briefly the process of nitrogen fixation and commercial significance and possible exploitation.

Historical Background

Since time immemorial, rotation of crops has been practised by the ancient Greeks. They observed that rotation gave better yields, if a legume crop was followed by corn or any other crop. The beneficial effects of growing leguminous plants on the soil are so obvious that these have been recognized for hundreds of years. Theophrastus, who lived in the third century B.C. has recorded that the Greeks used crops of broad beans to enrich the soils.

It has been known since Roman times that legumes such as beans, peanuts, alfalfa, soyabeans, peas, ,cover and lupine enhanced the fertility of the soil. Soil from the fields where legumes had been grown was added to fields that were to be assigned to legume cultivation for the first time. Such a transfer of soil to improve productivity was perhaps the simplest possible biotechnological operation. The Romans, however, could not have known that the underlying reason for the success of this practice is the presence of bacteria of the genus Rhizobium which infect the roots of some legumes and fix atmospheric nitrogen. The relationship is an intimate one in which the bacteria actually enter the roots of the legume and form visible growths called nodules. For the Romans a population of Rhizobium bacteria in the soil ensured root nodulation and subsequent nitrogen fixation in new fields of legumes. This also forms the basis of the traditional practice of crop rotation because fixed nitrogen left in the soil by a crop of legumes can be taken up by other crop plants which do not form nodules. Nobody, however, realized the actual biological significance of the rotation of crops. In modern agriculture, it is a common practice to rotate a nonleguminous crop, such as corn, with a leguminous one such as alfalfa. The leguminous plants are then either harvested for hay, leaving behind the nitrogen-rich roots or ploughed back into the field. A good crop of alfalfa that is ploughed back into soil may add as much as 450 kg of nitrogen to the soil per hectare. Symbiotic organisms, i.e., (Rhizobium) associated with legumes were reported by Wilfarth and Hellriegel in 1887. In 1888, Beijerinck succeeded in isolating Rhizobium from leguminous plants and showed that root nodule bacteria are required for healthy plant development. The legumes are by far the largest group of plants that enter into this nitrogen-fixation partnership with Rhizobium. In 1893 a German botanist, Winogradsky, discovered nitrogen fixation in a free living heterotrophic bacterium Clostridium pasteurianum. This organism grew luxuriantly in a culture medium containing a simple sugar, fresh water enriched with trace metals and CaC03 as a buffer. Another historically important discovery was made by Beijerinck in 1901 when he isolated aerobic free living nitrogen-fixing bacterium Azotobacter. Since then, the research on biological nitrogen fixation has developed rapidly. Later, nitrogen-fixing cyanobacteria were also isolated from fresh and marine water, hot springs, Arctic and Antarctic soil and rice fields. Now, it is quite well established that nitrogen fixation is exclusively confined to Prokaryotic cells. However, there are certain other nonleguminous plants which fix nitrogen and are described in the following section.

Nitrogen—Molecular, Oxidized and Reduced

Nitrogen is an important constituent of several cellular substances such as amino acids, proteins, enzymes, purines, pyrimidines, nucleic acids, porphyrins, vitamins and alkaloids.

Nitrogen is about one-third as abundant as carbon and occurs principally free in the diatomic state in the atmosphere and combined as Chile saltpetre (NaNO3). The N_2 molecule contains a triple bond (N = N). The nitrogen bond has the shortest bond length of 1.095 A, the highest ionization potential (15.58 eV) and the highest stretching frequency. Conversion of molecular nitrogen to atomic nitrogen requires at least 160 kcal of energy. The actual fixation steps in which the two atoms of N_2 combine with six atoms of HZ to form two molecules of ammonia (described in the following chapter) releases 13 kcal. Therefore, the whole process of nitrogen fixation requires 147 kcal.

Bond Energy (kcal/mol)

Bond energy	Bond energy
C—C 82	N—N 38
C—C 146	N—N 98
C—C 192	N—N 225

Nitrogen, although it dissociates into single atoms stably, is unstable with respect to oxidation by O_2 in the presence of water to nitrate ions (NO_3-). It is fortunate that this reaction is very slow, otherwise, atmospheric N_2 and O_2 would combine with the oceans to form solutions of nitric acid.

Nitrogen might be highly oxidized (NO_3-) or highly reduced (NH_3). The oxidation number of NO_3 is + 5 and that of NH_3 is - 3. The oxidation number (oxidation state) is essential to keep track of electron shifts in oxidation-reduction reactions and is defined as the charge which an atom appears to have when electrons are counted according to some rather arbitrary rules. The first of these rules is that electrons shared between two unlike atoms combine with the more electronegative atom. The second rule is that electrons shared between two like atoms are divided equally between the sharing atoms. NH_3 occurs to a slight extent in the atmosphere also, primarily as a product of the putrefaction of nitrogen-containing animals or vegetable matter.

Symbiotic and Asymbiotic Nitrogen-fixing Systems

In nature, nitrogen-fixing organisms occur independently (asymbiotic) or in association with other lower or higher plants (symbiotic). It is rather surprising that the property to fix nitrogen is widespread in symbiotic systems ranging from altogether different and unrelated plant groups such as lichens. Bryophytes, Pteriodophytes, Gymnosperms and Angiosperms. The examples include symbiotic association of cyanobacteria with fungi (lichen), cyanobacteria with Bryophytes (such as Riccia, and Anthoceros), cyanobacteria with Pteriodophytes (such as Azolla), cyanobacteria with Gymnosperms (such as Cycas), bacteria with legumes (such as Rhizobium) (Tables). There are several genera of Angiosperms, other than legumes, that have also established a nitrogen-fixing partnership with symbiotic bacteria, mostly actinomycetes (Table): Besides, one of the most recently described associations is that between Brazilian grass Digitaria decumbens and the bacterium Azospirillum lipoferum, which occurs in the inner cortex of the root, but also fixes nitrogen when grown in a culture with appropriate nutrients. Associations between Azospjrillum lipoferum and corn (Zea mays) roots have also been produced recently in Brazil, and one strain of corn achieved a yield of fixed nitrogen comparable to that of Rhizobium in nodules on the roots of soyabean (Glycine max).

Table

Rhizobium Species and Plants Nodulated

Rhizobium Species	*Plants Nodulated*
R. meliloti	Medicago saliva (alfalfa)
	Melilotus sp. (sweet clover)
	Medicago sp. (burr and barrel medic annuals)
	Trigonella foenum gracum (fenugreek)
R. trifolii	Trifolium sp. (clovers)
R. leguminosarum	Pisum sativum (garden and field peas)
	Vicia faba (broad bean)
	Lens esculenta (lentils)
	Lathyrus sp. (peavine)
R. Phaseoli	Phaseolus vulgaris (common, field, haricot, kidney, pinto, snap, beans, etc.)
	P. coccineus (runner bean, scarlet runner)
R. lupini	Lupinus sp. (all lupins)
	Ornithopus sativus (serradella)
R. japonicum	Glycine max (soybean)
Rhizobium sp.	Vigna unguiculata (cowpea)
(Cowpea type)	Arachis hypogae (peanut, groundnut)
	Vigna radiata (mung bean)
	Phaseolus lunatus (lima bean)

Rhizobium Species	*Plants Nodulated*
	P. acutifolius (tepary bean)
	Psophocarpus tetragonolobus (winged bean)
	Sphenostylis sp. (African yam bean)
	Pachyrhizus sp. (jicamus)
	Cenlrosema sp. (centro)
	Mucuna deeringiana (velvet bean)
	Canvavalia ensiformis (jack bean)
	Lablab purpureus (hyacinth bean)
	Phaseolus aconitifolius (moth bean)
	Cyamopsis letragonoloba (guar)
	Voandzeia subterranea (Bambara groundnut)
	Cajanus cajan (pigeon pea)
	Desmodium sp.
	Cassia sp.
	Lespedeza sp.
	Indigofera sp.
	Crotalaria sp.
	Pueraria sp.
Rhizobium sp.	Cicer arielinum (chick-pea, garbanro)
	Coronilla varia (crownvetch)
	Onobrychis vicisefolia (sanifoin)
	Leucaena leucocephala (ipil-ipil)
	Petalostemum sp. (prairie clover)
	Albizzia julibrissin
	Lotus sp. (trefoils)
	Anthyllis vulneraria (kidney vetch)
	Sesbania sp.

Table

Nitrogen Fixed by Various Rhizobium-Legume Associations

Plant	*Approximate Ranges of Nitrogen Fixed (kglhalyr)*
Alfalfa, Medicago sativa	100-300
Sweet Clover, Melilotus sp.	125
Clover, Trifolium sp.	100-150
Cowpea, Vigna unguiculata	85
Faba bean, Vicia faba	240-325

Plant	Approximate Ranges of Nitrogen Fixed (kglhalyr)
Lentils, Lens sp.	100
Lupines, Lupinus sp.	150-200
Peanuts, Arachis hypogea	50
Soybeans, Glycine max	60-80
Mung bean, Vigna radiata	55
Velvet bean, Mucuna pruriens	115
Pasture legumes, Desmodium sp. Lespedeza sp.	100-400

Table

Nitrogen Fixed by Various Genera and Species of Nodulating Non-Leguminous Trees and Shrubs

Species	Nitrogen kglhalyr
Alnus crispa	362
	115
	40
Alnus gluiinosa	125
	56-130
Alnus rubra	325
Casuarina equiserifolia	58
Ceanothus sp.	60
Coriaria arborea	129-192
Dryas drummondii	12
Nippophae rhamnoides	15
	179
Myrica gale	9

There are more than 15 nonleguminous angiosperms which fix nitrogen including, Alnus, Myrica, Purshia and Ceanothus. Besides, there are certain tree legumes such as Robinia and Lupinus arboreus which have been reported from New Zealand to fix nitrogen. The symbiotic organism involved is primarily an actinomycete which forms nodules.

In free living or asymbiotic systems, nitrogen fixation has been extensively studied in heterotrophic bacteria, photosynthetic bacteria and cyanobacteria. The important systems include Azotobacler (aerobic), Clostridium pasteurianum (anaerobic),

Klebsiella pneumoniae (facultative aerobe), Rhodospirillum rubrum (photosynthetic), Nostoc muscorum and Anabaena variabilis (cyanobacteria).

Leghemoglobin and Nitrogen Fixation in Nature

Leghemoglobin is the red pigment containing protein, present in nodules of legumes and is concerned with the process of nitrogen fixation. The red pigment is very similar to the hemoglobin of red blood cells. It has now been established that leghemoglobin is synthesized in the host cell and localized within the membrane envelope which surrounds the bacteroids in the nodules. Neither legumes nor the bacterium Rhizobium, in isolation, can synthesise leghemoglobin. Therefore, the molecule is definitely a product of symbiotic interactions at the genetic level. Leghemoglobin differs from hemoglobin present in red blood cells in several ways: (1) Heme moiety attached to a peptide chain, which represents the globin part, has a molecular weight of 16,000-17,000 daltons. The blood hemoglobin, on the other hand, has a higher molecular weight of 66,000 dalton. (2) Leghemoglobin has only one peptide chain linked to heme moiety. In contrast, blood hemoglobin has four peptide chains (2α 2 /3), each with one heme moiety.

It is now well-established that a number of host genes are induced following infection of the plant by Rhizobium. Leghemoglobin, the major host gem product, appears in nodules several days prior to the nitrogenase and its mRNA sequences are detected in all ineffective nodules tested. This indicates that the induction of leghemoglobin is independent of nitrogenase function. The level of induction, however, appears to depend upon the specific strain of Rhizobium. Now, evidence suggests that a number of host genes may be involved in the process of symbiotic nitrogen fixation. It is also suggested that in addition to leghemoglobins, these genes encode a group of nodulespecific proteins, nodulins. The role of the nodulins is not very well understood. These may be structural components of the nodule or play an important role in the establishment of symbiosis.

The influence of the host plant is observed at various levels including recognition and acceptance of the rhizobial strain, temporal regulation of nodule development, the number, size and gross morphology of nodules, the intracellular organization and structure of nodule cells and also the activity of the functional nodule. These processes are regulated in part by a number of host genes and several plant mutants have also seen isolated from various species. However, the exact number and function of these genes is not known.

Nitrogen fixation is a reductive process and is extremely sensitive to the presence of oxygen. Leghemoglobin enhances the transport of oxygen at a low partial pressure to the nodules and maintains a steady supply of oxygen at lower concentrations of the nodule. Therefore, leghemoglobin is essential for effective nitrogen-fixing legume symbiosis and without it, fixation is slow or absent. Recent work in Australia has shown that leghemoglobin has a very strong affinity for oxygen and prevents the local partial pressure of oxygen from reaching concentrations damaging to nitrogenase (enzyme for nitrogen fixation). However, leghemoglobin has not been recorded either in the cyanobacterial symbiotic system or in other higher plants such as Frankia and Parasponia which fix nitrogen without leghemoglobin. Thus, leghemoglobin is now considered a helpful but non-essential requirement.

Nitrogen fixation has been found to occur in fresh water, marine water, sand dunes, salt marshes, esturine muds, Arctic and Antarctic regions, and high Alpine regions besides agricultural and noncultivable lands. Quite frequently, Trichodesmium, a cyanobacterium which occurs in the Pacific Ocean, might cover an area up to 20,000 square miles and is an efficient nitrogen fixer in the marine environment. Roughly speaking, about 200 million metric tons of nitrogen are added to the earth's surface every year by the biological method. Man produces only about 28 million metric tons in the form of fertilizers. Simple organisms such as Azotobacter and Clostridium have been estimated to add about 7 kg of nitrogen to a hectare of soil per year. Rain water also brings down about 71 kg of nitrogen per hectare per year. Finally, legumes fix about 100-200 kg of nitrogen per hectare per year.

Nitrogenase-the Key Enzyme

Nitrogenase is the main enzyme responsible for the biological conversion of molecular nitrogen to ammonia. Till now, there has been no generally accepted nomenclature to describe nitrogenase and its components or subunits. Complicated technical names such as 'azoferredoxin' and 'molybdoferredoxin' have emerged. Likewise, the fraction or sequence of the elution of nitrogenase has been isolated and appropriately known as 'fraction I' (protein I or component I) and 'fraction II' (protein II or component II). Besides, games like 'MoFe protein' and 'Fe protein' are also used to describe the two components of nitrogenase.

Structural Subunits and Chemical Properties

For the last few years, several new techniques including Mossbauer spectroscopy, electron paramagnetic resonance (EPR), rapid reaction kinetics and classical steady-state kinetics have been successfully employed to unravel the complexities of nitrogenase. Sufficiently, highly purified material has become available for mechanistic studies at the enzyme level. The basic action pattern seems to be similar whether the enzyme system is obtained from the strict anaerobe Clostridium pasteurianum, the facultative anaerobe Klebsiella pneumoniae, or the aerobe Azotobacter vinelandii. Nitrogenase preparations from a number of organisms have been fractionated into two components and both of these are required for activity. John Posgatê's group at Sussex had devised a simple abbreviation to connote 'Fe protein' from Klebsiella pneumoniae as Kp2 and Clostridium pasteurianum and Azotobacter vinelandii as Cp and Av, respectively.

In general, methods for the purification of nitrogenase from different sources are almost similar. In order to prepare a crude extract for purification, the cell has to be broken to release the contents. General methods employed are ultrasonic disruption or high pressure disruption with a French pressure cell. Cells are also broken osmotically by treatment with lysozyme (an enzyme used to digest prokaryotic cell wall).

The component I or MoFe protein is a tetramer with a molecular weight ranging from 200,000 to 250,000. It contains two molybdenum (Mo) atoms, 28 to 34 nonheme iron (Fe) atoms and 26 to 28 acid labile sulphides (Tables).

Table

Physical and Chemical Properties of Nitrogenase from Various Sources

Composition	*Component I, Molybdoferredoxin (MoFe protein)*	*Component II, Azoferredoxin (Fe protein)*
Molecular weight	200,000-250,000	55,000-73,000
Number of subunits	Four	Two
Mol. wt. of subunits	Two-59,500	Both with
	Two-52,000	27,500
S^2 per molecule	18-19	4
Fe per molecule	30-34	4
Molybdenum per molecule	2	Nil
Mg^{2+}, Ca^{2+}	Present	Nil

Table

A Comparative Account of Properties of MoFe Protein from Photosynthetic and Nonphotosynthetic Microbes

		Kp	*Cp*	*Av*	*Rr*	*Ac*	*Rj*
Molecular wt.		229,000	220,000	270,000	215,000	216,000	2000,000
Subunit		51,000	50,700	54,000	56,000	56,600	50,000
Composition		59,600	59,500	60,600	Same	55,000	One type
Metal and	Mo	1	2	2	2	2.2	1.3
S^{2-} contents	Fe	18	22-24	32-36	25-30	20.4	29
gm^{-1} atom mol^{-1}	S^{-1}	18	22-29	28	19-22	20.1	26
EPR spectrum		432,363	427,328	4.3,3.65	—	—	—
		2.009	2.01	2.01			

Kp-Klebsiella pneumoniae; Cp-Clostridium pasteurianum; Av-Azotobacter vinelandii; Rr-Rhodospirillum rubrum; Ac-Anabaena cylindrica; and Rj-Rhizobium japonicum.

Amino acid analysis of both types of subunits has demonstrated that they are very similar to each other. Homogeneous preparations of component I have been obtained from C. pasteurianum, K. pneumoniae, A. vinelandii, soyabean nodules and some photosynthetic bacteria (Chromatium) and cyanobacteria (Anabaena). All the common amino acids are present in components I and II. However, there is a high proportion (more than 20%) of aspertic acid and glutamic acid.

Component II or azoferredoxin has a molecular weight ranging from 55,000 to 73,000 and is composed of two copies of a single Subunit. It has four nonheme Fe atoms and about four acid labile sulphides. Fe protein is a very labile protein and is difficult to obtain in the pure state. It's extremely sensitive to oxygen (within minutes of exposure) and forms multiple types of aggregates. Consequently, it becomes difficult to measure its physical and chemical properties. Certain properties of cyanobacterial nitrogenases have been studied in detail (Table).

Although nitrogenase complexes from various species appear to be similar, there is insufficient structural information about the proteins from the different organisms to make comparisons of the proteins. Indeed, there are some significant differences

Table

Biochemical Characteristics of Nitrogenase from Cyanobacteria

Organism	***Property***
	Oxygen sensitivity
Anabaena cylindrica and Plectonema boryanum	Completely inhibited by 10 per cent O_2 for 5 minutes
Gloeocapsa, LB 795	Stable in air in crude extracts Cold sensitivity
A. cylindrica	60% loss of activity at 0°C after 12 hrs
Gloeocapsa, LB 795	Not cold labile ATP dependent Hz evolution
A. cylindrica	Present

Organism	***Property***
Gloeocapsa, LB 795	Present
Anabaena cylindrica	N_2
	C_2H_2
Plectonema boryanum	Cyanide
	C_2H_2, cyanide
	Electron-donating system
Anabaena cylindrica	(i) $Na_2S_2O_4$
	(ii) Glucose-6-phosphate
	Ferredoxin/NADPH system
Plectonemu boryanum	$Na_2S_2O_4$
Gloeocapsa, LB 795	$Na_2S_2O_4$
	Ferredoxin-dependent pyruvate
	Cleavage
	ATP source
Anabaena cylindrica	Creatine phosphate/
	Creatine phosphokinase
Pleclonema boryunum	Creatine phosphate/
Gloeocapsa, LB	Creatine phosphokinase
	C reatine phosphokinase
	pH requirement for best reduction of substate
A. cylindrica	
P. boryanum	7-7.5
Gloeocapsa	

in the nitrogenase proteins from various species-here could be a difference of-as much as 5000 between the molecular weights of Cp2 and Kp2. Moreover, there are large variations in the reported number of cysteinyl residues per subunit for Kp2 (nine residues) and Av2 (two residues). There is also a considerable difference in the temperature stability of the Fe proteins from various sources. Finally, there appear to be differences among species in the biological regulation of the Fe protein as noted in a bacterium Rhodospirillum rubrum in which the Fe protein on isolation, undergoes activation by a peptide factor which requires ATP and divalent metal ions.

Vanadium Nitrogenase

Recently, workers from England and Canada have demonstrated that the conventional molybdoprotein can be replaced by a vanadoprotein (Vanadium as a cofactor instead of molybdenum) and the component II or iron containing protein replaces the one normally found with molybdoprotein.

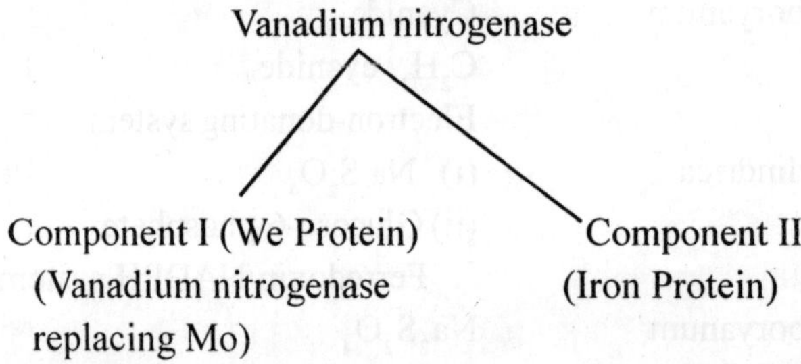

The vanadium nitrogenase has been demonstrated in Azotobacter chroococcum and A. vinelandii that are both obligately aerobic diarotrophic bacteria. It has been observed that purified V-nitrogenase catalyses the reduction of acetylene not only to ethylene, but also to ethane which Monitrogenase does not. Ethane formation has also been observed in cultures of Clostridium pasteurianum. WS suggesting that V-nitrogenase is not restricted to the Azobacteriaceae.

This alternative system of nitrogen fixation has been discovered in mutants lacking nif HDK (genes coding for MoFe protein (nif D and nif K) and Fe protein (nif H). These mutants could still fix nitrogen and function under conditions of Mo deficiency without the structural genes for the Mo nitrogenase. Nitrogen fixation is restored by supplying vanadium to these deletion mutants indicating that the conventional molybdoprotein of nitrogenase is replaced by a vanadoprotein.

$$\begin{array}{c} CH \\ \| \! | \\ CH \end{array} \longrightarrow \begin{array}{c} CH_2 \\ \| \\ CH_2 \end{array} \longrightarrow 2\,CH_4$$

V-Nitrogenase

Control of Biosynthesis of Nitrogenase

Nitrogenase synthesis is strongly inhibited by the presence of ammonia. Consequently, the entire process of nitrogen fixation is also brought to a stand still. Nitrogenase synthesis begins again when ammonia is eliminated. It is known that certain amino acids, particularly aspartic acid, stimulated the formation of nitrogenase after ammonia had been exhausted from the medium supporting nitrogen-fixing bacteria. In general, the formation of the nitrogenase system occurs in the presence of ammonia and when the ammonia supply is exhausted, nitrogenase synthesis is induced by molecular nitrogen before cell growth can take place. When nitrogen-fixing bacteria were grown on ammonia in a 20% O_2 to 80% helium atmosphere, nitrogenase activity could be detected 45 minutes after the exhaustion of the ammonia. Further, only high concentrations of ammonia would repress the synthesis of nitrogenase. Under a limited supply of ammonia, nitrogenase synthesis has, however, been observed in several organisms including A. chroococcum and Rhodospirillum rubrum. It has been further investigated that for N_2 to act as an inducer of nitrogenase, it would have to do so at a concentration of about 10-3 times that required for the induction of galactosidase by the nonmetabolizing substrate IPTG (Isopropylthiogalactoside) in E. coli. Nitrogen-fixing organisms possess a small intracellular pool of ammonia which can be continually used as a source of nitrogen for various biosynthetic processes requiring nitrogen. The concentration of ammonia within the cells checks or acts as a brake on the nitrogenase synthesis. In cultures grown with excess of ammonia, nitrogenase synthesis is completely repressed and, when ammonia is present in a small quantity, nitrogenase synthesis is only partially repressed. In a chemostat (a device for growing microorganisms with an automatic control of nutrient supply) with a limited ammonia supply, all the ammonia in the medium is utilized by the cells and there is no internal ammonia pool. Under these conditions, nitrogenase synthesis occurs at its maximum rate and the concentration of nitrogenase per cell is three times that of a normal N_2-fixing culture. Under inert gas (like argon, helium), the nitrogenase concentration per cell is maintained at this level. In the presence of molecular nitrogen, however, the ammonia pool will increase due to nitrogen fixation and the concentration of nitrogenase, on a per cell basis, is brought to normal.

Excess ammonia acts as a repressor of synthesis rather than as an inhibitor of nitrogenase. This is strongly indicated because excess ammonia, when added to an actively nitrogen-fixing culture,. prevents further synthesis of nitrogenase, but nitrogenase which is already present in the cells is fully active. It has also been observed that carbamyl phosphate, a product of early ammonia incorporation inhibits acetylene reduction and nitrogen-fixing activities in vitro and in vivo. At 1 mm carbamyl phosphate, a 30070 inhibition of activity occurred but even at concentrations of upto 20 mM, the total inhibition was only 50070. Further evidence is provided which suggests that carbamyl phosphate binds itself to the molybdoferredoxin of the nitfogenase. However, this binding occurs when the nitrogenase system is intact and operating.

Besides, the addition of methylamine and 2-methylalanine to nitrogen-fixing cultures also prevents nitrogenase synthesis. These have, however, no effect on the activity of preformed nitrogenase.

Substrate and Hydrogen-Deuterium Reactions

Considering there are several reducible substrates (including protons), apparently only nitrogen reduction is strongly inhibited by hydrogen. Earlier, it was thought that N_2O, H_2, 14CN and N_3 were strong inhibitors of nitrogen fixation. However, now it is clear that these inhibitors of N_2 fixation, are actually reduced by nitrogenase and are termed as 'nonphysiological' substrates. Only H_2 is a real inhibitor of nitrogen reduction. Acetylene (HC = CH) has turned out to be an important substrate and is reduced by nitrogenase. Acetylene is very similar to N_2 (N = N) in properties. The product of acetylene reduction is ethylene ($H_2C = CH$). The general requirements for acetylene reduction are same as those of nitrogen reduction including a reductant ($Na_2S_2O_4$), magnesium and ATP. Extracts from ammonia-grown cells cannot reduce acetylene. Acetylene reduction to ethylene has been used extensively as a sensitive assay of N_2 fixing activity. The ethylene produced is a hydrocarbon which is readily separated by gas chromatography and can be measured in low concentrations with a hydrogen flame ionization detector. The reduction of acetylene to ethylene requires two electrons, whereas the reduction of nitrogen to ammonia requires six electrons. Therefore, the rate of N_2 fixing activity is approximately one third that of acetylene reducing activity. Besides, several other substrates can be reduced by nitrogenase (see Table). The different products formed require different numbers of electrons and these vary from substrate to substrate. The products formed possess 2, 4, 6, 8, 10, 12 or 14 electrons and this has strongly indicated that reduction by nitrogenase only occurs in two electron steps irrespective of the substrate. All the substrates used are molecules longer than N_2 and are linear (with the exception of certain acetylene analogues). The size of the substrate molecule has a bearing on the rate of reduction. Nitrogen is more readily reduced than acetylene. Inturn, acetylene is more readily reduced than nitrous oxide, followed by azide and cyanide.

Table

Various Substrates used by Nitrogenase

Substrate	*Products formed*
N_2	NH_3
N_2O	H_2O, N_2
N_3	NH_3, N_2
C_2H_2	C_2H_4

Substrate	***Products formed***
HCN	CH_4, NH_3, CH_2NH_2
CH_3CN	C_2H_6, NH_3
CH3NC	CH_3NH_2, CH_4
	C_2H_4, C_2H_6
	C_3H_6, C_3H_8
CH_2CH CN	C_3H_6, C_3H_8, NH_3

A strong correlation exists between H_2 metabolism and N_2 fixation. H_2 is a competitive inhibitor of N_2 reduction in cell-free extracts as well as in intact cells. Nitrogenase catalyzed the formation of H-D in the presence of D_2 and this is called H-D (hydrogen-deuterium) exchange reactions. Inhibition of N2 reduction by HZ and H-D exchange reaction are of particular interest. In principle, H-D formation resulted from the decomposition of abound diimide by D, according to the reaction:

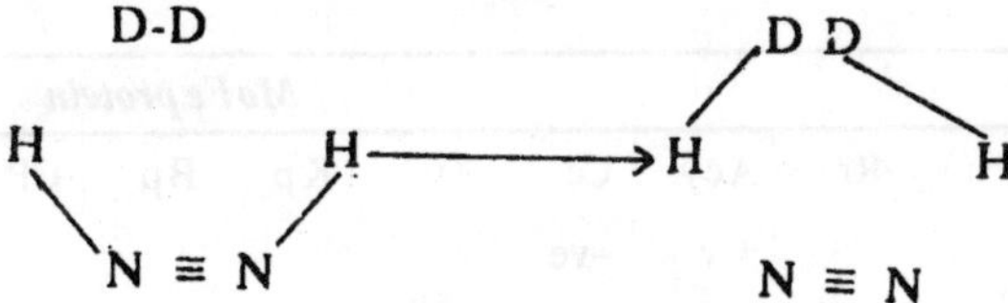

This reaction is analogous to the disproportionation of two molecules of diimide to hydrazine and nitrogen. The inhibition by hydrogen would then result from a decomposition of the diimide intermediate.

The H-D exchange reaction suggests the involvement of a bound diimide intermediate in N_2 fixation.

Cross-reactions and hybridization Studies

Nitrogenase is composed of two components and both the components together constitute the activity of nitrogenase. Since the nitrogenase components from different organisms have certain similarities, it was of interest to see if component I (MoFe protein) from the organism would complement component II (Fe protein) from another organism. Such hybrid nitrogenase enzymes have been able to successfully carry out the reduction of molecular nitrogen (Table).

Of particular interest with respect to these crosses was the fact that the nitrogenase components from physiologically similar organisms complement one another. For example, components of Bacillus polymyxa and Klebsiella pneumoniae are interchangeable and both are facultative anaerobes. Components from physiologically dissimilar organisms such as Azotobacter vinelandii, a strict anaerobe, do not complement each other. Combinations of components from facultative anaerobes with those of either strict aerobes or anaerobes cross-react sporadically. Even a complementation between components of a cyanobacterium (blue-green bacteria), Anabaena cylindrica and a photosynthetic bacteria Chloropseudomonas ethylicum is partially successful. The MoFe protein of the cyanobacterium complements the Fe protein from the bacterium. However, the reciprocal cross was negative. Considerable success achieved in enzymatically active hybrid molecules has further suggested a conservation of the molecular structure of nitrogenase between organisms which might be placed at a considerable distance on an evolutionary scale.

Table
Hybridization Reactions and Activity of Reconstituted Nitrogenase from different Microbes

Fe protein.	***MoFe protein***									
Anabaena cylindrica	Rr	Acy	Cc	Av	Kp	Bp	CP	Rj	Ac	Mf
(Acy)		+ v	–ve							
Rhodospirillum rubrum (Rr)	+ ve									
Chloropseudomonas ethylicum (Ce)		+ve								
Bacillus polymyxa (Bp)	50%			+ ve	+ ve	+ve				50%
Clostridium pasteurianum (Cp)	-ve			-ve	ve	-ve	-ve	-ve		-ve
Klebsiella pneumoniae (Kp)	+ve			+ve	+ve	+ve	+ve	+ve	+ve	+ve
Azotubacter vinelandii (Av)				+ve	+ve	-ve	-ve	+ve		
Azotobacter chroococcum (Ac)					+ ve	50%			+ ve	50%
Mycobacterium flavum (Mf)	+ ve				+ ve	+ ve			+ ve	+ ve
Rhizohium japonicum (Rj)				+ ve			+ ve			

Hydrogenase

The transfer of electrons from nitrogenase to molecular nitrogen has an energetically wasteful side reaction. Many of the electrons recombine with protons before they reach the nitrogen; they are released as molecular hydrogen gas (HZ). Certain strains of Rhizobiunt synthesize hydrogenase, an enzyme that converts molecular hydrogen back into electrons and protons for reuse by nitrogenase. Hydrogenase could therefore serve as a kind of afterburner that would make the nitrogen-fixing bacteria more energyefficient. The enhanced bacterial efficiency would enable the plant to direct its energy more to seed yield than to the support of its symbiotic bacteria. Investigators at Oregon State University have recently demonstrated the effectiveness of hydrogenase in the open field. They showed that field-grown soybeans inoculated with a strain of Rhizobium that makes hydrogenase have a greater yield than soybeans inoculated with a strain lacking the enzyme.

Workers at the John Innes Institute in Britain have shown that the hydrogenase made by certain strains of Rhizobium is coded for by a single gene and that the gene is found on a plasmid, a loop of DNA separate from the bacterial chromosome. It should be possible to transfer the hydrogenase gene to Rhizobium strains that lack the enzyme but have other characteristics making them desirable as nitrogen fixers.

Not all symbiotic nitrogen fixation is conducted by Rhizobium bacteria, and not all nitrogen-fixing bacteria attach themselves to legumes. The key to engineering microorganisms that fix nitrogen for cereal crops may be in understanding the details of such natural symbioses. There is only one well-ducumented instance in which Rhizobium has been found to nodulate a plant other than a legume; it was reported by an investigator in Western Australia. A number of other nitrogen-fixing bacteria, however, have been identified and isolated. The bacterium Frankia alni fixes nitrogen for the alder tree and certain other nonlegume plants. Alder can, therefore, be employed in crop rotation as legumes. Alder seedlings are mixed into forests in order to enrich the soil for commercially valuable trees such as Douglas fir and poplar. The enzyme hydrogenase primarily catalyzes the activation of molecular hydrogen:

$$H_2 \rightleftharpoons 2H^+ 2e^-$$

The enzyme was originally described by Gest and Kamen in 1949 in a photosynthetic bacterium Rhodospirillum rubrum. Since then, it has been purified from a number of bacteria including Clostridium, Azotobacter and Desulfovibrio. In these organisms, a conventional reversible hydrogenase is present. Recently, however, unidirectional hydrogenase has also been reported in C. pasteutianum. Cyanobacterial hydrogenase also seems to be unidirectional and constitutive, inhibited by CO and $C_2 H_2$. Low levels of reverse hydrogenase activity has been detected in cyanobacterium Anabaena. Hydrogenase is constitutive in photosynthetic anaerobes such as Rhodospirillum rubrum and Rhodopseudomonas capsulata. Hydrogenase is present in nonleguminous nodules also. Two extractable hydrogenases, one stimulated by ATP, have been reported in soybean and lupin bacterioids. Hydrogenase is formed by free living rhizobia. Purified hydrogenase has molybdoflavoprotein which participates in electron transfer from hydrogen to a variety of oxidants. The enzyme also has a bivalent cobalt which probably links the enzyme to hydrogen. It is through this hydrogenase-catalyzed reaction that electrons are provided for reduction of nitrogen to ammonia. In case the substrate (nitrogen) is not available, then these electrons combine with protons to release hydrogen gas. ATP-dependent H_2 evolution is a property of the nitrogenase enzyme complex and is quite distinct from ATP independent evolution of H_2 catalyzed by the conventional hydrogenase enzyme which may coexist in nitrogen-fixing cells. It is true that all nitrogen-fixing microbes possess hydrogenase, but not all organisms which have hydrogenase fix nitrogen.

The Hydrogen-donating System

The hydrogen-donating system will make protons and electrons available for the reduction of nitrogen. Pyruvic acid and NADPH generated in the dark reactions of photosynthesis have frequently been mentioned as electron donors. In nitrogen fixing organisms grown on ammonia as a source of nitrogen, the nitrogen-activating system is absent, whereas the hydrogendonating system can still be detected. This suggests that the hydrogen donating system is independent of the nitrogenase complex. However, it alone cannot fix nitrogen although it may be present in the cell. Hydrogen gas can also serve as a source of electrons. Therefore, when HZ gas and ATP are supplied from outside, the need for the hydrogen-donating system is eliminated. It is still not clear if more than one type of hydrogen-donating system can function simultaneously. Other potential sources of electrons are water that is photolyzed, carbon that is fixed during photosynthesis, and reduced sulphur compounds. Recently, it has become clear that electrons from the photolysis of water or from aerobic respiration are not essential for nitrogenase activity. This is supported by the fact that DCMU (an inhibitor of photosystem II) apparently does not inhibit the nitrogenase activity. Electrons from water are made available during photolysis on photosystem II, it is also known now that electron donation to nitrogenase could also be from (i) photosystem I-dependent reactions in light, (ii) the oxidative pentose phosphate pathway in light or dark and (iii) glycolysis and the Krebs cycle.

Ferredoxin and Flavodoxin

Ferredoxin, a very important electron carrier, was discovered in 1962 by Mortenson and Carnahan at the Du Pont Laboratories in Delaware. It has now been isolated from nonphotosynthetic bacteria, cyanobacteria, photosynthetic bacteria and higher plants. It is an iron-sulphur protein with acid labile sulphur. It was recovered for the first time from an anaerobic nitrogen-fixing bacterium Clostridium pasteurianum. The protein has an extremely low redox potential at a slightly alkaline pH. Its properties are given in Table.

Table

Physical Properties of Ferredoxin

Molecular weight	10,000-40,000
Standard oxidation-reduction potential	0.24-0.49
Number of iron-sulphur atoms per molecule	2-4

The inorganic sulphide group is liberated as hydrogen sulphide upon acidification. Ferredoxin loses its characteristic absorption spectrum immediately after iron and inorganic sulphide are removed. The protein plays a major role in nitrogen fixation by connecting the hydrogen-donating and nitrogen-activating system:

$$CO_2 + \text{acetyl-COA} + \text{ferredoxin}_{(red.)} \rightarrow \text{pyruvic acid} + \text{COA} + \text{ferredoxin}_{(oxa.)}$$

Irrespective of the source of electrons, reduced ferredoxin reduces the azoferredoxin and then molybdoferredoxin, which is ultimately activated by ATP through a postulated coupling enzyme:

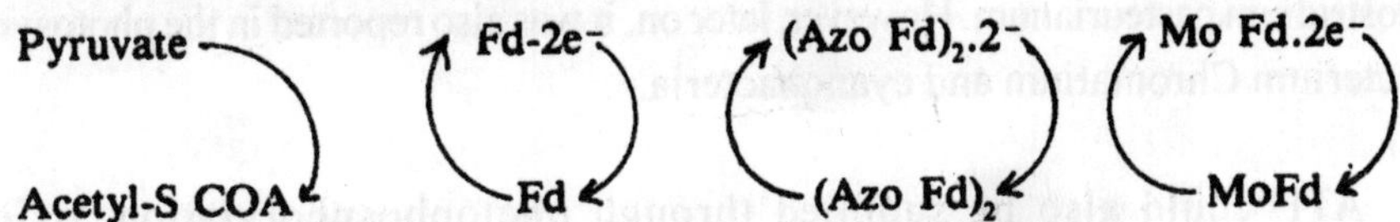

The bacteria which grow on a limited iron supply, produce a flavoprotein in place of ferredoxin, with the functional characteristics of ferredoxin. This protein is called flavodoxin. As with ferredoxin, flavodoxin has been subsequently detected in several organisms including photosynthetic bacteria and cyanobacteria. It has also been proved that ferredoxins from variable sources can function with different nitrogenases; Clostridium ferredoxin, especially, can couple with nitrogenases from Chromatium, Chloropseudomonas and Anabaena. In some cases, ferredoxin and flavodoxin both operate in the same cell, and it becomes very difficult to ascertain which of these proteins actually donates electrons to nitrogenase in vivo. However, both ferredoxin and flavodoxin have been shown to donate electrons in vitro to nitrogenase.

Phosphoroclastic Dissimilation

The major requirements for nitrogen fixation include ATP and an electron source. As mentioned in the previous sections, pyruvic acid is an important requirement in nitrogen fixation. A specific reaction called Phosphoroclastic dissimilation operates during the process of obtaining electrons and ATP formation. In this reaction, the keto acid is dissimilated to acetylphosphate, CO_2 and H_2.

$$\begin{matrix} CH_3 \\ | \\ C=O \\ | \\ {}^{*}COOH \end{matrix} + COA \xrightarrow{\text{pyruvate dehydrogenase}} \underset{\text{AcetylCoA}}{CH_3CO\text{-}SCoA} + CO_2 + H_2$$

$$H_2 \xrightleftharpoons{\text{Hydrogenase}} 2'I^+ + 2e^-$$

Consequently, the acetyl-CoA so formed will undergo phosphorylation to form acetyl phosphate:

CH_3 CO-SCoA + H_3PO_4 Phosphotransacetylase acetyl-PO_4 + CoA Acetyl phosphate could eventually become a source of ATP:

$$\text{acetyl-}PO_4 + ADP \rightarrow \text{acetate} + ATP$$

Phosphoroclastic dissimilation was initially reported in the anaerobic bacterium Clostridium,pasteurianum. However, later on, it was also reported in the photosynthetic bacterium Chromatium and cyanobacteria.

ATP could also be supplied through photophosphorylation, oxidative phosphorylation and by the breakdown of polyphosphates which are abundant in cyanobacteria. Oxidative phosphorylation presumably supplies ATP when nitrogen fixation occurs in the dark.

Six-Electron Requirement

Nitrogen fixation is a reductive process. The first inorganic product formed during nitrogen fixation is ammonia. This is indicated by using a heavy isotope of

nitrogen $^{15}N_2$ which appears in NH_3 formed within growing cells. $^{15}N_2$, can be measured in a mass spectrometer. The first organic product formed which has a high concentration of $^{15}N_2$, is glutamic acid. The oxidation number of N_2 is nil, whereas that of ammonia .formed is —3. On the following scale, NO_3 is highly oxidized (oxidation number + 5) and ammonia is highly reduced:

NO_3	NO_2	$N_2O_2^{2-}$	N2	NH_2OH	NH_3
+5	+3	+1	0	−1	−3

Therefore, proceeding from N_2 onward and ending with ammonia, a continuous reduction with a change in oxidation number from 0 to - 3, has to be achieved:

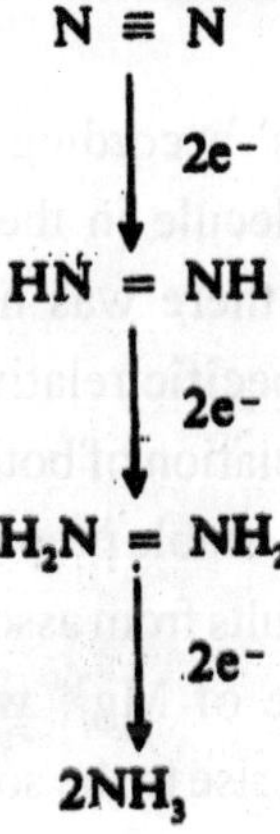

In general, a total number of six electrons are required to reduce N_2 to ammonia.

Recently, a proposal of eight-electron requirement has been suggested to explain H_2 evolution during nitrogen fixation:

$$N_2 + 8H^+ + 8e^- \rightarrow 2NH_3 + H_2$$

No concrete evidence, however, has been presented to support this explanation which does not seem to be a correct representation of the enzyme stoichiometry.

Oxygen Sensitivity

Both MoFe and Fe proteins of nitrogenase are irreversibly damaged by O_2 and, special anoxygenic techniques have been devised for their purification. MoFe protein is relatively less sensitive to O_2 than Fe protein. Clostridium pasteurianum needs barely five minutes to inactivate Fe protein in the presence of O_2 with the concomitant oxidation of two sulphydryl groups and one S^{2-}. Practically, all cyanobacterial nitrogenases are sensitive to O_2. Oxygen-inactivated proteins have no electron paramagnetic spectrum (EPR), and the Mossbauer of O_2 - inactivated K. pneumoniae proteins resembles that of reduced ferredoxin. The initial changes during oxygen inactivation render the protein incapable of accepting electrons. The a-helical content of Fe protein decreases on exposure to an increase in oxygen.

Despite the O_2 lability of nitrogenase proteins in vitro, the capacity to fix N_2 is found in obligate aerobes, oxygenic photosynthetic prokaryotes as well as in obligate and facultative anaerobes. For all N_2 fixing organisms, O_2 is potentially inhibitory to growth and different protective mechanisms are applicable to different systems:

(1) 'Conformational Proteins'-according to which there is a reversible change in the shape of the enzyme molecule in the presence of oxygen that renders it insensitive to oxygen. Initially, there was no evidence whether conformational protection resulted from some specific relative orientation of the two nitrogenase proteins themselves or from association of both with one or more stabilizing factors. Studies with partially purified O_2 - stable preparations from Azotobacter have shown that conformational protection results from association with a third protein. Protection also depends upon the presence of Mg^{2+} which may also induce considerable aggregation of nitrogenase giving rise to the so called particulate form of the enzyme observed in crude extracts.

(2) 'Respiratory; protection is perhaps confined to Azotobacter which has the highest respiratory rate known in living organisms (5000 ml O_2 taken mg dry wt^{-1} hr^{-1}) and, thus, excess of oxygen is utilized for the oxidation of various substrates in the cell. The protective response partly depends upon subtle adjustments in the composition and functioning of the respiratory chain which itself is very complex.

Although, short-term protection against O_2 has been attributed to the formation of a complex between riitrogenase and soluble 2Fe-2 protein in Azotobacter. Such a protein has not been isolated from any cyanobacterium. Heterocystous cyanobacteria have a special protection mechanism because heterocysts lack PS II and do not evolve oxygen during photosynthesis. Consequently, heterocysts provide an O_2-protected environment for nitrogenase. Besides, the constituents of the heterocyst envelope may provide an oxygen diffusion barrier which protects the oxygen labile nitrogenase system.

Hydrogen, Evolution and biotechnological operation

The first indication that group nitrogenase evolved hydrogen in vivo was obtained by a research group from Wisconsin in 1957. It was later observed that several species of nodulated plants wasted a substantial proportion of their energy for N_2 fixation as H_2 evolution. As described in the previous sections, ATP-dependent hydrogen evolution is significant during nitrogen fixation. All nitrogenase preparations isolated have been found to catalyze an ATP-dependent H_2 evolution even in the presence of N_2.

BIOCHEMICAL SHORT CIRCUIT leads to a wasteful release of hydrogen gas (H_2) as a byproduct of the enzymatic reaction by means of which Rhizobium bacteria convert molecular nitrogen (N_2) into ammonia (NH_3). The hydrogen has no known value to the plant or to the bacteria; instead it seems merely to waste energy derived from photosynthesis. The energy serves to segregate free protons (H^+) and electrons (e^-), which then drive the reactions of nitrogen fixation. If the protons and electrons recombine to form hydrogen, however, the energy is dissipated. In some strains of Rhizobium bacteria, a gene carried by a plasmid circumvents this inefficiency. The gene, designated hup codes for the synthesis by the bacteria of the enzyme hydrogenase: Hydrogenase catalyzes the breakdown of hydrogen gas into its constituent protons and electrons for reapplication in nitrogen fixation. The introduction of the hup gene into a Rhizobium inoculant has been shown to give the inoculated plant higher seed yields than those of plants infected with strains of Rhizobium bacteria that do not synthesize hydrogenase.

Evidence has been presented to indicate that this reaction also occurs in vivo in various nitrogen-fixing organisms. Hydrogen evolution persists in the presence of alternative substrates such as cyanide or isocyanide but not in the presence of acetylene. As a matter of fact, H_2 evolution accompanies nitrogen reduction by the purest nitrogenase preparations and even under 6 atm. of N_2. This has been clearly demonstrated in the photosynthetic bacterium Rhodospirillum rubrum. ATP-dependent H_2 evolution, however, is a property of the nitrogenase enzyme complex and is quite distinct from ATP-independent evolution of H_2 catalyzed by the conventional hydrogenase enzyme which may coexist in N_2 fixing cells. Hydrogenasecatalyzed hydrogen is independent of N_2 gas and requires methyl viologen (an electron donor). It is completely inhibited by carbon monoxide in contrast to ATP-dependent H_2 evolution due to nitrogenase. It should be noted that the reasons for HZ evolution by nitrogenase are least understood.

Aerobic and Anaerobic Systems

Although nitrogen fixation is a reductive process and is highly sensitive to oxygenic conditions, yet in nature, there are both aerobic and anaerobic systems present which operate smoothly and fix nitrogen. There are strict aerobes and anaerobes along with the facultative anaerobic system in free living microbes (Table).

In a symbiotic system, membrane envelopes compartmentalize rhizobia in the plant cell metabolism. Surrounding the packets of bacteria are large quantities of leghemoglobin. The function of leghemoglobin is to bind O_2 and prevent free oxygen from reaching the nitrogenase within the bacteria. The plant codes for this globin and the bacteria seem to produce the heme moiety of leghemoglobin.

Table

A Comparative Account of Various Aerobic and Anaerobic Nitrogen-fixing Systems

Organism	***Status***	***N2- fixing property***
Clostridium pasteunanum	Nonphotosynthetic bacteria	Strict anaerobe
Desu(fovibrio desulfuricans)	Nonphotosynthetic	Anaerobic
Klebsiella pneumoniae	Nonphotosynthetic	Facultative anaerobe
Bacillus polymyxa	Nonphotosynthetic	Facultative anaerobic
E. coli C-M 74	Nonphotosynthetic	Facultative anaerobe
Azotobacter chroococcum	Nonphotosynthetic	Aerobic system
Azotobacter vinelandii	Nonphotosynthetic	Aerobic system
Nostoc muscorum	Cyanobacteria photosynthetic	Anaerobic in cells, heterocystous, microaerophilic
Anabaena cylindrica	Cyanobacteria photosynthetic	Anaerobic in cells, heterocystous, microaerophilic aerobic
Gloeocapsa	Cyanobacteria photosynthetic	Anaerobic
Chromatium vinosum	Photosynthetic bacteria	Anaerobic
Rhodospirillum rubrum	Photosynthetic bacteria	Nonheterocystous
Plectonema boryanum	Photosynthetic cyanobacteria	Nonheterocystous, microaerophilic

The reasons for the existence of both aerobic and anaerobic systems are only speculative. Considering that cyanobacteria were primarily responsible for filling up our atmosphere with O_2 during the initial stages, an anaerobic system might have developed gradually. The genetic basis for this has not yet been understood.

Energetics

Almost two decades have passed since it was first discovered that ATP is involved in nitrogen fixation. In general, nitrogenase-catalyzed reductions have an overall negative free energy change in spite of which ATP is required for the reaction. The participation of ATP in the reduction process was inferred by experiments in which nitrogenase was inhibited by arsenate or by glucose in a cellfree system containing pyruvate, the auxiliary enzymes of the phosphoroclastic reaction, and hexokinase. Arsenate is known to inhibit acetokinase and thus also ATP formation from acetyl phosphate. The Dupont group reported a requirement of two ATP per electron pair transferred by the nitrogenase of C. pasteurianum. However, other laboratories reported a stoichiometry of 4-5 ATP per 2e- for the C. pasteurianum and this was found to be pH-dependent.

The action of ATP requires the presence of a divalent metal. In order of decreasing effectiveness, Mg^{2+}, Mn^{2+}, Co^{2+} and Fee+, and Fe'+ have been employed. More specifically, Mg-ATP binds itself to the Fe protein and causes a conformational change in the protein. This binding renders the iron more labile to removal by iron-binding reagents, resulting in a substantial lowering of its redox potential, causing oxidation of the Fe protein and reduction of the MoFe protein. Extensive work with fluorescence labels, spin labels, NMR and specific inhibitors have almost proved that there may be at least two ATP sites and that a pair of free sulphhydryls and possibly some of the iron groups in close proximity to the site of substrate reduction area at the ATPase site. Recently, the Sussex group has reported the catalysis of exchanges of terminal phosphate groups of ATP and ADP by purified nitrogenase proteins.

Still question of the stoichiometric relationship between the molecules of ATP hydrolyzed and the number of pairs of electrons transferred by nitrogenase to an acceptor remains unresolved. (Table).

Table

In vivo ATP Requirement Per N^2 Molecule Reduced

Organism	*No. of ATP molecules*
Clostridium pasteurlanum	20
Klebsiella pneumoniae	25-29
Azotobacter chroococcum	4-5

Stoichiometric observations are, generally, expressed in terms of the ATP: $2e^-$ ratio. It may be essential to describe glucose oxidation or fermentation pathways before the actual ATP requirement is determined. For example, K. pneumoniae yields 2 ATP per glucose while fermenting glucose through EMP pathways. Valentine's group calculated 21-25 ATP/N_2 for five different mutant strains of K: pneumoniae (25°C, pH 7.3). The reduction of N_2 to NH^{4+} is also energydependent (in the form of low potential electrons) and apparently, 4.3 electron pairs are consumed per N_2 reduced (from the observed H_2/NH^{4+} molar ratio of 0.65). Based on their observations, the Davis group finally suggested a total of 3540 ATP/N_2 for the appropriate reactions catalyzed by nitrogenase. An operational schematic representation of nitrogen fixation, in general, is illustrated in Fig.

Organization of nif Genes

In symbiotic nitrogen fixation, it has been clearly established that plant genes are involved in nitrogen fixation. This is documented specifically by the leghemoglobin, a protein moiety, which is coded for by the plant. The information pertaining to the strict coding behaviour of nitrogenase genes, whether coded by the bacteria or plant, is not very clear, although evidence goes in favour of bacteria. However, the major hurdle in studying such a symbiotic system is that bacteria, in isolation, apparently do not fix nitrogen. Therefore, more attention has been diverted towards understanding the organization of the nitrogen-fixing gene complex in free living microbes. The initial credit goes to Karlssons and Barker (1948) who isolated, for the first time, conditional nonnitrogen-fixing mutants in Azotobacter. This mutant could not fix atmospheric nitrogen when grown on glucose, but could fix nitrogen when grown on ethanol. Similarly, several mutants with nonfunctional or defective nitrogenase have been reported in Azotobacter and Clostridium. Recently, Klebsiella pneumoniae has attracted attention and has been widely studied as far as the genetics of nitrogen fixation is concerned. The organism can grow under aerobic or anaerobic conditions but fixes nitrogen only under anaerobic conditions. It is distributed fairly well in soil and water. Bergerson has described some nitrogen-fixing strains from the leaf nodules of the plants. An interesting association of Klebsiella was recorded in the intestines of New Guinea natives (fed mostly on sweet potatoes). The implications of K. pneumoniae and its role as an efficient nitrogen fixer in the human gut remains to be determined.

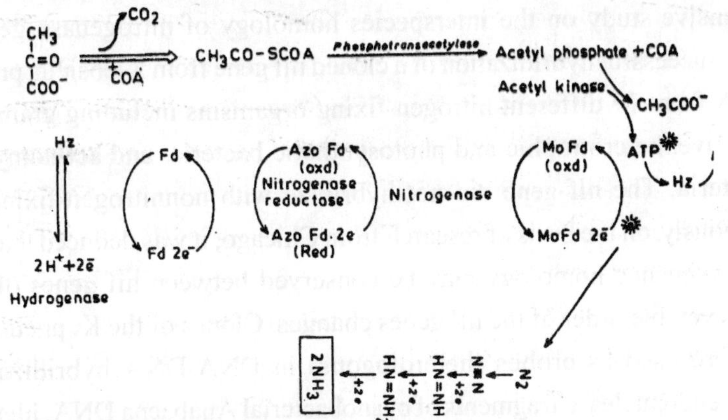

Fig. Schematic representation of current status of Nitrogen fixation process

Methodological studies on the fine genetic dissection of nitrogen fixation in K. pneumoniae demanded the isolation of a large number of mutants with varying nitrogen fixing abilities. The major question which confronts scientists, currently, is the whereabouts of the piece of DNA which codes for the structural and regulatory genes governing nitrogenase. Are these genes clustered together constituting a 'nif operon' or diffused over the entire chromosome? Valentine's group, in their experiments on nitrogen-fixing mutants, has reported generalized transduction employing, PI phage and the transfer of nitrogen fixation property from the wild type to a mutant of K. pneumoniae, unable to fix molecular nitrogen. It was suggested that most nif genes are clustered near the histidine (his) operon. They further reported (from a large number of recombination studies and transductional analysis) the region of the nif operon to be 15-20 gene length in size. Apparently, this would code for almost all the protein required during the entire process of nitrogen fixation. About 42 nif strains were isolated; of these, 39 were found to be genetically linked with his. They have also observed two mutations which are phenotypically similar to nif mutants and map close to his, but are not linked to his because no nif+ colonies were found among the his+ transductants. Now, it is accepted that nif genes are close to the operator site of the his operon. An extensive study employing deletions extending through the his genes suggested the order rfb and his nif shu (rbf is the locus coding for phage resistance and specifying the cell wall lipopolysaccharides; gnd codes for gluconate 6-phosphate dehydrogenase, and shu codes for utilization of shikimate). Streicher and coworkers had earlier reported eductants (mutants isolated from lysogens which have very large deletions) lacking genes of the his operon and gnd, unable to fix nitrogen and produce no nitrogenase.

The Harvard group (Now the Medical School, Boston) has made a comprehensive study on the interspecies homology of nitrogenase genes. They reported a successful hybridization of a cloned nif gene from Klebsiella pneumoniae with DNA from 19 different nitrogen-fixing organisms including gram negative, gram positive, heterotrophic and photosynthetic bacteria, and actinomycetes and cyanobacteria. The nif gene did not hybridize with nonnitrogen-fixing species. Simultaneously, on the basis of research from Chicago, it was deduced that although the DNA sequence homology may be conserved between nif genes of different organisms yet, the order of the nif genes changes. Clones of the K. pneumoniae nif cluster were used as probes (heterologous) in DNA-DNA hybridization with restriction endonuclease fragments of cyanobacterial Anabaena DNA. Identification and cloning of the nif gene from Anabaena was possible by using such heterologus probes. Very recently, Haselkorn's group has further determined the nucleotide sequence of the structural gene coding for nitrogenase reductase (component II or Fe protein of nitrogenase) from a cyanobacterium Anabaena. They have reported 194 bases of 5' -flanking sequence and 170 bases of 3' -flanking sequence. An astonishing similarity was observed between the amino acid sequence of nitrogenase reductase from Clostridium pasteurianum and the systeine-containing peptides of protein from Azotobacter vinelandii. They further observed that amino acid sequence around five systeines, located in the amino terminal (two-third of the proteins), are totally conserved in all the three organisms tested.

In Klebsiella pneumoniae, Genetic analysis of N_2 fixation has clearly indicated the involvement of at least seventeen genes which are linked and arranged in about seven or eight transcriptional units (Fig.). Genes specifically required for the fixation of dinitrogen (N_2) in K. pneumoniae are clustered on the genetic map near the operator end of the histidine (his) operon. The Harvard group has also constructed a map of nif genes, primarily using the amplifiable nif plasmids. The minimum size of the entire nif gene cluster has been found to be about 24 kilobase (1 kilobase is equal to 1000 nucleotide pairs). The maximum separation between insertions in two adjacent genes (nif M and nif V) was found to be 1.5 kb.

Assigning functions, nif H, nif D and nif K, are the structural genes for nitrogenase reductase. These three genes form one transcriptional unit with the promoter at the start of the nif H. These three genes have been highly conserved during evolution as proved by probing and cloning of these genes from a variety of nitrogen fixing

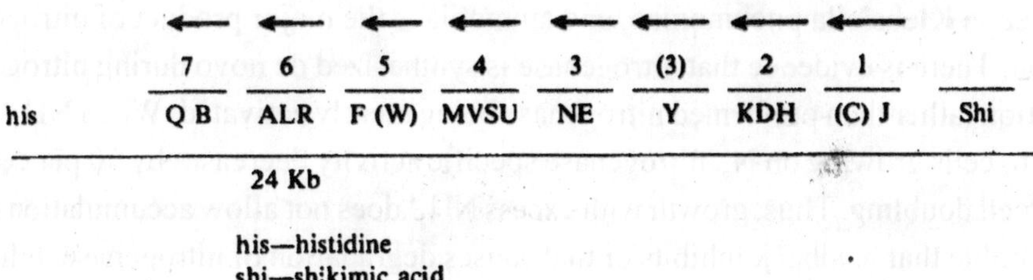

Mol.wt.: A66; L 52, F 26; M 27; V 38; S 42; U 28; N 50; E 46; K 60; D 56; H 35; J 120

Fig. Organization and Transcriptional Units of *nif* Region

microbes using Klebsiella nif HDK genes as probes. Although there is a fair organizational resemblance between the nif gene cluster of various organisms, yet in Anabaena (a cyanobacterium) nif H and nif D were rightly localized but nif K has been found to be 11 kb away from nif Din vegetative cells (In heterocysts, however, nif HDK operon is complete). Some exceptions of similar nature have also been discovered in Rhizobium species. The genes involved in the synthesis of FeMo protein include nif E, nif N, nif B and nif Q. Presumably, nif M, nif V and nif S are required to convert the nif H produced into an active nitrogenase reductase. Besides, the genes concerned with the electron transport include nif F and nif J. A positive regulatory protein product is thought to be produced by nif A, and this is apparently required for the expression of all other nif genes. W.J. Brill from Wisconsin has studied the relationship of gene products and its regulation of nif in K. pneumoniae. The expression of his and nif genes has been compared in Serratia marcescens, Erwinia herbicola and Proteus mirabilis. Restriction maps (deletions) have recently been prepared in the nif region of K. pneumoniae. The nif genes have also been isolated from Rhizobium gaponicum (from soyabean nodules) and introduced in E. coli using plasmid as vector. The structure and regulation of nif genes have been most extensively studied in free living Klebsiella pneumoniae and Rhizobium meliloti (the endosymbiont of alfalfa, Medicago saliva).

Regulation of nif Expression

There are multiple regulatory mechanisms underlying the synthesis of nitrogenase and these have been studied extensively in Klebsiella pneumoniae, Azotobacter vinelandii and K. aerogenes. Molecular N_2 first enters into the main stream of nitrogen fixation and many workers have reportedly studied the inducing property of nitrogen for nitrogenase synthesis. Due to the ubiquity of molecular N_2 and the failure to obtain absolutely nitrogen-free cellular conditions, however, this type of control system is extremely difficult to work out.

The most sophisticated control system for modulating nitrogen fixation has been recorded in Klebsiella pneumoniae, and ammonia is the major product of nitrogen fixation. There is evidence that nitrogenase is synthesized de novo during nitrogen starvation rather than preformed nitrogenase being merely activated. When NH_4^+ is added to cells growing on N_2, nitrogenase specific activity decreases by 50 per cent in one cell doubling. Thus, growth with excess NH_4^+ does not allow accumulation of a metabolite that feedback inhibits or that causes degradation of nitrogenase. It has been suggested that NH_4^+ in the medium prevents transcription of nif DNA. Earlier, it was reported that Rifampicin added to cultures just beginning to derepress nitrogenase synthesis mimics the effect of ammonia. It is not known if the components are derepressed coordinately. At present, there are two major questions confronting the regulation of nitrogen fixation by ammonium: (1) Is NH_4^+ the actual effector for the repression of nitrogenase synthesis? (2) Is a metabolite or an enzyme involved in NH_4^+ assimilation responsible for repression ? Apparently, NH_4^+ is not the sole effector for repression.

Briefly, the reactions can be summarized as follows (cells growing with limiting ammonia):

(i) NH_3 + α – ketoglutarate + NADPH + H + $\xrightarrow{\text{Glutamate dehydrogenase}}$ glutamate + NADP$^+$ + H_2O

(ii) Glutamate + ATP + NH_3 $\xrightarrow{\text{Glutamine synthetase}}$ glutamine + ADP + Pi

(iii) Glutamine + α – ketoglutarate + NADPH + H$^+$ $\xrightarrow{\text{Glutamate synthase}}$ 2 glutamate + NADP$^+$

Magasanik from MIT has expanded on the role of glutamine synthetase as a central point in nitrogen metabolism because this enzyme seems to control its own synthesis as well as the synthesis of glutamate dehydrogenase and enzymes of several pathways involved in the degradation of N-containing compounds such as histidine and proline. The fact that NH,+ represses histidase synthesis in K. aerogenes and nitrogenase synthesis in K. pneumoniae led several workers to speculate about whether nitrogenase synthesis also requires glutamine synthetase, which plays an active role. The Davis group isolated glutamine auxotrophs (unable to grow in the absence of glutamine) of K. pneumoniae and found that these strains do not produce active nitrogenase. The involvement of glutamine synthetase in nitrogenase synthesis has further been confirmed by transferring the F factor (an extra chromosomal genetic particle in bacteria) containing the nif genes of K. pneumoniae to a non N_2-fixing strain of K. aerogenes having a mutation that causes the formation of constitutive levels of glutamine synthesis. This hybrid strain synthesizes high levels of nitrogenase in the presence of excess NH_4+. The glutamine requirement was lost when this strain was grown with excess NH_4+. Therefore, the mutation is probably in a gene that controls glutamine synthetase. It is suggested that this synthetase is needed to activate the synthesis of nitragenase. The involvement of glutamine synthetase in nitrogenase regulation is the most difficult and unsolved question because of an obvious paradox which seems to exist where the inhibitors of glutamine synthetase derepress nitrogenase synthesis and nitrogenase synthesis is repressed in glutamine auxotrophs. The problem has not yet been fully resolved.

If glutamine synthetase does exert a direct effect on the expression of nif, then transgenosis of repressible nif genes from Klebsiella to plants would require homologous glutamine synthetase structural and regulatory genes to be transferred along with nif. Further evidence that implicates glutamine synthetase as a direct participant in the genetic control of nitrogen fixation has been given by Valentine and coworkers. They isolated glutamine-requiring auxotrophs (gln A-) lacking glutamine synthetase which are also unable to synthesize rutrogenase. It was assumed that the mutations which block the catalytic activity of glutamine synthetase simultaneously destroy the regulatory properties of the glutamine synthetase.

The nif-specific regulation is mediated by the products of nif LA operon. The nif A product (Nif-A) is a transcriptional activator required for the expression of all K. pneumoniae nif operons (except for its own). nif L has a regulatory role. It appears to antagonize the action of nif A in the presence of oxygen or intermediate concentrations of fixed nitrogen. The regulation of nif is in general mediated by a centralized system (ntr system) which controls the expression of various nitrogen assimilatory genes in enteric bacteria. Under conditions of ammonia starvation Ntr C (product of ntr C) combined with Ntr A (product of ntr A) activates nif LA operon and also activates other genes involved in nitrogen assimilation such as but (histidine utilization) and put (proline utilization). Under conditions of ammonia excess, Ntr C (product of ntr C) is present in an inactive form.

Recently, the existence of an enzyme system known as the 'adenylylation enzyme cascade' had been revealed. In short, the addition of NH,+ (or any other nitrogen compound that represses nitrogenase synthesis) is thought to trigger the enzyme cascade system that catalyzes the covalent modification of glutamine synthetase by the attachment of adenyl moieties on specific tyrosine residues of the protein. The

Davis group has further suggested that the cascade might be regarded as the 'biochemical' part of the model in contrast to the genetic part, dealing with the activation of the expression of the nif genes. It was suggested that adenylylation blocks the binding of glutamine synthetase to the nif promoter or deadenylylation leads to the binding of glutamine synthetase and the subsequent activation of the nif transcription. A strong correlation between glutamine synthetase and DNA/RNA polymerase, sticking to the nif promoter region, is believed to exist.

BIOTECHNOLOGICAL PROGRAMME AND CURRENT STATUS

All over the world today, agricultural scientists in many developing countries are concentrating on rhizobial technologies and the culture of other organisms that have superior biological nitrogen fixation capabilities. With the advent of techniques for the production of strains in which nif genes are constitutive or derepressed, the prospects of improving the fertility of soil have increased. Some of these strains are able to excrete large amounts of fixes) nitrogen as NH_4^+ (upto 20.2u mole of NH_4^+ are produced mg^{-1} of cell protein 24 h^{-1}). The strains with a derepressed system continue to synthesize nitrogenase in the presence of NH_4^+, in amounts upto 65 per cent of those produced in the absence of NH_4^+. Cyanobacteria afford an excellent tool for obtaining such derepressed mutants for producing ammonium fertilizers because of their ubiquity and also for the derivation of energy for nitrogen fixation directly from the Sun.

How can the nitrogen-fixing ability of Rhizobium on legume roots be enhanced, thereby increasing the crop yield? A straight-forward approach is through plant breeding, which does not require microbiological techniques at all. For example, if the rate of photosynthesis were increased by selective breeding, the bacteria in the nodules might be able to fix more nitrogen. By combining plant breeding with microbiological techniques for modifying the Rhizobium, however, even high yields of plant protein might be obtained.

In view of the great successes that have been achieved in securing a whole plant from a single cell protoplast, the possibility of employing gene cloning to transfer desirable nif genes in the plant cells is fast developing. However, there are multiple barriers including DNA uptake, DNA replication and expression. In vitro, construction of small amplifiable plasmids, which carry some of the K. pneumoniae nif genes, has already been achieved. Plant protoplasts have proved to be an ideal system for the induced transfer and uptake of cell organelles. It is surprising, therefore, that serious attempts have not been made to transfer nitrogen-fixing bacteria directly into protoplasts.

Now that both conceptual and technological breakthroughs have been made for gene cloning in bacteria, the benefits of the technology directed towards the localization of some of the potential benefits for society in the agricultural areas are overdue. Efforts should extend the gene cloning technology to nitrogen-fixing bacteria and cyanobacteria of agricultural importance, and the establishment of a gene cloning system in plant cells for the purpose of obtaining potentially nitrogen-fixing crop plants. Nature has been liberal in donating the nif gene pool and its greatest homology to the nitrogen-fixing organisms. The amino acid composition of nitrogenases, from several species, are very similar. Moreover, purified MoFe protein from one bacterial species reconstitutes an enzymatically active hybrid nitrogenase with Fe protein from another species. On the basis of ribsomal DNA sequence data, it has been proposed that evolutionary divergence among gram negative species such as K. pneumoniae, gram positive species such as Clostridium pasteurianum and cyanobacteria such as Anabaena variabilis occurred as long as 3 × 109 years ago and these still conserved the nif DNA sequence.

The mechanisms of oxygen repression of nitrogenase is another exciting area to be explored. This may be even absolutely essential for a successful gene transfer to a higher plant cell with oxygenic metabolism as is the case with all the crop plants. There is no information whether the O_2 repression is related to or mediated by mechanisms identical to NH repression. How the perception of O2 takes place in the cell is not at all known. Perhaps a solution may have to be obtained through genetics and O_2 sensitivity might be a genetically determined property. The lack of mutants insensitive to O_2 repression is hampering the work. Constitutive mutants might also be extremely helpful in such investigations. A recent finding from Sussex has suggested that nif L may be involved oxygen regulation nitrogenase synthesis in K. pneumoniae.

More recently, the role of glutamine synthetase in the regulation of nitrogenase is gaining importance. It has been shown that Rhizobium strains possess two distinct GS activities and can be separated by isoelectric focusing. GS I is apparently a dodecameric enzyme and its catalytic activity can be modulated by reversible adenylylation. In contrast, GS II is thermolabile and does not seem to be adenylylated. GS I is definitely involved in the regulation of nitrogenase gene expression, both in cultured and in root nodule symbiosis. GS has also been purified from cyanobacteria and each molecule is composed of 12 subunits arranged in two superimposed hexagonal rings.

As described in the previous sections, hydrogen is the by-product of the nitrogenase reactions in all the organisms that fix molecular nitrogen and dissipate upto one-third of the energy flux through nitrogenase. As an energy saving device, some nitrogen fixing organisms including certain strains of Rhizobium possess an active hydrogen uptake (hup) system permitting hydrogen to be recycled. Consequently, these systems (hup+) are certainly more energy-efficient. The Davis group has already shown that hup genes are located on plasmids which can be transferred by conjugation. Consequently, it may be possible to introduce these hup+ plasmids into systems lacking these, resulting in the development of genetically improved strains of Rhizobium. These hup+ strains can recycle the H_2 and recover two moles of ATP per mole of H_2, reducing the energy waste by about 12.5 per cent.

The genetic construction of new nitrogen-fixing organisms must be a very carefully determined process. Multiple requirements for N_2 fixation are obligatory and include (i) nitrogenase; (ii) a strong reductant; (iii) adenosine triphosphate; (iv) NH,+ assimilation pathway; (v) protection system for oxygen denaturation; and (vi) regulatory system for nif genes. Fixation of nitrogen by an organism calls for a heavy energy demand and this may be a serious hurdle in searching for new recipients for nif genes. Purified nitrogenase requires at least I S moles of ATP to reduce per mole of nitrogenous substrate to NH_4+. How a potential recipient for nif genes could continuously supply this enormous amount of energy is a problem yet to be solved. Even higher plants such as pea that use sunshine show that 32 per cent of the total carbon fixed by the plant is transported to the nodules and the nodules, in turn, return 45 per cent of the carbon as amino compounds for the plant growth while using the remaining 55 per cent for their own growth and to supply energy for the fixation of nitrogen. Therefore, it is imperative that any genetic modification leading to nitrogen fixation in a higher plant cell must account for a very liberal energy budget. E. coli is like any nitrogen-fixing bacteria in possessing the genetic machinery supplying most of these requirements including a powerful reductant (during the fermentation of carbohydrates). Besides, the NH_4^+ repression system is operative as the synthesis of glutamine synthetase is repressed by NH_4

Currently, plasmids are proving to have a bright future in genetic engineering of nitrogen fixation as these play a role in the synthesis of large quantities of nitrogenase needed for nitrogen fixation. Cannon and coworkers from England have recently constructed amplifiable plasmids carrying 14 out of 18 Klebsiella pneumoniae nitrogen fixation nif genes, including those that code for the MoFe and Fe proteins of nitrogenase. Gene amplification is a very useful procedure concerned with the 'amplification' of genes fused with the bacterial plasmids. The possibility of enzymatically inserting nif DNA into a fast replicating plasmid from E. coli (Col E 1) might increase the synthesis of nitrogenase and, consequently, the production of NH 4^{+}.

If genes can be introduced into plant cells at will, one important application may be the insertion of nitrogen fixation genes (nif genes) into cereal plants. (Table). Workers at the University of Sussex have assembled a bacterial plasmid that includes all 17 of the known nif genes from the nitrogen-fixing bacterium Klebsiella pneumoniae. When the plasmid was transferred to E. coli, normally incapable of fixing nitrogen, it became a nitrogen-fixing microorganism.

Table

Selected Nitrogen Fixing Microbes Useful for nif Gene Transfer Programme

Donor	***Receptor***	***Expression***
Klebsiella pneumoniae Strain M 5al	E. coli	Successful
	Klebsiella aerogenes	"
	Salmonella typhimurium	"
	Serratia marcescens	"
	Erwinia herbiocola	"
	Pseudomonas sp.	"
Rhizobium trifolii Strain Ti	Agrobacterium tumefaciens	"
Rhizobium trifolii Strain T1K	Klebsiella aerogenes	"
	E.coli	"
Klebsiella pneumoniae	Yeast	Unsuccessful

Even more promising have been the recent successes of groups of investigators at Cornell University, the Pasteur Institute and the University of Paris. They introduced the 17 nif genes of K. pneumoniae into yeast (Fig. 10.3). Yeasts, being eukaryotic against bacteria which are prokaryotic, are closely related to the higher plants than to bacteria. Hence, the introduction of nif genes into yeast cells marks the crossing of a significant biological barrier.

Nevertheless, the yeast cells carrying the nif genes were not able to express the inserted DNA: they were not able to fix nitrogen from the atmosphere. The failure illustrates the complexity associated with the genetic engineering of biological functions embodied in more than one gene. The transferred DNA must first be transcribed correctly into RNA by the yeast.

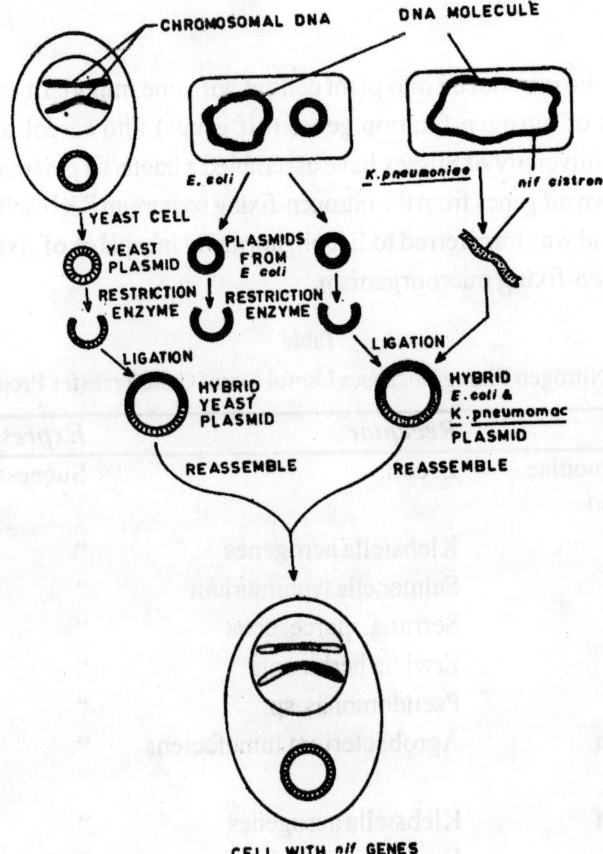

Fig. Genetic engineering of nif gene in yast cell

Correct transcription cannot be assumed as a matter of course, because the yeast must correctly interpret the bacterial signals to start and stop the transcription. The RNA must then be exported from the nucleus and recognized by the ribosomes as a messenger RNA for translation into protein. The 17 proteins that express the nif genes must then function together in the foreign cytoplasm of the yeast cell. There may be impediments to such functioning. For example, the nitrogenase molecule has many iron atoms in its structure. Enough iron is evidently available in nitrogen-fixing bacteria, but it is not certain whether the heavy demand for iron can be met in a plant cell without jeopardizing the synthesis of other enzymes essential to the plant.

Doubts of this kind can arise even when the expression of a particular plant function calls for only one gene. The enhancement of the protein quality of food is a major goal of agricultural research. The stored proteins in some grains and other seeds, for instance, are deficient in amino acids essential to human and animal nutrition. It was once thought that the insertion of a single gene, coding for a stored protein with a better balance of amino acids, would be sufficient to improve the quality of the protein. It is now understood that major seed proteins are often collections of numerous related proteins, each of the latter being encoded by a separate gene. In order to make a significant improvement in the overall quality of the stored protein many of the genes might have to be individually modified.

Although such technical problems may be difficult to overcome, they are probably not the limiting factors in the application of microbiology to agriculture. In the long run, more stringent constraints will be the comparatively small financial commitment to basic research in agriculture and the loss of irreplaceable genetic resources from the totai gene pool.

Genetics and genetic regulatory mechanisms of nitrogen fixation afford a very bright and exciting future and this field would emphasize a research pursuit which is likely to be a conglomeration of multiple disciplines and of direct utility to humanity at large. The crucial technical problem which an expert geneticist is facing is the successful establishment of nitrogen-fixing genes in a higher plant cell and an efficient replication system. Whether the oxygen sensitivity of nitrogenase would also turn out to be a genetically controlled phenomenon is not clear and may be difficult to envisage in the absence of suitable mutants. This remains another technical hitch in a successful genetic modification of nif genes. A fine genetic dissection with individual functions of nif operon is likely soon and it would be a prerequisite for further genetic modification.

SECTION - 3

ROLE IN INDUSTRY AND AGRICULTURE

EIGHT

Metals and Mining

The release of elemental carbon, nitrogen, sulphur, phosphorus and other trace elements by soil microorganisms from organic materials, is known as mineralization and one of the important biochemical properties of microbes. Interest in geomicrobiology has been growing in recent years, with the development of new insights into the role of microbes in a number of geological processes. Microbes are now recognized as important geologic agents, playing a role in processes such as mineral formation, mineral degradation, sedimentation, weathering and geochemical cycling.

Mineral Cycling

The most important vital cycles of nature include carbon, sulphur and nitrogen cycles.

The earth is composed of a defined amount of carbon that must constantly be recycled to allow the formation of organic compounds of which all living things are made. Photosynthesis is the basis of the total carbon fixed and ultimately rendered available to the living systems; carbon is released back to the atmosphere in the form of carbon dioxide. Through decomposition of dead organisms and the waste products of plants and animals, some carbon dioxide is returned to the atmosphere. In addition, the burning of coal and other fossil fuels releases a considerable quantity of CO_2. In this process carbon may be stored for some time before it is returned to the cycle via the atmosphere and water (Fig.).

In recent times, a considerable amount of CO_2 as well as highly toxic compounds such as SO_2 are being added to the environment by industry and automobiles. The consequences of this are quite dramatic in the smok-filled atmospheres of most major cities. In fact, even CO_2 may be hazardous in excessive amounts since many control centres in living systems are dependent on a balance between CO_2 and O_2.

Oxygen is important not only as a structural component of sugar, fat and protein molecules but also as an agent for the process called oxidation, by which energy is released from foods. Oxygen is primarily released from the atmosphere during photosynthesis of green plants or from water in the case of aquatic forms (Fig.).

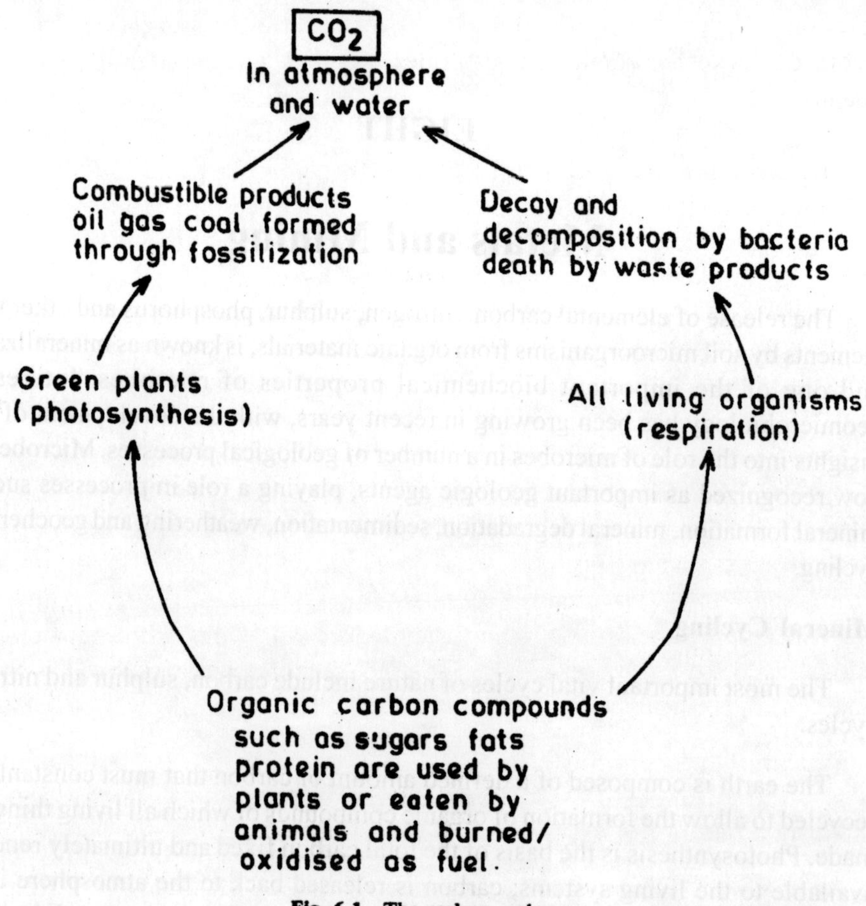

Fig. 6.1 The carbon cycle

O_2
In atmosphere
and water

Green plants
Algae
(photosynthesis)

← H_2O —

Living organisms
(Respiration)

Fig. The oxygen cycle (simplified)

The process by which nitrogen is made available to living organisms is somewhat more complex than the oxygen and carbon cycles. Nitrogen is an essential element in compounds such as amino acids, purines, vitamins, nucleic acids, enzymes and proteins.

These compounds are usually produced first in green plants and then made available to animals. Although nitrogen makes up 78 per cent of the atmosphere, it is not directly available to the majority of living organisms since most of the nitrogen used by green plant is in the form of nitrates. A small percentage of atmospheric nitrogen is directly fixed (converted to ammonia) by several nitrogen fixing microbes including cyanobacteria and bacteria (besides legumes and other higher plants). Lightening and electrical discharges also fix a small amount of nitrogen.

Decomposition to ammonia follows rapidly and some nitrogen is then converted to nitrite and then to nitrate, which is readily used by the plant to synthesize new amino acids. Ammonia is converted to nitrite by the bacteria of the genus *Nitrosomonas*. Nitrite is further oxidized to nitrate by *Nitrobacter*. The ammonia, nitrites, and nitrates are usually present as ions (NH_4+, NO_2-, NO_3-) which can unite with other ions of opposite charges to produce salts. Other sources of ammonia are also available to nitrite forming bacteria such as the ammonia produced in animal wastes (such as urine containing urea).

Decomposing organisms
N_2—fixing bacteria NH_3 ———— NO_2 ———— NO_3
Waste products

A significant portion of the nitrogen is released to the atmosphere from the nitrate and ammonia and consequently, it becomes useless to the plants and animals. This process known as denitrification completes the nitrogen cycle. (Fig.)

Phosphorus

The major reservoir of phosphorus is the rock formations of the earth's crust. Slowly, through the processes of weathering and leaching, phosphates are returned to the soil and the sea. Dissolved phosphate can be incorporated into living organisms, usually into plants and then into animals. Phosphorus, in this form, is essential to the functioning of the metabolic processes that convert food to usable energy, energy transfer, and the structure of the hereditary material. In addition, it is a major component of the bones and teeth. Since phosphorus is eroded away, phosphates are often added to lawns and farmland when the soil has become depleted. Following the decomposition of organisms, phosphates are recycled. Some phosphate, however, is deposited in sediments, and the deep, rather insoluble sediments are lost to living organisms until an upheaval occurs. Eventually, these will be returned to the cycle, although some concern has been expressed over the fact that gradually phosphorus is permanently lost to the deep sediments. Land dwelling organisms contribute to the terrestrial phosphate deposits and thereby, assist in maintaining the cycle. Fish eaten by man and birds help to return phosphorus from the sea to the land. (Fig. 6.4).

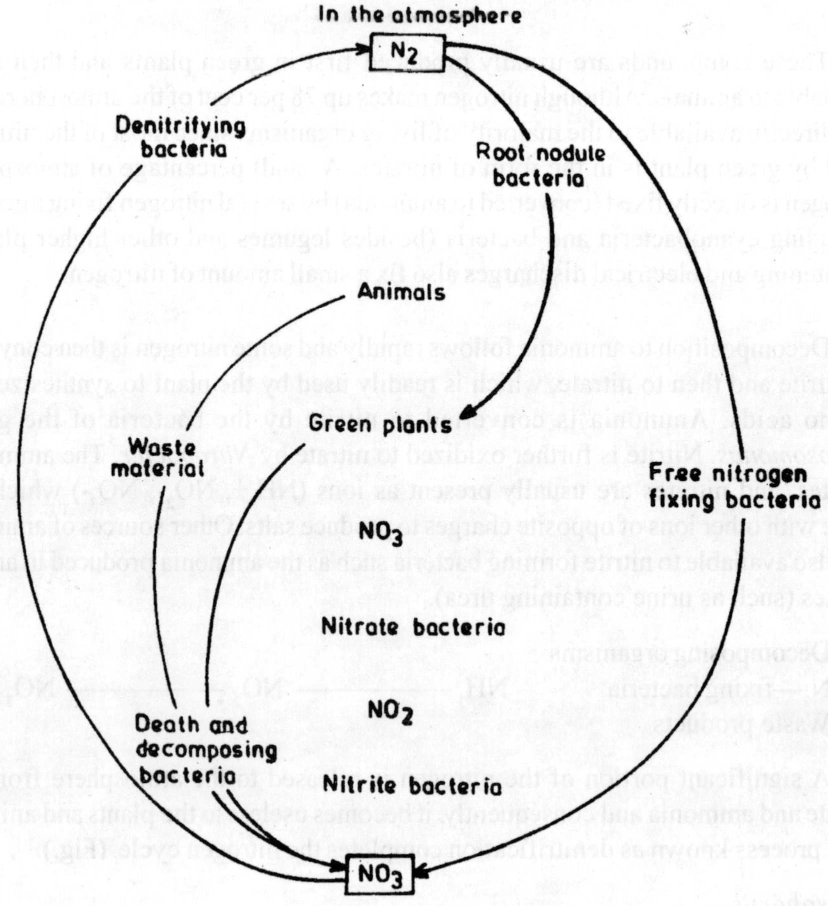

Fig. The nitrogen cycle

Sulphur

Sulphur is made available to plants (which cannot utilize elemental sulphur) by chemoautotrophic bacteria such as Thiobacillus which oxidize elemental sulphur to sulphates.

$$2S + 2H_2O + 3O_2 \longrightarrow 2H_2SO_4$$

Sulphates are accumulated by plants and their sulphur is incorporated into proteins and present in amino acids including systeine, methionine. With the death of organisms, bacteria first decompose the protein and the sulphur containing amino acids to yield various compounds including H2S. Sulphur is also released in the form of sulphate molecules which are commonly found in organic matter. Anaerobic bacteria such as those of the genus Desulfovibrio convert this sulphur to hydrogen sulphide. Other bacteria such as Thiobacillus, Beggiatoa and Thiothrix release the sulphur from hydrogen sulphide and convert it into sulphate (Fig. 6.5) Man is releasing large quantitites of sulphur into the atmosphere by burning fossil fuel at a rate that will equal the emission from natural sources by the year 2000.

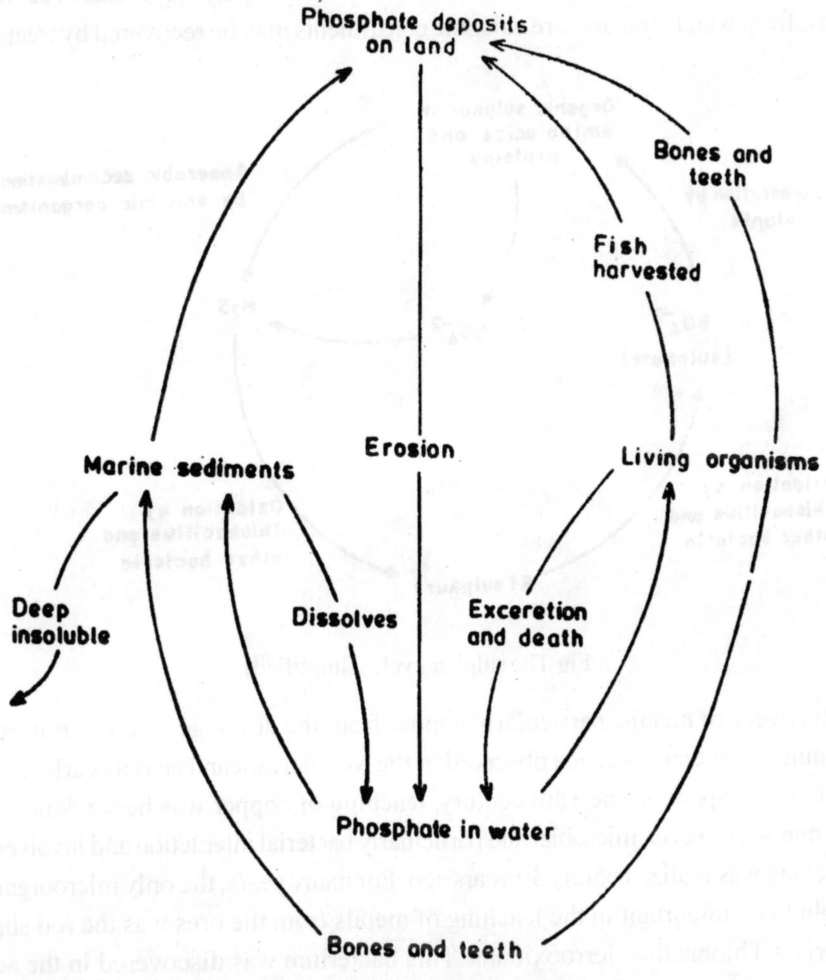

Fig. The phosphorus cycle

Microbial Transformation of Metals

The term metal is applied to any substance that has a silvery lustre and is a good conductor of electricity and heat. Some metals, of which the alkali metals are examples, also are relatively soft, malleable and ductible.

Almost all microorganisms interact with metals. Two most important sections that are of potential economic and industrial interest are leaching metals from their ores and concentrating metals from wastes or dilute mixtures. In the first case, metal is extracted from large quantities of lowgrade ores and the second reaction would provide methods for recycling precious metals and controlling pollution caused by toxic metals. An ore is a solid and naturally occurring mineral aggregate of economic interest from which one or more valuable constituents may be recovered by treatment.

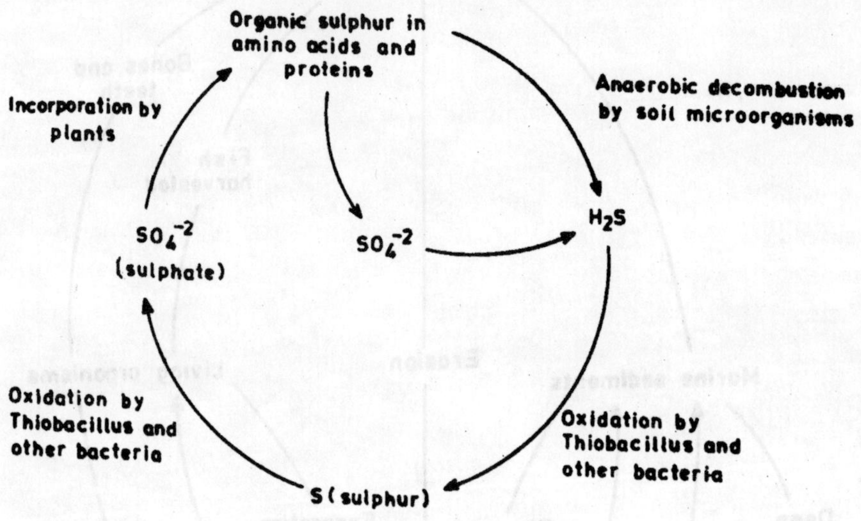

Fig The sulphur cycle (simplified)

Recovery of metals, particularly copper from the drainage water of mines, has been quite an ancient practice observed in the Mediterranean Basis as early as 1000 B.C. Even in Spain, in the 18th century, leaching of copper was being done at Rio Tinto mines. However, microbial and particularly bacterial interaction and involvement in leaching was realized barely 30 years ago. For many years, the only microorganism thought to be important in the leaching of metals from the ores was the rod shaped bacterium Thiobacillus jerrooxidans. This bacterium was discovered in the acidic water draining coal mines and by 1957 a correlation was established between the presence of this bacterium and the dissolution of metals in copper leaching operations.

Leaching in general, begins with the circulation of water through large quantities (often hundreds of tons) of ore. The bacteria, which are naturally associated with the rocks, then cause the metals to be leached by one of the methods: (i) either the bacteria act directly on the ore to extract the metal or (ii) the bacteria produce substances such as ferric iron and sulphuric acid which then extract the metal. It seems that simply adding acid is not as efficient as using live bacteria. Although acid certainly plays a role in metal extraction, it is possible that direct bacterial attack on some ores is also involved. As a matter of fact, several bacteria which help in mineral leaching have been shown to bind tenaciously to these minerals.

Thiobacillus ferrooxidans gets energy for growth from the oxidation of either iron or sulphur. The iron must be in the ferrous or bivalent form (Fe^{++}) and it is converted by the bacterium into ferric (Fe^{+++}). The bacterium also attacks several forms of sulphur including both soluble and insoluble sulphides, elemental sulphur and soluble compounds that incorporate either the thiosulphate or dithionite(S_2O_3--) or the tetrathionite (S_4O_6--). In either case, the product of the transformation is a substance in which the sulphur atom has fewer valence electrons, culminating in the formation of sulphate ions (SO_4--). Thiobacillus thiooxidans is another organism resembling T. ferrooxidans and grows on sulphur and some other soluble sulphur compounds. Don Kelley from Warwick studied in detail the joint activity of two bacteria and proved that both the organisms, if put to work together, can result in a far better extraction of metals from ores. The effect seems to be more or less synergistic and not merely additive. Similarly, the combination of Leptospirillum ferrooxidans and T. organoparpus can degrade pyrite (FeS_2) and chalcopyrite ($CuFeS_2$) neither of which can be extracted by the bacterium alone.

Other interesting leaching microorganisms are the moderately thermophilic and acidophilic bacteria designated TH (meaning thermophilic and *Thiobacillus* like). Strains of TH bacteria grow on ferrous ions or on minerals such as pyrites, chalcopyrite, covellite (Cus) and pentlandite (Fe Ni_9S_8). An organic supplement to the mineral diet is apparently required by these organisms for growth. TH strains have been isolated from acidic hot springs and from leaching environments where copper sulphide is present. Their actual mode of action and commercial utilization still remains to be worked out.

Sulfolobus is a very well studied thermophilic and acidophilic organism (Archaebacteria) to exploit for mineral leaching. It grows in acidic hot springs and volcanic fissures near 60°C and some strains grow even at very high temperatures. *Sulfolobus acidocaldarius* and S. *brierlevi oxidize* sulphur and iron for energy relying on either carbon dioxide or simple organic compounds for carbon. Oxygen serves as the ultimate acceptor of the electrons removed in the process of chemical oxidation. It can also grow anaerobically and molybdenum (MO^{6+}) and ferric iron (Fe^{+++}) can serve as electron acceptors in the absence of air. Minerals such as chalcopyrite and molybdenite (MoS_2) are readily attacked by this organism. In sharp contrast to thiobacilli, S. brierley is highly resistant to molybdenum (upto 750 mg per litre). Sulfolobus species are now under intensive experimentation to be utilized in extraction of several resistant minerals.

Leaching and Extraction

The application of the leaching process to uranium mining is of particular interest because of the possibility of in situ mining. Instead of using conventional techniques to haul uranium ore to the surface, microbial suspensions can extract the metal from its source. Water is percolated through underground shafts where the bacteria dissolve the metals. The solution is then pumped to the surface where the metal is recovered. The bacteria do not directly attack the uranium minerals; instead they generate ferric iron from pyrite and soluble ferrous ions. Ferric ions readily attack minerals incorporating quadrivalent uranium (U^{4+}), converting this ion to hexavalent uranium (U^{6+}) which is soluble in dilute sulphuric acid. The procedure described above is already being used in Canadian uranium mines. It has been suggested that extending this practice to most mines would have significant environmental benefits because of the minimal disruption of the land surface. With this method, the deeper and lower grade deposits could be mined more readily.

Although bacterial leaching is currently exploited only for the recovery of copper and uranium, the bacteria can also leach out other metals as well since these are not specific. The organisms readily degrade other sulphide minerals, yielding zinc from sphalerite (ZnS) and lead from galena (Pbs). The leaching bacteria can, therefore, be considered for the extraction of many other metals.

Bacterial leaching is likely to find its greatest application in the controlled treatment of vast quantities of solid materials. From the solutions, the microbes can remove metals in three steps: (i) the absorption of metal ions into the surface of a microorganism (ii) the intracellular uptake of metals and (iii) chemical transformation of metals by biological agents. Microbes possess a negative electric charge owing to the presence of negatively charged groups of atoms on the cell membrane and the cell wall. The charged groups include carboxyl (CoO^-) phosphoryl (PO_4---), sulphhydryl (HS^-) and hydroxyl (OH^-). Consequently most positively charged metal ions from the solution are absorbed on the bacterial surface. The common yeast Saccharomyces cerevisiae and the fungus Rhizopus arrhizus have been shown to adsorb uranium from waste water. It has been observed that upto 18 per cent of the dry weight of the cell represents uranium concentration. Other microbes also deposit metals such as oxides of manganese (*SphoerotilusLeptothrix group and Hyphomicrobium*) and iron (SphaerotilusLaptothrix group and *Gallionella*). Besides several microbes transport and accumulate other trace substances such as Mg^{++}, Ca^{++}, K^{+}, Na^{+} SO_4^{-}, CrO_4--, SeO_4-- (selenate), VO_4-- (Vanadate), WO_4-- (tungstate) and MoO_4— The common soil water bacterium Pseudomonas aeruginosa accumulates nearly 100 mg of uranium per liter of solution in less than 10 seconds.

The mechanism of leaching of metals by microbes is not very well understood. However, the direct leaching involves an enzymatic attack by the bacteria on components of the mineral that are susceptible to oxidation. In the process of obtaining energy from the inorganic material, the bacteria cause electrons to be transferred from iron or sulphur to oxygen. In principle, the inorganic ions do not enter the bacterial cell; the electrons released by the oxidation reaction are transported through a protein system in the cell membrane and, therefrom, the oxygen atoms (in aerobic systems) forming water. The transferred electrons give up energy, which is coupled to the formation of adenosine triphosphate. The other method is the indirect leaching which does not proceed through a frontal attack by the oacteria on the atomic structure of the mineral. Instead, the bacteria generate ferric irons by oxidizing soluble, ferrous iron. The ferric iron can react with other metals because it is a powerful oxidizing agent. The control of acidity is of utmost importance in leaching, because an acidic environment must be maintained in order to keep ferric iron and other metals in solution. Acidity is controlled by the oxidation of iron, sulphur and metal sulphides, by the dissolution of carbonate ions and by the decomposition of the ferric iron through reaction with water. Increasing the leaching rate of the bacteria would certainly be a desirable improvement. The major drawback of bacterial leaching has been the slowness of the biological process compared with some chemical methods of extraction.

Bioextractive Metallurgy—a Future Discourse

The microbiological processes currently exploited by the mineral industry are fairly simple in engineering design and their effectiveness is sensitive to seasonal changes and sudden alterations in the chemistry of the system. Current developments in the study of the uptake of metals by *S. cerevisiae, R. arrhizus,* and *P. aeruginosa* make it probable that these microorganisms can be utilized in precise processes for the recovery of metals from waste-water streams. The accumulation of metals by microbes, by whatever method, is nonspecific. Many organisms have cellular components that are highly metalspecific. One of the best understood metalbinding agents, is the protein metallothione in which is rich in sulphur containing amino acids forming a sulphhydryl (HS) chelation site. In the marine cyanobacterium *Synechococcus* a comparatively small cadmium binding metallothionein can bind an average of 1.28 atoms of cadmium per molecule of protein. Recent progress in the genetic manipulation of microorganisms for industrial purposes is likely to scale up the leaching processes in metal bearing ores as well as metalcontaminated waste waters.

The important future programming in bioextractive metallurgy include (i) microbial genesis of sulphur from sulphate or of methane from organic residues in natural environments, (ii) immobilization or volatilization of polluting toxic elements such as arsenic or mercury (iii) microbial desulphurization of coal (iv) microbial removal of methane from coal mines (v) Use of aliphatic hydrocarbonutilizing bacteria in detecting petroleum deposits including microbial processes for obtaining petroleum products from oil shale and tar sands (vi) microbial processing and removal of pyritio impurities from pottery clay (the problem is quite serious in Kerala in India). (vii) Immobilization of manganese and recovery of manganese and radium from the tailings through the use of microorganisms using bioaccumulation or accretion or bioconversion.

The natural property of several microbes to accumulate high intracellular concentrations of several metals is being investigated for water purification. Schist Lake in Manitoba which receives metal contaminated tailings from a mining and smelting operation and sewage effluent from a small town is receiving attention. Nutrients in the sewage promote algal blooms and the algae in turn accumulate metals. Such natural cleaning symptoms are attracting serious attention for microbial extraction of metals from highly polluted streams and dump fields. Therefore, microbial extractive metallurgy has a potential role to play.

NINE

Microbial Use in Industry and Agriculture

Micro-organisms for large scale industrial processes have long been used. Although it has assumed renewed emphasis in recent years. Centuries ago, people in Asia and Africa learned to make wine, beer, vinegar and saki with bacteria and yeast, without knowing the scientific basis of such productions. The technology related to microbial production of metabolites such as ethanol, lactic acid, butanol, riboflavin, etc. and enzymes such as protease, amylase and invertase was also developed as early as first few decades of the 20th century. Large scale production of the antibiotic penicillin was perfected during World War II and the production of many other antibiotics, amino acids, nucleotides, enzymes, etc. had been successfully accomplished in the 1950's and later. Following this history, in recent years, micro-organisms have found their application not only in the production of a variety of metabolites but also in the bio-transformation of several chemicals. The genetically engineered micro-organisms are also being used for the commercial production of some nonmicrobial products such as insulin, interferon, human growth hormone and viral vaccines. Microbes are also being used to meet effectively the crisis in both environment and energy sectors. They can reduce environmental pollution through a variety of processes and other means including the following: (i) recovery of metals from polluted waterways; (ii) elimination of sulphur from metal ores and coal-fired power and (iii) use of biofertilizers and biopesticides. In the energy sector, they can be used for production of single-cell proteins (SCP) to meet food and fodder problems, and for biogas production to provide energy to electrify villages. Through the use of biofertilizers and bioinsecticides, they can also add to crop productivity. Similarly, they can provide cheap and cost effective methods of mining and metallurgy. Thus; microbial biotechnology will have a great impact on industry in the 21st Century. Some of these aspects of microbiological applications including the techniques of isolation and mass scale culture of micro-organisms are described in this chapter.

Isolation and Culturing of Micro-organisms

Isolation methods

Micro-organisms are though ubiquitous in their occurrence, common sources for their isolation are soils, lakes and river muds. Common techniques used for the isolation of industrially useful micro-organisms include the following: (i) direct sponge of the soil; (ii) soil dilution; (iii) gradient plate method (pour plate and streak plate techniques); (iv) aerosol dilution; (v) flotation; (vi) centrifugation.

Although for the detailed description of the above techniques, the readers are advised to consult a book on practical microbiology, they should recognize that the technique of isolating micro-organisms also varies according to the nature and physiological properties of the microbe to be isolated. For example, isolation of fungi from a mixture of fungi and bacteria can be achieved easily by incorporating in the growth medium, antibiotics to which the bacteria are sensitive. Similarly, from a mixture of sporulating (spore forming) and non-sporulating bacteria, spore forming bacteria can be isolated by heating the mixture to 70 - 80° C for 5-10 min, leading to the death of all vegetative forms, so that the remaining spores will belong to only sporulating bacteria. Small samples of microbial cultures can also be obtained from permanent cultures, which are maintained by government as well as private agencies in most countries.

Sterilization and disinfection methods

For any microbiological culture, firstly, it is necessary to have a contamination free seed culture, and secondly the equipment and media that are used, should be fully sterilized. In large scale installations, heat (dry or steam) is the most important means for sterilization, since vegetative cells of most micro-organisms are rapidly eliminated by heating the culture. The equipment is generally sterilized at temperatures that are not very high, although for destroying spores (specially of bacteria) temperatures as high as 121°C are needed. Other sterilization techniques involve use of radiation (UV, X-rays), chemicals or mechanical removal of organisms.

In most micro-biological operations, the air passing through the reactor culture vessel should also be sterilized, since outdoor air contains approximately 5-2000 micro-organisms m^{-3}, with 50% of them in the form of fungal spores and 40% being gram negative bacteria. Among various methods available for sterilizing gases are filtration, gas (ozone) injection, gas scrubbing, radiation (UV) and heat. In most cartridges of filter systems, containing membranes made up of cellulose ester, polysulfone or nylon, which can filter off most of the micro-organisms.

Culture methods

Most of the industrially useful bacteria and fungi are cultured on either a solid medium or a liquid medium with a support. In either case, the medium should contain all essential ingredients needed for optimum growth of that particular micro-organism. Solid medium is generally prepared by melting agar at 100°C (in water) and then solidifying it at about 45°C. The medium, which is still in a molten stale, is poured into a suitable container and is allowed to cool and solidify. It is then inoculated by dipping a needle into the suspension of bacteria or fungi and drawing this needle right across the surface of the medium. After a few days of inoculation, colonies of micro-organism appear on the surface of the medium. In liquid cultures, all essential nutrients are dissolved in water. Agar or any other solidifying agent is not needed.

Table
Sterilization time and sterilization temperatures of various groups of organisms

Celts	*Sterilization lime (minutes)*	*Sterilization temp. (°C)*
Vegetative cells	5-10	60
Fungal spores/yeast spores	15	80
Streptomyces spores	5-10	60-80
Bacterial spores, general	5	121
Spores of Bacillus stearothennophilius	15	121

Culture maintenance

The pure culture of a specific organism is to be maintained for its continuous use. The easiest method is to store them at low temperature (2-6°C) in agar or liquid medium. But the most common method is to store them frozen either at -18°C or at -80°C in freezers or at -196°C in liquid nitrogen. Frozen cultures can be stored for several years without any loss of viability. The best method for culture is, of course the lyophilization (freeze drying). The addition of protective agents such as skimmed milk or sucrose reduces the lethality during lyophilization process. The cell suspensions are placed in small vials, which are then frozen by immersing in a mixture of dry ice and acetone or liquid nitrogen. The vials are then evacuated and dried under vacuum, sealed and stored allow temperature. In this way, the cultures can be satisfactorily sealed and stored allow temperature. In this way, the cultures can be satisfactorily maintained for essentially indefinite periods.

In molecular biology experiments, bacterial strains (transformed with recombinant DNA) are stored in glycerine stabs, where 0.85 ml of cell suspension is added to 0.15 ml of sterile glycerol, mixed thoroughly and stored at -20°C or -70°C. Some spore forming bacteria and fungi are often stored in sterile soils. Spore suspensions are added to sterile soils and the mixture is dried at room temperature and stored in a refrigerator. Bacterial cultures maintained by this procedure have been found viable even after 70-80 years.

Production of Organic Compounds by Microbial Fermentation

A variety of organic compounds, used as organic feedstocks for many chemical industries, are produced industrially by fermentation process. Microbial production of organic compounds that can be used as substitutes I for traditional fuels also received major attention in 1970's, when petroleum and natural gas became scarce. For both feed and fuel, generally the carbohydrate rich plant products are used as substrates. It is estimated that about 95% of the total 180×10^9 tonnes of plant material produced annually on this earth remains unutilized as human or animal feed. By evolving suitable biotechnology, such a large amount of biomass can be converted into a variety of useful products.

Ethanol production

Ethanol for human consumption has been manufactured as a component of alcoholic fermentation since prehistoric limes. In recent years, it has also found its use as an important chemical feedstock and as a fuel supplement. In Brazil, in the 1980s, 20% of the petroleum imports have been replaced by ethanol produced from sugarcane. In India, in the year 1988, approximately 910 million litres of ethanol was produced against a consumption of 826 million litres. The production is estimated to increase to 1150 million litres by the year 2000. This calls for a policy to export surplus alcohol and also to create demand for additional usage of alcohol in India.

Ethanol is generally produced by fermentation of some sugar rich products with the help of yeast, Saccharomyces cerevisiae or sometime with Kluyveromyces fragilis. Several other organisms, fungi and bacteria are also known to produce small quantities of ethanol. However, in then later cases in addition to ethanol, other undesirable products are also produced so that often they are not used for alcohol production. Fermentation is often most active under anaerobic conditions, when carbohydrate is converted into ethanol via pyruvate and acetaldehyde. Under aerobic conditions on the other hand very little alcohol is produced, most of the carbohydrate getting oxidized to CO_2 + H_2O. Furthermore, ethanol y inhibitory for yeast growth at high concentration, although yeast strain differ in their tolerance to ethanol's inhibitory action. The steps involved in ethanol formation by this method are shown below

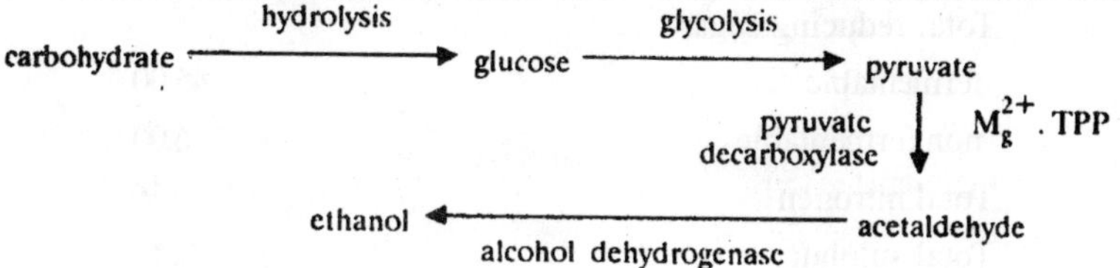

Fig. Different steps involved in the manufacture of ethanol from a carbohydrate via pyruvate and acetaldehyde.

Commercial production of ethanol involves three steps: (i) preparation of substrate, (ii) fermentation and (iii) distillation. These steps are shown in and are described in the following text.

Preparation of substrate— For commercial ethanol production, following three substrates are used as renewable raw materials: (i) roots, tubers or grains that are rich in starch, (ii) molasses or juice derived from sugarcane, palm or sugarbeet, and (iii) wood or waste products (rich in cellulose) from processed wood. In India, ethanol is produced mainly by fermentation of molasses, which are produced as a by-product in sugar factories and are quite rich in fermentable sugars.

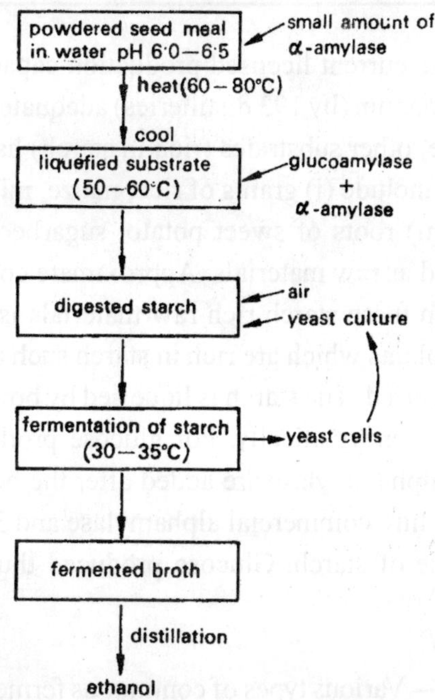

Fig. A flow diagram showing various steps involved in the manufacture of ethanol from starchy seed.

Chemical composition (approximate) of the molasses.

Component	*Percentage*
Total reducing sugar	
-fermentable	45.00
-non fermentable	6.00
Total nitrogen	0.36
Total sulphur	0.0.5
Volatile acidity (as acidic)	0.18
Ash content	11.00
Moisture	28.00

Approximate production cost of one litre of alcohol from different starch rich raw materials (based on "Fermentation Technology and Industrial Application of Biotechnology" published by Director General of Technical Development or DGTD, Ministry of Industry, Govt. of India).

	Raw material	*Cost (Rs)*		*Raw material*	*Cost (Rs)*
1.	Potato	15.30	4.	Rice	6.50
2.	Maize	10.50	5.	Sweet potato	12.30
3.	Tapioca	3.00	6.	Sugar beet	5.15

However, for the current licensed production capacity of about 1500 million litres of alcohol per annum (by 193 distilleries) adequate quantity of molasses is not available. Therefore, other substrates (rich in starch) have often been used. These alternate substrates include (i) grains of rice, maize, millets and barley containing usable starch and (ii) roots of sweet potato, sugarbeet and cassava. Potato and tapioca are also used as raw materials. Approximate cost of the production of one litre of alcohol from these starch rich raw materials is given elsewhere. Starting with those parts of plants which are rich in starch such as seed, the material is first ground, filtered and dried. The starch is liquefied by boiling under pressure, cooled and then hydrolysed enzymatically. For glucose production as an intermediate, glucoamylase and alpha-amylase are added after the boiling and cooling process. Approximately one litre commercial alphamylase and 3. 5 litres glucoamylase are added for one tonne of starch. Glucose produced thus can be used for ethanol production.

Fermentation— Various types of continuous fermentation plants are available for producing ethanol, where sometimes several fermenters are connected in a series. In a typical fermenter using molasses as a substrate, diammonium phosphate (as a source of nitrogen for the yeast) is mixed up with the substrate. The pH is maintained at 5.0 and fermentation is carried out at 30-35°C. After fermentation, yeast cells are separated by centrifugation or sedimentation and used again.

Distillation— Ethanol from the fermented mixture is recovered by distillation process. The first distillation yields about 80% pure ethanol but by repeated distillation, ethanol concentration of 99.4% or more can be achieved. Final dessication can be achieved by chemical dessicants and by absorption.

In recent years, cellulose also seems to have become a distinct possibility as a renewable source of alcohol. Bagasse, sulphite waste, liquor of paper industry and agriculture biomass (hay) are the sources of cellulose. The biomass is first pretreated through physical or chemical processes leading to reduction in size. This helps in hydrolysis of cellulose, and hemicellulose by enzymatic or chemical treatment. The soluble sugar obtained after hydrolysis are subjected to fermentation in the usual manner. Cellulolytic microorganisms can also be used for degradation of biomass, so that direct fermentative production of alcohols or acids will be possible. Biochemical Engineering Research Centre (BERG) at I.I.T., Delhi is conducting research for developing improved methods for cellulose pretreatment.

Acetone/butanol production

Acetone, butanol, butyric acid and isopropanol are obtained through fermentation of starch, molasses, sucrose, wood hydrolysates and pentoses using bacteria belonging to the genus Clostridium. The relative proportion of different products varies with the species and the strain of *Clostridium* used. The biosynthetic steps involved in the production of each of these compounds are interrelated, all of them being derived from acetyl CoA produced from either glucose or fatty acid degradation.

In a typical fermentation process, the substrate includes molasses plus maize steep liquor, which are taken in a suitable container which is gassed with CO_2, The fermentation is carried out at 33-35°C, with initial pH of 5.8 to 6.0. However, the pH drops in the first 18 hours to about 5.2 due to the production of acetic and butyric acids, but in the next 18 hours increases again due to the production of acetone and butanol from these acids. Contamination due to bacteriophages and lactobacilli is a common problem in such a fermentation and therefore absolute sterilization is necessary. In India, alcohol produced from molasses is used as a feedstock for production of acetone and butanol. 1.5-1.9 MT of alcohol is used per MT of butanol and 2.2 MT of alcohol is used per MT of acetone.

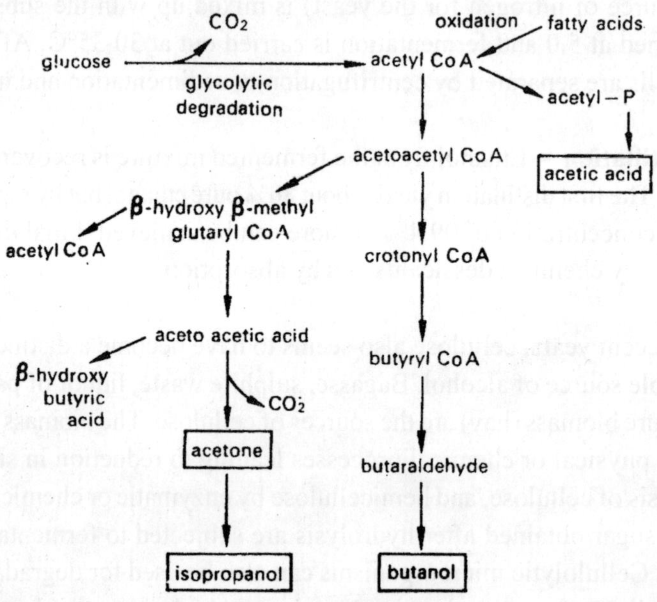

Fig. An outline of the pathway of biosynthesis of acetic acid, butanol, acetone and isopropanol.

Acetone, butanol and ethanol are recovered through continuous distillation and fractionation. Carbon dioxide produced is recovered and converted into liquid CO_2 or dry ice. The residue left after the distillation may be dried and utilized as animal feed.

Gluconic acid production

Gluconic acid and its salts are widely used in pharmaceutical food, feed, detergent, textile, leather, photographic and concrete industries. During glucose oxidation, gluconic acid is produced by most Aspergillus spp. and several species of other fungal genera like Penicillium, Mucor, Fusarium, Pullutaria, etc. In the industry, Aspergillus niger seems to be the most widely used micro-organism, since it contains very active form of the enzyme glucose oxidase. Glucose is the precursor for the synthesis of gluconic acid as follows:

$$\beta\text{-D-glucose} + V_2O_2 \xrightarrow{\text{glucose oxidase}} \text{gluconic acid}$$

A typical fermentation medium for producing gluconic acid using Aspergillus sp. contains the following : glucose 150g/l, maize steep liquor 3.7g/1, KH_2PO_4 0.2g/I, $MgSO_47H_2O$ 0.17g/1, urea 0.10g/1, $(NH_4)2HPO_4$ 0.4g/l. For some strains of Aspergillus niger, other nitrogen sources such as $NaNO_3$, NH_4NO_3 or $(NH_4)_2SO_4$ and peptone are also added in the medium. Maize syrup supplies the trace elements required for the fungus. The fungus is incubated in such a medium which is kept well aerated at pH 5.5 to 6.0. Optimum yield is obtained at 28-30°C. Often the accumulation of free gluconic acid in the medium inhibits further fungal growth, with a poor yield of the acid. To avoid this, $CaCO_3$ or NaOH is also added in the medium, so that the gluconic acid secreted from the fungus is neutralized to Ca/Na gluconate.

Production of other organic compounds

A variety of other important organic compounds are also produced by microbial fermentation and it will not be possible to discuss the production of each of these compounds under the limited scope of this book. Some such compounds, their most common substrates, and the organisms employed for producing them are listed in below.

Some organic compounds produced by microbial fermenlalion

Compound	*Common substrate*	*Common microorganism used*
Amino acids		
DL-alanine	Glucose	Microbacterium aminoniaphilum
L-glutamic acid	Glucose	Micrococcus glutamicus
L-leucine	Glucose	Brevibacterium lactofermentum
L-lysine	Acetate	Brevibacterium flavum
L-threonine	Surose	Escherichia coli K12
Nucleotides		
5' IMP	Starch hydrolysate	Bacillus subtilis
5' IMP	Starch hydrolysate glucose	Brevibacterium ammnoniagens
5' GMP	Glucose	Bacillus subtilis
Organic acids		
Acetic acid	Sugarcane juice	Acctabacter spp.
Citric acid	Beet molasses	Aspergillus niger
Citric acid	N. alkane	Saccharomyces cerevisiae
Lactic acid	Whey	Lactobacillus prueckii
Lactic acid	Glucose	Lactobacillus balgaricus
Vitamins		
Vitamin B_{12}	Sugarcane	Pseudomonas denitrificans
Vitamin B_{12}	Glucose	Propionobacterium sphermanii
Riboflavin	Maize steep liquor	Ashbya gossypii

Production of Enzymes by Micro-organisms

As described in an earlier chapter, enzymes have found a variety of applications in medicine, food industries, textile and leather industries and also in analytical processes. The first enzyme produced industrially in U.S.A. as early as 1894, was **'fungal taka-diastase'** employed as a pharmaceutical agent (for digestive disorders). Similarly, microbial 'rennin' is being used instead of calf's rennin since 1967. Although, the enzymes can also be extracted from the tissues of higher organisms, it is easier and often cheaper to obtain them from micro-organisms.

Production of alpha amylases

Alpha amylases are the most important group of enzymes, which hydrolyse starch. They are produced and excreted out into the medium by many bacteria and fungi. The constitution of the culture medium and other conditions (e.g. pH and temperature) would vary with the organism utilized. For industrial production of cx - amylases, mainly batch fermentation by bacteria (*Bacillus subtilis*) is utilized.

Bacteria and fungi producing alpha amylase

Bacteria	*Fungi*
Bacillus amyloliquefaciens	*Aspergillus*
B. caldolyticus	*Candida*
B. coagulans	*Cephalosporium*
B. coreus	*Mucor*
B. licheufonuii	*Neurospora*
B. subtilis	*Penicillium*
B. stearothenuoPltihrs	
Escherichia spp..	
Lactobacillus spp.	
Micrococcus spp.	
Pseudomonas spp.	
Protects spp.	
Serratia spp.	

Production of proteases

Proteases are the second most important group of industrial enzymes after alpha amylases. They are used primarily in the detergent and dairy industry. In the market, they are available as alkaline, neutral and acid proteases.

Alkaline proteases— They are produced and secreted out by many bacteria and fungi. The most important producers are species of Bacillus (e.g. B. licheniformis, B. amyloliquefaciens, B. firmus, B. megaterium, and B pumilus), species of Streptomyces (e.g. S. fradiae, S. griseus, and S. rectus) and some fungi (e.g. Aspergillus niger, A. sojae, A. oryzae and A flavus). The cultures are maintained at 30-37°C in fermenters or flasks, where production of a protease is chiefly regulated by the composition of culture medium. High partial pressure of oxygen is generally necessary for optimal protease production. In most cases, the enzymes are collected, purified and immobilized before their use in the detergent industry.

Neutral proteases—They are also produced and excreted out by both bacteria and fungi including Bacillus subtilis, B. cereus, B. megaterium, Pseudomonas deruginosa, Streptomyces griseus, Aspergillus oryzae, A. sojae and Percularia oryzae'. Neutral proteases are used in leather industry and food industry. They are relatively unstable.

Acid proteases— They include rennin, which is an important enzyme used in cheese production. Many other acid proteases are used in medicines. The important fungal genera producing rennin are Aspergillus, Candida, Coriolus, Endothia and Mucor. Rennin produced by Endothia parasitica was the first rennin marketed commercially in 1967. The enzyme is excreted out by the fungus in the medium. It is concentrated and precipitated out in an evaporation process.

Production of lipases

Lipases hydrolyse fats (triglycerides) into diglycerides, monoglycerides, fatty acids and glycerol. They are used primarily in medicines as digestive enzymes to supplement pancreatic lipases. Lipases are produced and secreted out by many fungi and bacteria such as Penicillium chrysogenum, Pseudomonas fragi, Rhizopus delamar, Aspergillus niger, Mucor sp., Staphylococcus aureus, etc.

Production of Antibiotics by Micro-organisms

Antibiotics are produced by a number of micro-organisms and inhibit the growth of other micro-organisms even at very low concentrations. As such, the antibiotics have found wide application in chemotherapy, plant pathology, food preservation, veterinary medicine and as research tools in biochemistry and molecular biology. At present, about 7000 antibiotics are known and about 100 of these are produced commercially by microbial fermentation process. The first antibiotic to be isolated was penicillin which was discovered by Alexander Fleming in 1929 and was produced on large scale using cultures of Penicillium notatum during 1940's. During World War II, the demand for chemotherapeutic agents to treat wound infections led to the development of a production process for penicillin and the beginning of an era of antibiotic research. This continues to be the most fascinating area of microbial biotechnology even today.

Fungi, bacteria and actinomycetes are important antibiotic producing organisms. Most species of Streptomyces arc quite active in the production of a variety of antibiotics. Some micro-organisms and the antibiotics producted by them are listed below. The status of the production of different antibiotics in indict is summarized on the next page.

Some micro-organism and antibiotics produced by them

Ogranism	*Antibiotic*
Penicillium spp.	Penicillin
Bacillus licheniformis	Bacitracin
Cephalosporium acremonium	Cephalosporin
Nocordia uniformis	Nocardins
S. caespitosus	Actinomycins
Strepimyces antibiotieus	Mitomycin
S. erythreus	Erythromycin
S. griseus	Streptomycin, cycloheximide
S. virginae	Virginiamycin

Status of the production of different antibiotics in India (based on data from DGTD, 1989)

		Quantities (annual)		*Production*	
		Capacity (1989-90)	*Production (1985)*	*Import (or excess demand)*	*units in India*
1.	Penicillin*	637 MMU	481 MMU	600 MMU	HAL., IDPL,
	Ampicillin	150 t	110t	470t	Alembic Chemicals
	Amoxicillin	37 t	1.4 t	35.6 t	—
	Cloxacillin	Total demand in 1990=32 t			—
	Cephalaxin	10 t	3 t	37 t	Standard Pharmaceuticals
2.	Streptomycin	445 t	200 t	70 t	IDPL, HAL. Symbiotics

		Quantities (annual)		*Production*	
		Capacity (1989-90)	*Production (1985)*	*Import (or excess demand)*	*units in India*
3.	Chloramphenicol	185 t	86 t	214 t	—
4.	Tetracyclin	220 t	219 t	-	IDPL
5.	Erythromycin	88 t	30 t	60 t	Alembic
6.	Gentamycin	1.0 t	437 kg	2.5 t	HAL
7.	Rifampicin and neomycin	140 t	No production	100 t+16 t	—
8.	Kanamycin	1000 kg	"	17 t	Alembic

* Penicillin is used to produce 'semi-synthetic penicillin - 6 APA (ampicillin, amoxicillin and cloxacillin) and 7ACDA (Cephalexin). The use of 6APA and 7ACDA has increased due to their lower side effects, and they are manufactured from Penicillin G,

* HAL = Hindustan Antibiotics Ltd., Pune
IDPL = Indain Drugs and Pharmaceuticals Ltd., Rishikesh

The most important aspect of microbial antibiotic production is the selection of a high yielding strain and a suitable fermentation medium. Since antibiotics are the secondary metabolites of an organism, it is necessary to limit the growth of the micro-organism and to divert the organism from primary to secondary metabolism. This is usually achieved by designing the medium in such a way that a key nutrient is exhausted at the appropriate time, producing the desired metabolic switch or diversion. Following are the different steps in commercial production of antibiotics using microbial fermentation.

Steps for commercial production of antibiotics

Inoculum and media preparation— A small sample of the cultured stock is taken and the amount of material is amplified by growth on solid surfaces or in liquid culture. Spores of the micro-organism are collected and retained in a suspension form. This suspension is transferred into a container suitable for attachment to the fermenter.

Different media are required at different stages of antibiotic production. Manufacturers spend considerable amount of time and money in developing media both for increasing antibiotic productivity and for decreasing costs. The medium must supply to the organism, source of energy, carbon, nitrogen, phosphate, trace elements and in some cases specific precursors for the biosynthesis of antibiotics (e.g. phenyl acetic acid for the production of penicillin G). Composition of a typical medium used in the final stage of penicillin production is shown below. In many cases, small amounts of yeast extracts or similar other substances are added to increase the rate of fermentation.

Composition of fermentation medium for the production of penicillin.

Components	*Per cent (approximate, w/v)*
Sugar (e.g. glucose/sucrose/lactose)	5—10
Maize steep liquor	5—9
Pharma media	3—5
Chalk	0—1.0
$(NH_4)_2SO_4$	0—1.0
Phosphate	0—0.5
Edible oil	0.3—0.8
Phenyl (or phenoxy) acetic acid	0.3—0.6

The fermenter— A typical antibiotic fermentation plant is illustrated on the previous page. The medium may be batched (made up) and sterilized in the fermenter itself. However, most of the fermenters are attached to a batching vessel and to sterilizers as shown earlier. The feed vessels connected to the fermenter are used to supply nutrients or precursors during fermentation process. The seed and the final stage fermenters must be capable of operating under aseptic conditions. The fermenter vessels are usually stainless steel containers with a capacity of 30 to 300 m^3. Agitation of the broth is facilitated via two or three flat bedded impellers. Air which is sterile, filtered and compressed is passed into the vessel immediately below the bottom impeller. Heat generated by the agitation system and by the metabolic activities of the micro-organisms is dissipated by using cooling water columns, since optimum temperature for most of the antibiotic producing organisms is 24-28 °C. For continuous sterilization of the system, steam is used. The steam delivery system is fitted in such a way that only the desired section of the system can be sterilized.

The production stage— All antibiotics are produced by batch or fed batch fermentation methods. Fed batch techniques are used to extend the duration and productivity of fermentation. By continuous addition of glucose or other sugars, antibiotic can be produced with the same system for a period extending upto two weeks. Addition of sugars and of other nutrients (if required) may be achieved by metering pumps, low meters, or burettes. Sometimes, considerable amount of foam is generated during fermentation, specially when the media are proteinaceous and they are vigorously agitated and aerated. The foam reduces the vessel capacity and, therefore, it is undesirable. To prevent it, often anti-foam chemicals such as vegetable oils are added in the media. The fermentation is generally carried out at 24-28°C.

Recovery of antibiotic— A variety of physico-chemical methods are used for the removal of antibiotic from the fermenter. When penicillin is the antibiotic, most of it (about 99%) is excreted out by the fungus into the medium. In such a case, the mycelium is separated by filtration and the antibiotic is recovered from the medium by two stage continuous countercurrent extraction of the fermenter broth with amyl or butyl acetate. Streptomycin and many other aminoglycoside antibiotics are also secreted out and are removed by ion exchange columns. Cell bound amino glycosides such as gentamycin, sisamicin and fortimicin are released from the cell by acidification of the culture broth with H_2SO_4 to achieve a pH of 2.0-2.5.

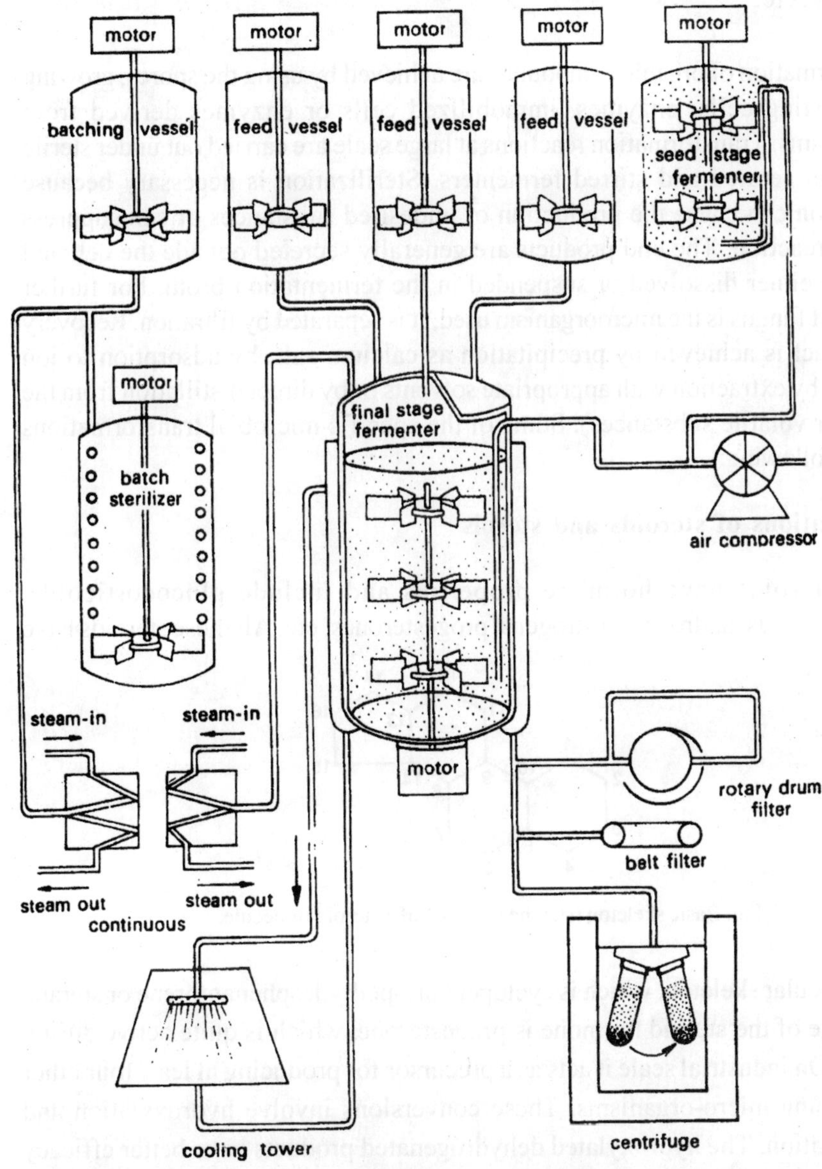

Fig. A typical antibiotic fermentation plant.

Microbial Transformations

Micro-organisms have the ability to convert one organic molecule into another. Many complex bioconversions that are achieved by micro-organisms, can not be achieved by normal chemical means and are achieved with great difficulty by using organisms other than microbes. Microbial transformations of the molecules which have found industrial applications involve oxidation, reduction, isomerization, hydrolysis, condensation, etc.

Transformation of organic compounds are achieved by using the spores, growing cultures, resting cells, enzymes, immobilized cells or enzymes derived from microorganisms. Transformation reactions at large scale are carried out under sterile conditions in aerated and stirred fermenters. Sterilization is necessary because contamination can cause the production of undesired substances or can suppress the desired reaction. The end products are generally secreted outside the cell and they remain either dissolved or suspended in the fermentation broth. For further processing, if fungus is the microorganism used, it is separated by filtration. Recovery of the product is achieved by precipitation as calcium salt, by adsorption to ion exchangers, by extraction with appropriate solvents or by direct distillation from the medium (for volatile substances). Some of the specific microbial transformations include the following:

Transformations of steroids and sterols

Most steroids have hormone properties and include glucocorticoids, mineralocorticoids, androgens, estrogens, progesterones, etc. All these steroids have

1 2 3 4 5 6 7 8 9 10 11 12 13 14 15 16 17

Fig. Basic skeleton (sterane nucleus) of a steroid molecule.

a basic molecular skeleton, which is cyclopentano-perhydrophenanthrene or sterane nucleus. One of the steroid hormone is progesterone, which is quite active during pregnancy. On industrial scale it acts as a precursor for producing at least four other products, using micro-organisms. These conversions involve hydroxylation and dehydrogenation. The hydroxylated dehydrogenated products have better efficacy as hormone. For example, the oxygen at C-11 is essential for the anti-inflammatory effect of cortisone (another steroid hormone). The cortisone itself can be produced from progesterone and then modified. In 1949, 1.Og of cortisone was produced at a cost of about 200 US dollars, but as a result of introduction of the microbial process for the C-11 -hydroxylation of progesterone, the cost had decreased to under 1 US dollar by 1979. Currently, almost all positions of the steroid molecules can be specifically hydroxylated by different micro-organisms.

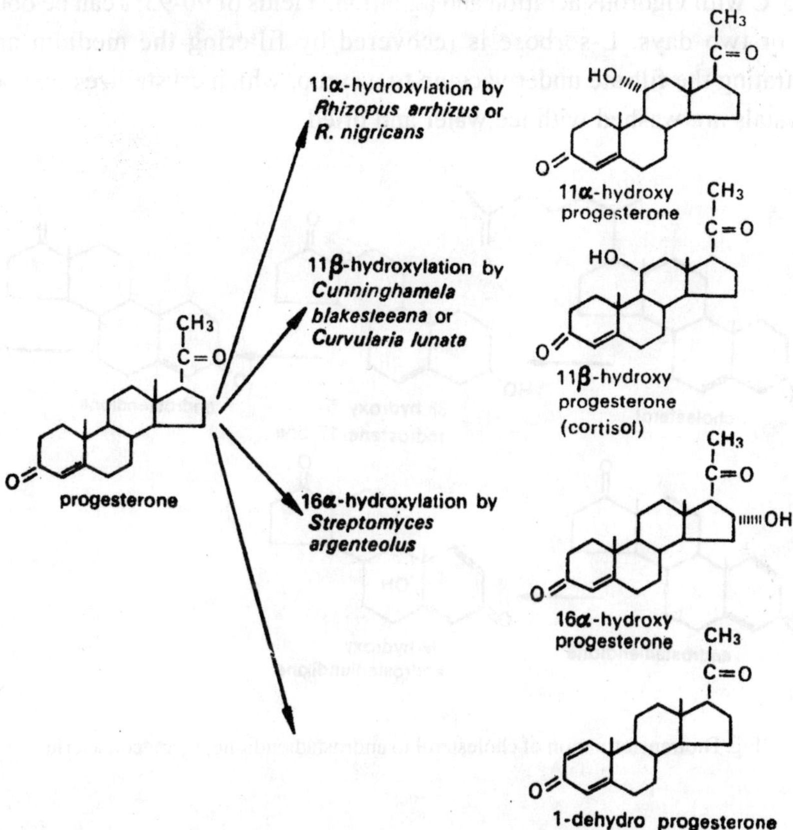

Fig. Microbial conversions of provesteron.

Steroid hormones can also be produced from sterol precursors which are quite inexpensive. For instance, diosgenin, β-sitosterol, cholesterol, etc. are often used as hormone precursors. These sterols have the same sterane nucleus and one or more side chains. The basic approach in their microbial bio-conversion is to remove the side chain. Production of some steroid hormones using cholesterol as primary molecule with the help of mycobacteria is diagrammatically represented below.

Transformation of sorbitol to sorbose

L-sorbose is a chemical precursor of L-ascorbic acid (vitamin C), which is used in vitamin preparations and also used as antioxidant in food manufacture. L-sorbose is produced from dehydrogenation of D-sorbitol using Acetobacter suboxydans.

Sorbitol is added at an initial concentration of 20% in the nutrient medium consisting of 0.5% yeast extract or maize steep liquor and $CaCO_3$. Fermentation is carried out at 30-35°C with vigorous aeration and agitation. Yields of 90-95% can be obtained in one or two days. L-sorbose is recovered by filtering the medium and by concentrating the filtrate under vacuum to a syrup, which crystallizes on cooling. The crystals are washed with ice water and dried.

cholesterol → 3β hydroxy-5-androstene-17-one → androstendione → androstadiendione → 9α-hydroxy androstadiendione

Fig. Biotransformation of cholesterol to androstadiendione; by mycobacteria.

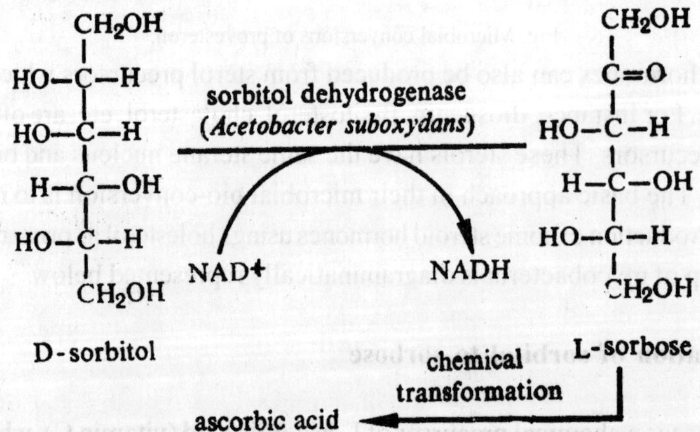

Fig. Microbial conversion of sorbitol to sorbose.

Transformation of antibiotics

Studies in recent years have also been conducted to produce new antibiotics that are more effective than the existing ones, using micro-organisms. Several such bio-conversions have been achieved with remarkable success. For example, Bacillus megaterium IFO 12108 has been used to convert antibiotic lankacidin C (+ methyl butyrate) into lankacidin C butyrate, which has an improved antimicrobial activity with reduced toxicity.

Single Cell Proteins (SCP) from Micro-organisms

The term single cell protein (SCP) was coined at Massachussetts Institute of Technology (MIT) by a group of scientists in 1966. The term, as used today, is rather misleading, since it refers to not only the isolated cell protein but to "any microbial biomass from both uni-and multicellular bacteria, yeasts, filamentous fungi or algae which can be used as food or feed additives". On an average, the microbial biomass contains about 45 to 55% protein, although in some bacteria the protein content is as high as 80%. The biomass also contains other essential nutrients as well, and as such it is an ideal supplement to conventional food supply. However, in many countries, use of microbial biomass as a supplement to diet has met with scepticism because of certain psychological barriers. But even in these countries, it will play a major role via SCP feeding to animals which will be then consumed by the humans. As compared with traditional methods of producing proteins for food or feed, large scale production of microbial biomass has the following advantages: (i) Microorganisms in general have a high rate of multiplication; (ii) Microbes have a high protein content; (iii) They can utilize a large number of different carbon sources, some of which are waste products; (iv) Strains with high yield and good composition can be selected or produced relatively easily; (v) Microbial biomass production is independent of seasonal and climatic variation.

In India, at CFTRI (Central Food Technology Res. Institute), Mysore, research is being conducted on the use of blue green alga, Spirulina as a supplement to diet (food and feed). The alga is cultured , dried, powdered and then used in the from of one gram tablets. It contains 60% protein, essential vitamins and unsaturated fatty acids. When fishes and children (3-12 years old) were fed on this fond, encouraging results were obtained. It is, therefore, hoped that Spirulina will become an important supplement to the diet of humans and also of marine and fresh water life.

However, there are certain disadvantages also, of using microbial biomass as supplement to diet. Many micro-organisms produce toxic substances and it has to be made sure that the biomass does not contain any such substance. Sometimes the microorganisms when taken in, lead to indigestion and allergenic reactions. The high nucleic acid in many microbial biomass products is also undesirable.

The SCP process

Regardless of the type of substrate or organism employed, the production of SCP involves following basic steps: (i) preparation of suitable medium with suitable carbon source; (ii) prevention of contamination of medium and the plant; (iii) production of the desired microorganism; (iv) separation of microbial biomass and its processing.

The medium for SCP production varies according to the microorganism. Among other things, the medium must contain a carbon source for cultivating the heterotrophic micro-organisms, although green algae (Chlorella, Scenedesmus, Spirulina, etc.) can be cultivated autotrophically without a dissolved carbon source. The carbon sources for heterotrophic organisms can be divided into two groups, namely fossil and renewable. Fossil carbon sources include n-alkanes, gaseous hydrocarbons, methanol, ethanol, etc. and the renewable sources include CO_2 molasses, whey, polysaccharide hydrolysates, effluents (of breweries, distilleries, confectioneries industries, potato and canning industries and wood pulp industries) or other solid substrates. In most of the industrial establishments, n-alkanes appear to be preferred carbon source. In addition to the alkane fraction, salts of PO_4^{3-}, K^+, Mg^{2+}, Mn^{2+}, Zn^{2+}, Fe^{2+} and gaseous ammonia (as source of nitrogen) are also added for cultivating many micro-organisms including *Saccharomycopsis lipolytica.*

Production of mushrooms (fruiting bodies of higher fungi) has also acquired importance in several countries in recent years, since it is considered to be a delicacy for human food. These are cultured on solid substrates which range from straw and composted straw to pieces of wood and sawdust. World production of the common type of mushroom exceeds 106 tonnes per year. At the University of Waterloo (Canada), success has been achieved in initiating a new project where the fungus Chaetomium cellulyticum is cultured on a variety of agricultural and forestry wastes.

In most of the industrial systems for biomass production, it is desirable to keep the medium and the plant free from contamination. The circulating air and the gaseous components of the medium (NH_3, CO_2, etc.) are sterilized by passing through filters. Other components are sterilized by steam, in most cases. The fermentation equipment is also sterilized.

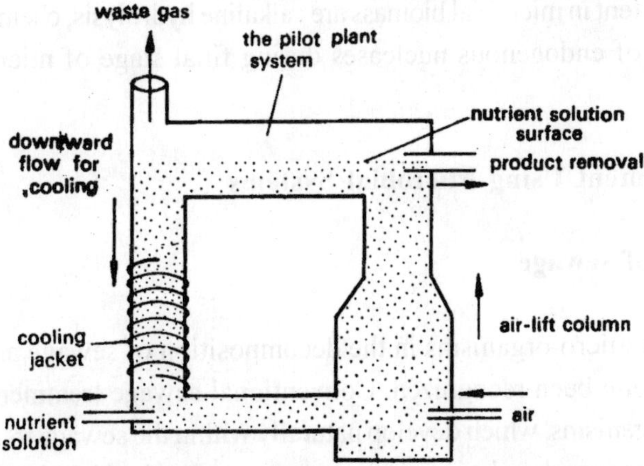

Fig. An outline of the ICI fermenter to produce single cell protein (SCP) from methanol.

The desired micro-organism is cultured on the medium under clean conditions. The organism to be cultured on the medium must have the following basic properties: (i) it should be non-pathogenic to plants, animals and to humans; (ii) it should have good nutritional value; (iii) it should be usable as food or feed; (iv) it should not contain any toxic compound; (v) its production cost should be low.

The choice of the fermenter (cultivation vessel) depends upon the micro-organisms. Aeration is an important operation in the cultivation. Heat is generated during cultivation, and it is to be removed by using a cooling device. The production is usually continued for indefinite period for maximum economy. A simple fermenter, used by Imperial Chemical Industry (ICI), to produce SCP from Pseudontonas methylotrophus grown on methanol, is illustrated earlier. Using such a fermenter, it is possible to produce about 4-30g/l of microbial biomass at 38-40°C and pH 6.8.

The microbial biomass can be harvested by a variety of methods. Single cell organisms like yeast and bacteria are normally recovered by centrifugation. Filamentous organisms are recovered by filtration. It is important to recover as much water as possible prior to final drying. The whole operation is to be done under clean and hygienic conditions to keep the product and the broth, that leaves the plant free of bacterial contamination. The final dried products are normally bacteriologically stable if handled properly. In some cases, an after-treatment of the biomass is desirable to reduce the unwanted compounds in the product or to isolate the protein. One of the important tasks is to reduce nucleic acid content, which is high in microorganisms (4-6% in algae, 10-16% in bacteria, 6-10% in yeasts and 2.5-6% in fungi) and can be hazardous to health. Some of the methods employed for reducing nucleic acid content in microbial biomass are : alkaline hydrolysis, chemical extraction, and activation of endogenous nucleases during final stage of microbial biomass production.

Sewage Treatment Using Microbial Systems

Degradation of sewage

The role of micro-organisms in the decomposition of sewage and other waste materials has long been recognized. Conventional sewage treatment involves the use of micro-organisms, which develop naturally within the sewage treatment system. In some newer approaches, however, the sewage is inoculated with a specific micro-organism, which has been specially selected for that particular sewage treatment process. Such organisms might be called 'starter cultures'. The use of starter cultures increases the efficiency of sewage degradation. Following are some of the examples of sewage treatment by using starter cultures : (i) Bacteria have been developed, which degrade alkanes and aromatic compounds at 0-15°C, in saline habitats. These could be useful in the degradation of oil spills in the ocean; (ii) Starter cultures of mixed micro-organisms are available, which metabolize DDT and polychlorinated diphenols or phenols and which possess high protease, lipase or cellulase activity; (iii) Starter cultures have also been used to deodorise animal excrements; (iv) A strain of Pseudomonas putida containing plasmids has been developed, which can degrade octane, xylene, metaxylene and camphor (see later for more details).

The degradation of sewage by micro-organisms requires large amount of oxygen, so that in order to provide room for oxygen most sewage treatment plants are bulky. For improving the efficiency, tubular loop reactors and air lift fermenters have been developed. They both have better oxygen transfer efficiency. Aeration by pure oxygen has also been tried in U.S.A. and West Germany. Commercial sewage treatment plants that use pure oxygen, are marketed by several companies.

Methane production

Methane is produced during anaerobic decomposition of sewage and other organic wastes by bacteria. The methane thus produced is collected and used as fuel in many countries. The methogenic bacteria are able to utilize acetate, methanol, formate, and H_2+CO_2 of the organic wastes for the production of methane gas. The starting materials in the wastes are, however, complex organic molecules such as cellulose, starch, fats and proteins. These are first broken down to simpler substrates such as acetate and $H_2 + CO_2$ by other micro-organisms. The methogenic bacteria then act on these to produce methane, as shown below.

$$\text{organic macromolecules} \xrightarrow{\text{anaerobic fermentation}} \text{acetate and } CO_2 + H_2 \xrightarrow{\text{methogenic becteria}} \text{methane}$$

Fig. The production of methane from organic macromolecules found in the organic wastes.

In the sewage treatment system, the degradation of macromolecules (cellulose, protein, fats, etc.) to smaller molecules is called liquefaction phase. A large number of micro-organisms can perform this degradation. However, only a restricted group of organisms are able to produce methane from the degraded products.

The composition and the amount of the gas (called biogas), produced from the degradation of biological products, is related to the quality of the stubstrate and the temperature. In anaerobic sludge digestion process, upto 600 litres of methane is produced per kg dry organic matter. The biogas generated in such a process contains 60-70% methane, 25-30% CO_2 and the rest is $H_2 + N_2$. In our own country, small and inexpensive installations have been set up in the villages to convert cow-dung into biogas. About 200 to 1100 litres of methane per kg dry cow-dung is produced at 55°C, in such installations. In most systems, the amount of biogas generated increases with the increase in temperature. The time for the generation of gas is also reduced at high temperatures.

Biotechnology in Paper Industry

Manufacture of paper from wood involves (i) wood processing; (ii) pulping; (iii) bleaching and (iv) sheet formation. Pulping of wood (preferably from softwood with 3-5 mm long fibres, but rarely from hardwood with 1.5 mm long fibres) requires separation of the wood fibres from each other, which are then reformed into a sheet. The wood fibres are glued together with the help of lignin and separation of these fibres is described as chemical pulping, when lignin is removed by degradation and is described as mechanical pulping, when fibres are mechanically feared apart. Both these methods of pulping are used. Mechanical pulping gives higher yields and is cheap, but the quality of paper produced is relatively poor, turning yellow on exposure to sunlight. Further, the mechanical pulping requires lot of electrical energy. These difficulties can be overcome through the use of biotechnology. The chemical pulp is also subjected to bleaching, inorder to remove residual lignin leading to satisfactory increase in the brightness of paper. This bleaching step creates numerous toxic derivatives of lignin that constitutes environment hazard. The use of biotechnology in pulping will be discussed in this chapter and that in reducing the environmental impact of bleaching will be discussed later on.

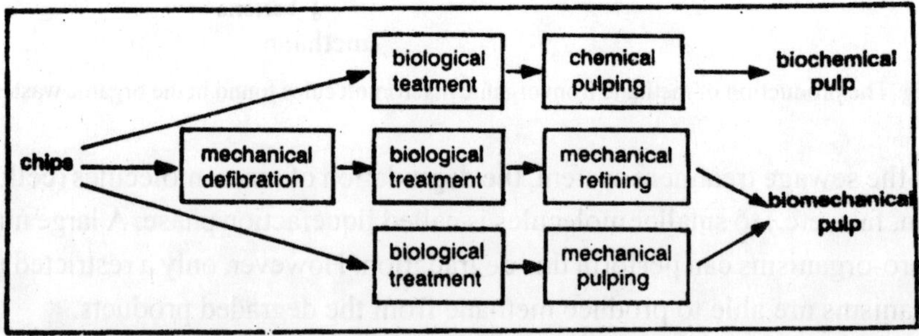

Fig. Three ways in which a biological step can be integrated in pulping process leading to lignin degradation.

Biological lignin degradation

In the pulping process, degradation of lignin can be achieved through treatment with microbes, of which lignolytic fungi are the most important. A biological step can be integrated, in pulping process, both in chemical pulping as well as in mechanical pulping. Three ways have been suggested for this purpose, which are outlined in.

It may be seen that the biological step may be a pre-treatment or a post- defiltration to remove lignin. However, the use of a biological step in lignin degradation, is still at the level of research and experimentation. Its use at industrial scale is seen as a distinct possibility due to successful results already obtained in several experiments. To support research in this area, a Biopulping Consortium (funded by 20 companies) was established in USA in 1987.

Biochemical pulping— In a recent report it was shown that a treatment of aspen chips (aspen is a wood; wood is received by pulp mills in two forms, logs or chips, the latter being more popular) with Phanerochaete chrysosporium before craft pulping (sulphate process) gives improved tensile strength (resistance to rupture by a- force parallel to sheet) and burst strength (resistance to rupture by force perpendícular to sheet), but decreased tear strength (resistance to elongation under transverse shear), brightness and yield. Brightness may be improved later by bleaching.

More research and experimentation is needed before biochemical pulping becomes a reality and used in paper industry.

Biomechanical pulping— Mechanical pulping involves two steps: (i) primary refining or defibration and (ii) secondary refining. Biological step may be introduced after the defibration or before this step, and both have some advantages. Treatment of coarse pulp (obtained after defibration) with lignolytic fungi (e.g. Trametes versicolot; P. chrysosporium, Pleurotus ostreatus) has been shown to give increased tensile strength, and may reduce energy requirement for secondary refining, pre-treatment also leads to increased pulp strength.

For pre-treatment as above, wood chips (3-8 mm thick) are supplemented with a nutrient solution and sterilized by autoclaving. They are then inoculated with the mycelium of the fungus, pre-grown in agar or liquid medium, on wood chips or on cereal grains. The inoculated chips are incubated at high humidity and at a temperature (25-40°C) optimal for growth of the specific fungus being used. Aeration is achieved by diffusion or forced ventilation sufficient to replenish oxygen. The chips may be agitated intermittently for distribution of inoculum and moisture. Within 2-6 weeks, the fungus colonizes the wood leading to softening and swelling of the fibre walls. This treatment may lead to substantial energy savings and paper strength improvement. Different lignolytic fungi used for this purpose include the following: Tremetes versicolor, Dichomitus squalens, Pholiota mutabilis, Phlebia subserialis, P. brevispora, P. tremellosa, Pitanerochacte chrysosporium, and Ceriporiopsis subvermispora.

Before biological pulping is used in industry, considerable research in the following area would be needed and is being undertaken : (i) More lignolytic fungi (there are 2000 known species of lignolytic fungi) need to be tried, and search needs to be made for better strains, which act quickly, have high competitive ability and do not darken the wood. (ii) Fermentation technology needs to be developed, which includes optimization of conditions for fungal activity, designing of a reactor and scaling up of the fermentation step. (iii) Mechanism of lignin degradation by fungi needs to be studied. Energy savings and strength improvement generally do not correlate closely with bulk lignin removal or even with each other, and a better understanding of the mechanism may really help.

Biohydrometallurgy and Biomineralization

For some years now, bacterial activity has been implicated in the weathering, leaching and deposition of mineral ores. This has led to a marriage between biotechnology and metallurgy, creating a new discipline known as biohydrometallurgy. Similarly in many cases, the metals are deposited as insoluble oxides and sulphides due to large scale and long term activity of bacteria. This area of biotechnology research is described as biomineralization.

Extraction of metals and mining

In 1947, for the first time, U.S. microbiologists discovered that copper was leached from its ores due to the activity of a bacterium, Thiobacillus ferroxidans. These bacteria are 'acidophilic autotrophs', meaning that they are able to live solely on sulphides and thrive in an acidic environment. In recent years, it has been shown that T. ferroxidans and T. thiooxidans both can be used for extraction of copper and uranium from insoluble minerals by either of the following two activities : (i) by redox processes at the mineral surfaces : (ii) biosorption of metals on bacterial cell surfaces. These bacteria, as their names indicate, live by oxidizing the sulphur that binds copper, zinc, lead and uranium into their respective sulphide minerals. This releases the valuable metal. In contrast to this biometallurgy, the conventional metallurgy involves smelting ores at high temperature, a process which is far more polluting and energy intensive approach. In view of this, biometallurgy is an approach which is environment friendly, and therefore, its use should be encouraged. Its use also needs to be encouraged to extract metal from low grade ores as done in U.S.A., since conventional metallurgy requires high grade ores, which are gradually becoming exhausted.

Biohydrometallurgy was first used in 1970, in Canada for extraction of uranium and in South Africa for extraction of gold. In 1980s, when copper industry was in great difficulty, different firms started looking for technology, which may be more cost-effective and less polluting. This technology involving Thiobacillus is now responsible for 30 per cent of the copper produced in U.S.A., and for 10-20 per cent of copper supply on world basis. This technology also led to (i) reduction in cost of production of copper from 60-90 US dollars to only 30 US dollars per pound and (ii) elimination of the release of SO_2 in the atmosphere, which is estimated at the level of two tonnes of SO_2 from one tonne of copper smelted. The bacteria like T. ferroxidans can also be used for removing sulphur from coal to reduce the emission of So, from coal- fired power stations. This will bring enormous environment benefits for fast developing industry in India and China.

Biohydrometallurgy of copper involves the following steps: (i) the low grade ore and tailings, left from any earlier conventional mining, are piled up in an area, where the ground has been made permeable ; (ii) the pile is sprayed with a leaching solution that contains iron in the form of the Fe^{3+} ion, sulphate ions, $(SO_4)^{2-}$ and T. ferroxidans; the sulphate ions in the leaching solution form sulphuric acid giving a suitable acidic solution essential for growth and activity of bacteria; (iii) copper, released due to catalytic chemical reaction (facilitated by T. ferroxidans), is drained in the form of a solution; (iv) the metal (copper) is then removed from the solution with the help of another solvent; (v) the remaining leaching solution flows into an open pond, where T. ferroxidans catalyses a reaction that oxidises Fe^{2+} into Fe^{3+}ions and sulphide into sulphate ions, so that the leaching solution is now recharged; (vi) the recharged leaching solution is pumped back to the top of the pile for the cycle to begin again; (vii) the copper is eventually extracted from the solution in (iv) above as sheets through electrowinning process by passing electricity, so that copper is collected on negative terminals; this process is cosily and non-biological, but 'bioabsorption filters' such as algae will be used in future to make the process entirely biological.

Biohydromatallurgy is also used in mining. For instance, in Arizona (USA), in a mine five holes were created by detonating an explosive underground. A mixture of acidic solution and T. ferroxidans is pumped down the central hole, where bacteria do their work. The resulting solution, rich in valuable copper, is pumped from the other four holes, processed and recycled. However, mining by biological means is much slower and takes decades, where conventional methods take a few months/ years. If both methods are compared, biological method (though cheaper) delays cash flow due to slower production and therefore hindered its adoption. Research is, therefore, in progress to increase the speed as is being attempted at School of Biotechnology, University of New South Wales (Australia). Some progress using heterotrophic (not autotrophic like T. ferroxidans) bacteria for extraction of manganese has been made. Genetically engineered bacteria may be used in future to increase the speed of reaction. Metals can also be separated from waterways, polluted with heavy metals, using algae and bacteria. These microbes oxidise dissolved manganese which may be separated through process of bioabsorption mentioned earlier. More progress will by made in this area in future.

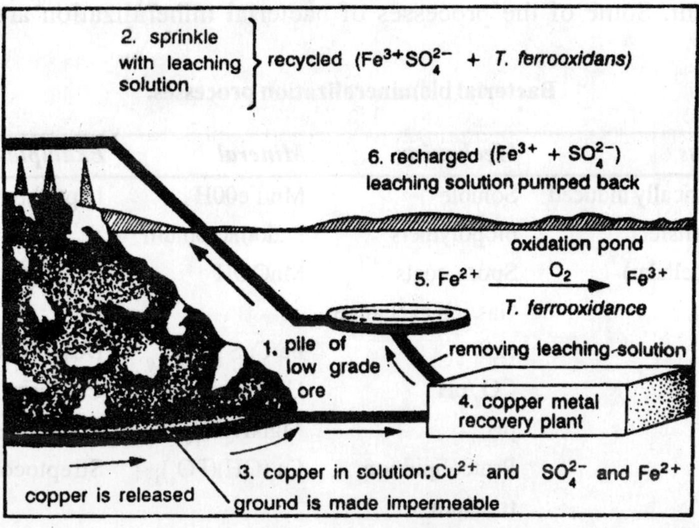

Fig. Different steps involved in mining of copper with bacteria.

Microbialty induced metal deposits

It has also been shown that some bacteria which are common constituents of soils may play an important role in the formation of some gold and silver deposits in sediments laid down by rivers - alluvial deposits. During recent years, several examples of microbially induced chemical deposition have been brought to light. For instance, it has been suggested in several separate studies that in Venezuela and Alaska, gold deposits have arisen through such a process. It is suggested that negatively charged polymer on the outside of the bacteria attracts the positively charged gold particles in the soil, so that they clump together as grains and eventually as gold nuggets. It has also been demonstrated that metals like gold continue to accumulate on the negatively charged polymers even after the bacteria have died, so that the process can continue indefinitely giving gold nuggets. Recently (John Watterson, 1992; Geology 20:315-318), association of lace like gold-decorated microbial structures with gold particles was demonstrated from nine Alaskan sites. In this study, preliminary evidence was also available for the presence of specific bacteria-Podomicrobium like budding bacteria, which are also involved in iron and manganese oxide deposition processes. Chlorella vulgaris and Bacillus cereus have also been shown, in the laboratory, to have the ability of nucleating metallic gold from dilute solutions having gold as Au(III) or as gold chloride.

Bacterial mineralization results from interaction of extruded metabolic products (iron, gases, polypeptides, electrons) with extraneous metal ions in the surrounding environment. Some of the processes of bacterial mineralization are listed in

Bacterial biomineralization processes.

Process	*Mechanism*	*Mineral*	*Examples*
1. Biologically induced	Soluble	Mn/Fe00H	Leptothrix
(biotransfer)	biopolymers	Pedomicrobium	
(extracellular)	Spore coats	MnOOH	Bacillus
	Gas/ion exchange		
	H_2S	Fe/CuS	Desulfovibrio
	CO_2/pH	$CaCO_3$	Calothrix
	pH	$MgNH_4PO_4$	Proteus mirabilis
	Proteolipid/ion transport	$Ca_5(OH)(PO_4)_3$	Streptococcus
	Phosphatase activity	$(UO_2)_3(PO_4)_2$	Citrobacter

Process	***Mechanism***	***Mineral***	***Examples***
	Electron transfer	Fe_3O_4	GS-15
		UO_2	GS-I5
	Nucleation proteins	H_2O(ice)	Pseudomonas syringae
	S-layer templates?	FeOOH	Leptothrix
2. Biologically controlled (biosynthesis) (intracellular)		Fe_3O_4	Aquaspirillum magnetotacticum
		Fe_3S_4/FeS_2	Wild types
	Ferritin micelles	$FeO(OH)/(PO_4)$	P. aeruginosa

this table. It is argued that genes responsible for mineralization in a variety of microbes can be isolated, cloned and transferred into E.coli, which may then help in metal deposition more efficiently. Improvement of strains of bacteria with increased rate of leaching for metal extraction are already being attempted. For instance, plasmids have recently been constructed, which when present with T. ferroxidans increases its resistance to aresenite and arsenate (these may not allow growth of bacteria), and enhances recovery of gold from arsenopyrite-pyrite ores. It is hoped that genetically engineered bacteria with enhanced bioleaching or nucleating capabilities will soon he available to companies engaged in mining and metal extraction.

Biofertilizers

Biofertilizers are defined as biologically active products or microbial inoculants of bacteria, algae and fungi (separately or in combination), which may help biological nitrogen fixation for the benefit of plants. Biofertilizers also include organic fertilizers (manure, etc.), which are rendered in an available form due to the interaction of micro-organisms or due to their association with plants. Biofertilizers thus include the following : (i) symbiotic nitrogen fixers- spp; (ii) asymbiotic free nitrogen fixers (.Azotobacter, Azospirillum, etc.); (iii) algae biofertilizers (blue green algae or BGA in association with Azolla); (iv) phosphate solubilising bacteria; (v) mycorrhizae ; (vi) organic fertilizers.

The need for the use of biofertilzers has arisen, primarily for two reasons. First, because increase in the use of fertilizers leads to increased crop productivity, second because increased usage of chemical fertilizer leads to damage in soil texture and raises other environmental problems. Therefore, the use of biofertilizers is both economical and environment friendly. The pragmatic approach will be to develop the integrated nutrient supply system involving a combination of the use of chemical fertilizers and biofertilizers. India is not self sufficent in fertilizer production. An estimated capital investment of Rs. 7,000 crores was needed by the end of Seventh Five Year Plan period to achieve self sufficiency. Realizing the importance of biofertilizers in supplementing the use of chemical fertilizers, the Government of India had launched the 'National Project on Development and use of Biofertilizers' during the Sixth Five Year Plan. Under this project, one national centre and six regional centres and 40 BGA production centres have been established. These centres will produce 800 tonnes of Rhizobia and 600 tonnes of BGA annually.

Location of national and regional biofertilizer production and development centres (after DGTD, 1989)

1. National Biofertilizer Development Centre, Ghaziabad, U.P.
2. Regional Biofertilizer Development Centre (North), Hisar, Haryana.
3. Regional Biofertilizer Development Centre (Central), JNKVV, Jabalpur, M.P.
4. Regional Biofertilizer Development Centre (West), MPKV, Pune, Maharashtra.
5. Regional Biofertilizer Development Centre (South), UAS, Bangalore, Karnataka.
6. Regional Biofertilizer Development Centre (East), OUAT, Bhubaneshwar, Orissa.
7. Regional Biofertilizer Development Centre (North-East) ICAR Research Complex or North Eastern Hill Region, Shillong, Meghalaya.

Rhizobia

Rhizobia (gram negative soil bacteria) are able to enter into symbiotic relationship with legumes (pulses, etc.). They fix atmospheric nitrogen and thus not only increase the production of the inoculated crops, but also leave a fair amount of nitrogen in the soil, which benefits the subsequent crop. Following seven groups of Rhizobia have been recognised for inoculating legumes in India. R. leguminosarum, R. melilotii, R. trifoli, R. phaseoli, R. lupinli, R. japonicum, and Rhizobium spp. The nitrogen fixing ability of legumes inoculated with these Rhizobia ranges from 50kg to 150 kg per hectare. Therefore, to increase this ability to fix nitrogen, following efforts are being made : (i) Efficient rhizobia for different crops and locations, which

Capacity and present production of Rizobium culture in the country* (alter DGTD, 1889)

	Organisation/manufacturer	*Capacity (in tonnes)*	*Present production (in tonnes)*
1.	NAFED	150.0	—
2.	Karnataka State	140	—
3.	Tamil Nadu State	113.0	118.0
4.	Kerala State	4.0	0.8
5.	Hisar Agricultural University	23.0	
6.	Indian Organic Chemicals Ltd., Bombay	360.0	5-4
7.	Vimpex Agencies, Bangalore	68.0	—

	Organisation/manufacturer	***Capacity (in tonnes)***	***Present production (in tonnes)***
8.	Bacfil Inoculants, Lucknow	750.11	118.0
9.	Micro Bac., India Shyamnagar, 24-Parganas	600.0	360.0
10.	Rallis Agro-Chemical Research Station, Bangalore	180	-
11.	Nitro Fix Industries, Calcutta	50.0	25.0
	Total	2.150.0	627.2

*Source: National Seminar on Development and Use of Biofertilizers (1985).

are tolerant to various stresses (drought, temperature, high or low pH) are isolated, maintained and used; (ii) A hup$^+$ gene (uptake hydrogenase) is used, which helps in recycling of H_2 produced during nitrogen fixation (N_2 $2NH_3$ + $3H_2$). This saves energy required for reduction. Rhizobin strains with hup$^+$gene have been shown to fix more nitrogen; (iii) Strains deficient in nitrate reductase have improved nitrogen fixing ability. Nitrate reductase catalyses the reaction NO_3 - NO_2, and NO_2 inhibits nitrogen fixation.

The demand for Rhizobia in India during Seventh Plan (@ 500g per ha) was estimated to be 7,150 tonnes. However, against a capacity of 2,150 tonnes, only 627.2 tonnes of Rhizobia are being produced. The new technology for industrial preparation of rhizobial cultures included the following: (i) seed pelleting with gum arabic (40%) or carboxy methyl cellulose (5%) as precoated seed; (ii) polyacrylamide entrapped rhizobia; (iii) freeze-dried or lyophilized culture of rhizobia; (iv) Rhizobium paste; (v) inoculants in liquid/frozen concentrate form; (vi) granular soil inoculants to be applied by aerial application; (vii) natural peat granule as soil implant inoculum.

Location of sub-centres for production and field multiplication of blue green algae (after DGTD, 1989)

	Name of the State	***Name of the organization/places***
1.	Andhra Pradesh	(i) PD & D Farm, Garikapad, District Krishna.
		(ii) Soil Testing Lab., Samalkota, Distt. East Godavari.
		(iii) Agriculture Farm, Amaravati.
		(iv) PD & D Farm, Amaravati.
2.	Assam	(i) Assam Agro-Industries Corporation, Guwahati.
		(ii) Assam Agriculture University Campus, Jorhat.

	Name of the State	Name of the organization/places
3.	Bihar	(i) Rajendra Agriculture University, Rice Research Institute, Mithapur, Patna.
4.	Gujarat	(i) Gujarat Agriculture University, Rice Research Station, Nawagam.
5.	Haryana	(i) Government Farm, Karnal.
6.	Himachal Pradesh	(i) Himachal Pradesh Krishi Vishva Vidyalaya, Palampur.
		(ii) Potato Research Station, Pekhubala, Una.
7.	Jammu & Kashmir	(i) Government Research Station, Jammu.
8.	Karnataka	(i) Government Seeds Farm, Gangavati,
		(ii) University of Agricultural Sciences, Bangalore, GKVK Campus, Bangalore.
		(iii) Regional Research Station, Mandya.
		(iv) Agricultural Development Centre. Tyavangi Chennai Taluk, Distt. Shimago.
9.	Kerala	(i) Kerala Agriculture University, Pattambi
10.	Maharashtra	(i) Punjab Rao Krishi Vidyapeeth, Akola Agricultural Research Station, Sindewahi.
		(ii) Mahatma Phule Agricultural University, College of Agriculture, Pune.
11.	Madhya Pradesh	(i) Jawaharlal Nehru Krishi Vishwa Vidyalaya, Jabalpur, University Centre, Raipur.
		(ii) Agriculture Farm, Jagdalpur, District Bastar.
12.	Manipur	(i) Government Rice Research Station. Wangabalm. Imphal.
13.	Meghalaya	(i) District Rice Research Station & Lab., Sangsangri. Dish. Garo Hills.
14.	Orissa	(i) Central Rice Research Station, Bhubaneshwar, District Puri.
		(ii) Regional Research Station. Chiplima.
		(iii) Regional Research Station, Ranital, Distt. Balasore.
		(iv) Regional Research Station, Keonjhar, Distt. Keonjhar.
15.	Punjab	(i) Punjab Agriculture University, Ludhiana.
16.	Tamil Nadu	(i) Tamil Nadu Agricultural University, Coimbatore.
		(ii) Paddy Experimental Station. Adduthurai, District Thanjavur.
		(iii) paddy Experimental Station, Tirur.
		(iv) Regional Station, Attiyandal.
		(v) Regional Station, Vandarayanpet.

	Name of the State		Name of the organization/places
17.	West Bengal	(i)	Bidhan Chandra Krishi Vishva Vidyalaya. Nadia University Farm, Kalyani.
		(ii)	Block Seed Farm, Singur.
		(iii)	Contai, A.R.F., Midnapore,
		(iv)	Berhampur Model Farm. Murshidabad.
18.	Uttar Pradesh	(i)	N.D.U.A.T., University Centre, Kumarganj.
		(ii)	Chandra Shekhar Azad Agricultural University, Chakri Farm, Kanpur.
		(iii)	Central State Farm. State Farm Corporation of India, Baharaich.

Asymbiotic nitrogen-fixers

Azotobacter and Azospirillum (also Bacillus polymixa, etc.), when applied to rhizosphere, fix atmospheric N2 and make it available to crop plants. They also synthesize growth promoting antibiotic substance, helpful to the plant. Most efficient strains of Azotobacter fix 30 kg of N2 from 1000 kg of organic matter. When applied to fields, positive responses by field crops were observed leading to saving of 10-25 kg/ha of N. Similarly, Azospirillum with farmyard manure (FYM) led to saving of 15-25 kg equivalent of N_2 per hectare in crops like sorghum and other millets. Azospirillum colonizes not only the roots, but also the above ground portions of the plant through associative symbiosis.

Algal fertilizers (BGA + Azolla)

Bluge green algae (BGA) and Azolla constitute a system, which is the main source of algal biofertilizer in south and southeast Asia, particularly for lowland paddy. BGA inoculation (without Azolla) with composite cultures of algal genera Anabaena, Nostoc, Plectonema, Aulosira, Oscil- Tolypothrix, etc. have been found to be more effective than single cultures. Production and multiplication of BGA cultures is done at centres listed in the previous table. Application of dried blue green algae flakes at the rate of 10 kg. /ha is recommended ten days after transplantation of rice. Besides being a source of N_2, BGA provides for the following other advantages: (i) algal biomass accumulates as organic matter; (ii) growthpromoting substances are produced, which stimulate growth of rice seedling; (iii) it provides partial tolerance to pesticides and fungicides; (iv) it also helps in reclamation of saline and alkaline soils.

BGA is also supplied with Azolla (an aquatic fern), which harbours Anabaena azolle (a BGA) in leaf cavities, providing symbiotic association. Azolla with Anabaena, has been used with some success, but there are following limitations : (i) Azolla, as a green manuring crop is labour intensive; (ii) raising of Azolla needs assured and adequate supply of water; (iii) damage of Azolla (in the nursery) is caused by pest diseases, heat, etc. (iv) optimum temperature is required for Azolla multiplication. Phosphate solubilising bacteria (PSB, e.g. Thiobacillus, Bacillus spp., etc.) and plant growth-promoting rhizobacteria (PGPR), including Pseudomonas fluorescens and P. putida are important new biofertilizers. PSBs convert non-available inorganic phosphates into soluble organic phosphates, which can be utilized by crop plants. PGPRs produce siderophores (ironchelating substances, e.g. pseudobactin), which chelate with iron and make it unavailable to harmful fungi (e.g. Eriwinia) in rhizosphere, leading to their death ; PGPRs have resulted in an increase in yields of potato (30%), radish (60-144%) and sugarbeet (4-8t/ha). These biofertilizers are yet to be commercially utilized in our country.

Mycorrhiza

Mycorrhiza is a symbiotic association of fungi with roots of plants, so that the nutrients absorbed from soil by the fungus are released to the host cells and in turn, the fungus takes its food requirements from the host. Mycorrhizae are of two types: (i) ectomycorrhiza, and (ii) endomycorrhiza.

Ectomycorrhizae are found on the roots of forest trees (e.g. pine, oak, beech, eucalyptus, etc.). They absorb nitrogen, phosphorus, potassium and calcium. They also convert complex organic molecules into simpler available forms, protect the roots from the pathogens, and produce growth promoting substances (cytokinins).

Endomycorrhizae are found in the roots of most fruits and other horticultural crops (e.g. coffee, pepper, cardamom, and betelvine). They particularly help in phosphorus nutrition. They also produce growth promoting substances and offer resistance against pathogens. The seedlings in nurseries are inoculated with endomycorrhizae and then transplanted in the field as done by citrus growers in USA. Fungus strains of Azotobacter, Aspergillus, Azospirillum, Beijerineleia, and Glomus are used as endomycorrhizae.

Organic fertilizers

India has a vast potential of organic waste resources, which include animal dung, animal urine, bone meals, slaughter house wastes, crop residues, oilcakes, urban garbage, sewage/sullage effluent, etc. Much of these organic wastes remain unutilized, leaving enormous scope for development of organic manures through recycling.

Production and use of organic wastes (per day) in India.

Source of organic fertilizer	*Production*	*Usage*
Urban waste	15 m.t.	6.7 m.t.
Rural waste	650 m.t.	250 m.t.
Sewage (sullage) effluent	8m gallons	2.5 m.gallons

Bioinsecticides

The use of chemical pesticides is being discouraged in recent years due to (i) spread of insect resistance; (ii) environmental pollution; and (iii) rising costs. Therefore, the use of bioinsecticides and an integrated approach to insect pest control have become necessary. Bioinsecticides are biodegradable (environment friendly), non-toxic and cost effective. Some of these bioinsecticides, introduced in the USA in 1950's, are based on Bacillus thuringiensis.

Bacillus thuringiensis

Bioinsecticides are generally composed of bacterial spores, crystal proteins and inert fillers. The presence of spores in these bioinsecticides is considered undesirable, since they are tolerant to most of the adverse conditions and survive in soil for long periods thus causing harm to soil microbial population and to beneficial insects like silk moth. In view of this, at Hindustan Lever Research Centre (HLRC) at Bombay and IARI at New Delhi, mutant strains of Bacillus thuringiensis (BT) have been isolated , which produce crystal proteins and form no spores. Bioinsecticides based on these mutant strains will be ecologically safe, and have been successfully tested on insects affecting cotton, maize, cabbage, sunflower, pigeonpea etc.

Some bioinsecticides

Class of bioinsecticides	*Organism*	*Status*
Microbial insecticides	(F) Verticillum lecanii	Commercial in Europe
	(F) Metarhizium artisopliae	Commercial in South Africa
	(F) Hirsutella thompsonii	Commercial in U.S.A.

Class of bioinsecticides	*Organism*	*Status*
	(B) Bacillus sphaericus	in development
	(B) Bacillus thuringiensis	4 strains commercial
	(B) Bacillus papillae	Commercial
	(F) Erynia neoaphidis	Field trials, U.K. & elsewhere
	(F) Beauveria bassiana	Field trials, U.S.A. & elsewhere
	(F) Nomurea rileyii	Field trials, U.S.A. & elsewhere
	(V) Heliothis nuclear polyhedrosis virus	Commercial, U.S.A.
Microbial herbicides	(F) Phytophthora palmivora	Commercial in U.S.A.
	(F) Colletotrichum	Commercial in U.S.A. gloeospriodes
	(F) Cercospora rodmanii	Experimental in U.S.A.
Microbial disease prevention	(F) Trichoderma viridie	Commercial in U.K., France
	(F) Agrobacterium radiobacter	Commercial in U.S.A., Australia

(F) Fungus; (B) Bacterium; (V) Virus. Source: SG Lisanshy (1982)

Commercially available BT formulations.

Product	*Producer**	*Trade Name*
Bacillus thuringiensis HD-I	1. Abbott	Dipel** HG & 4L
	2. Biochem products	Bactospeine, Novabac
	3. Sandoz	Thuricide ^ HP, HPCL, 32 B & 16 B
	4. Upjohn	SOK-BT
	5. Agriculture,& Nutrition Co.	Bactur TH
	6. Pennwalt, Holland NV	Tribactur

*There are several producers in the USSR which are not listed here.

** 1 kg. of formulated Dipel contains 165g of alkali extractable protein of which 90g is crystal protein. This implies that formulated product has been ten-fold diluted by inert filler materials.

^ Formulated product Thuricide is adjusted to 800 IU/mg of powder (lactose acetone coprecipitation of HD-1 yields about 12,000 IU/mg without dilution).

The bioinsecticide market makes 5% of the total insecticide market, amounting to one billion US dollars. Production of bioinsecticides is 1,000 tonnes per year in USA and 2,000 to 3,000 tonnes worldwide, selling at the retail price of 10-25 US dollars per kg. BT is registered for use against 90 different insects, as BT formulations listed in the last table. Eighty per cent (80 %) of BT is used to control insects of lettuce and cole crops, particularly cabbage looper (Trichoplusia sp.), tobacco bud worm (Heliothis virescens), tobacco horn worm (Manducta sexta) and several other forest pests. BT can be used at any time upto harvest, and is also used as a clean up spray before taking the produce to market. Bacillus thuringiensis israelensis (BTI) is another strain of BT isolated from Israel and extremely lethal to insects belonging to Diptera. This BTI strain will find increasing use in future. HLRC in Bombay is also working on BTI formulations.

However, for many menacing insects, no bioinsecticides are available and whenever available , their effect is slow. Therefore, they can form a part of integrated pest management programme.

Nuclear polyhedrosis virus (NPV)

NPV and some granulosis viruses also possess insecticidal properties. Boll worms cause extensive damage to cotton and can be controlled by NPV. Since in India, 50% of insecticide is used for cotton alone, NPV can be very useful as an insecticide. Sandoz is marketing a product called 'Elcar' , which is based on NPV.

Applications of Genetically Engineered Bacteria

A variety of micro-organisms, particularly bacteria have been modified through the techniques of genetic engineering to meet specific needs. These genetically engineered bacteria have found applications in areas including the following: (i) crop production and protection through biological control of insects, fungal diseases, frost damage, etc., (ii) biodegradation of xenobiotics (waste from non-biological systems) and toxic waste transformations, (iii) production of fuels and chemicals including antibiotics, enzymes and diagnostics and (iv) extraction of metals from ores. Application of this technology will ultimately release beneficial agents in the environment. It is hoped that if safely deployed, genetically modified bacteria should be able to provide significant benefits in the management of environment systems. Some examples of these genetically engineered bacteria will be briefly discussed in this section.

Crop production and protection

Several bacterial strains have been modified by introduction of foreign genes to control insects (by production of endotoxins), fungal diseases (by production of chitinases, which suppress fungal flora in the soil), or frost damage (through production of mutants for ice nucleation gene or Ice$^+$ responsible for frost damage). Deleterious bacteria and fungi or their effects have been controlled in some cases by using a variety of approaches, which include the following : (i) production of avirulent mutant of the pathogen and releasing it to compete with the virulent strains ; (ii) production of antibiotic producing strains of plant associated bacteria to control pathogen through antibiotic production or (iii) production of bacterial strains that will degrade the toxin produced by the pathogen.

There are also other positive measures, where the N fixing efficiency of Rhizobia can be increased by transfer of useful nif genes and other related genes involved in nodulation, nitrogenase activity and other related functions. Some examples of these genetically engineered useful bacteria for the possible increase in crop production and protection are listed below.

Table

Genetically engineered bacteria in crop production and crop protection

	Bacterium	*Altered Trait*	*Possible use*
1.	Clavibacter xyli	Transfer of B. thuringiensis delta endotoxin gene	Control of corn ear worm
2.	Pseudomonas fluorescens	Transfer of Serratia marcescens chitinase gene	Control of fungal disease
3.	P. fluorescens, P. syringae	Deletion of 'ice' gene	Control of frost damage
4.	P. fluorescens	Addition of lac ZY	Assessment of movement of bacteria for biological control ('take all disease' of wheat)
5.	A. radiobacter	Deletion of 'tra' gene of Agrocin 84 plasmid	Biological control of crown gall
6.	Rhizobium melilotii	Additional copies of 'nif' gene	Increased efficiency of N_2 fixation

Biodegradation of xenobiotics and toxic wastes

Bacteria can be modified genetically to develop catabolic pathways for degradation of xenobiotics (wastes from nonbiological systems) and other waste material. Bacterial genes for this purpose are isolated from bacterial strains (or their plasmids) found at waste sites. Most of these genes are available from gram-negative bacteria (e.g. Pseudomonas spp.), which are themselves not very efficient degraders, since multiple genes may sometimes be needed for efficient biodegradation. Therefore, for efficient biodegradation, efficient degraders have to be prepared through genetic engineering and they need to be established in the environment at a required density. Some of the genetically engineered bacterial strains for biodegrdation of waste material are listed on the next page.

A list of genetically engineered bacteria used for degradation of xenobiotics and toxic wastes.

	Bacterium	*Substrate that can be degraded*
1.	Pseudomonas capacia	2, 4, 5 - trichloro-phenoxyacetic acid
2.	P. putida & other spp (also E. coli)	2, 2, 5 -dichloropropionate; mono and dichloroaromatics
3.	Alcaligenes sp.	Dichlorophenoxyacetic acid, mixed chlorophenols; 1,4 - dichlorobenzene
4.	Acinetobacter sp.	4 - chlorobenzene

Production of chemicals and fuels

Recombinant DNA technology is also having a significant impact on microbial production of chemicals and fuels. Following are some of the examples of this utility: (i) genetically engineered strains of Bacillus amyloliquefaciens and Lactobacillus casei have been prepared for production of amino acids on a large scale; (ii) industrially useful bacteria can utilize cheaper feedstocks (substrates) like D-xylutose, lignocellulose or cellulose; for instance, Zymomonas mobilis (normally incapable of using lactose) carrying cellulase gene from Cellulomonas uda has six fold increase in cellulase activity; (iii) E. coli and Klebsiella planticola carrying genes from Z. mobilis could utilize glucose and xylose to give maximum yield of ethanol; (iv) genes for thermophilic (tolerant of high temperature) enzymes have been isolated from thermophiles like Bacillus stearothermophilus, B. licheniformis and Dictyglomus thermophilum and placed in E. coli, so that these modified strains can be commercially used for the synthesis of thermophilic enzymes.

TEN

Food Production

Today, the food processing industry has wide ranging processes for transferring agricultural products into edible products. The two most important methods are (i) to design microorganisms that transform inedible biomass into food for human consumption or animal feed, (ii) to design organism that aid in food processing either by acting directly or providing additional material. Traditionally microbes have been used to stabilize, flavour and modify various properties of food. Besides, attempts are always being made to process food such that it is free from microbes harmful to public health.

Single Cell Protein

The phrase "Single cell Protein" (SCP) was introduced for the first time in the 1960's to refer to the microbial proteins produced by yeasts, bacteria, cyanobacteria, fungi and algae and useful as human food or animal feed. Later on, several workers and public enterprises seriously explored the possibility of extracting SCP from microbes grown on hydrocarbons. The production of microbial proteins (SCP) has been practiced since First World War when Delbruck in Berlin, developed brewer's yeast on a large scale. The cells were mainly incorporated into soups and sausages. Even during the Second World War, food yeasts including Candida arborea and Candida utilis had been frequently used in Germany. Besides proteins, these microbial cells are also rich in carbohydrates, fats, vitamins and minerals. SCP is a potential protein source food for humans or animals. At present, SCP processes for the production of animal feed are the most attractive, since conventional animal feed stuffs such as soyabean meal and fish meal as being imported to many tropical and subtropical countries at the present international prices. In using SCP for animal feed, however, there is a huge loss of conversion efficiency as opposed to direct human use. For human use, microbes used are primarily Saccharomyces cerevisiae and Candida utilis and are fully accepted by public health authorities for human consumption. If these organisms are used as a major protein diet source, they need be processed further to minimise the nucleic acid content, to avoid the formation of kidney stones or gout.

The choice of organism as a source of SCP, however, depends upon several factors including economics of management. (Table). The idea of using SCP as an animal feed or human food is quite old and yeast has been used as food protein

Table

Products of Cellulose or Lignocellulose-Utilizing Microorganism and SCP Production

Microorganism	***Product***
Volvariella volvacea	Human food (mushrooms); animal feed
Lentinus elodes	Human food (mushrooms); animal feed
Pleurotus sp.	human food (mushrooms); animal feed
Thermoactinomyceters sp. and other thermophilic actinomycetes	Human food (SCP); animal feed
Phanerochaete chrysosporium	Delignified cellulose for use as feed, fiber, or further conversions
Trichoderma reesei	Cellulases for converting cellulose to sugars; animal feed (SCP)
Clostridium thermocellum	Cellulases for converting cellulose to sugar; ethanol, acetate, lactate, and HZ; animal feed (SCP)
Pseudomonas fluorescens var.	Animal feed; cellulases for converting cellulose to sugars
cellulose and similar bacteria	
Thermophilic Sporocytophaga	Animal feed; ethanol, acetic acid

since the beginning of this century. However, in recent years, a large number of plants have been introduced for the manufacture of SCP especially for the production of yeast. At present, the Soviet Union is a potential producer of SCP and proposes to produce enough fodder yeast from her internal sources to be self-sufficient in animal protein food stuffs by 1990. At several other places, SCP processes have also been practiced on a commercial scale, based on the growth of yeasts in deep tank-agitated and aerated cultures. Examples of raw materials used for these processes and yeasts that utilize them are molasses, Saccharomyces cerevisiae; n-paraffin hydrocarbons, Saccharomycopsis rapolytica and cheese whey, Kluyveromyces fragilis. Sweden has developed the Symba process utilizing starchy wastes by combining two yeasts, Saccharomycopsis fibuligera and Candida utilis.

Bacteria are also used as a source of SCP, particularly, species of Cellulomonas and Alcaligenes. Bacteria are preferred over yeast for production of SCP because of their fast growth rate and higher content of crude protein and certain amino acids particularly methionine. However, bacterial cells are smaller than yeasts and may be more costly to harvest unless the cells can be flocculated to give a higher solids slurry prior to centrifugation.

Fungi are also being exploited as a source of SCP. (Table). However, problems have been encountered with the use of fungal mycelium as a source of SCP, including slow growth rate. Therefore, fungi need to be maintained under sterile conditions, for a longer period of time to prevent overgrowth by bacterial and wild yeast contaminants. This consequently increases the production costs. The most important thermo-tolerant cellulolytic fungus Chaetomium cellulolyticum, has been found to be very useful in the conversion of cellulose to SCP.

Table

Raw Materials for Growing Fungi on a Commercial Scale

Raw materials	***Fungal Species***
Cane and beet molasses	Agaricus campestris
Maize syrups, dexture (D glycose)	Morchetta esculenta
Cheese whey, and canning	M. hortensis, M. crassipes
Coffee-processing wastes	Trichodorma Sp.
Maize wet-milling waste	Gliocladiurn Sp.
Maize and Pea-canning wastes	Trichoderma reesei.

Of all the microbes, algae are the most important source of SCP as these can grow profusely in open ponds and use CO_2 (as a carbon source) and sunlight. Spirulina species have been eaten for many years by inhabitants of the northern shores of Lake Chad in Africa and Aztec Indians in Mexico. Lot of work has also been done in India using Spirulina as a source of SCP because of its high nutritive value.

Recently several oil companies including British Petroleum have become interested in the production of microbial proteins from hydrocarbons. The Italian company Liquichimica built a plant for the production of yeast proteins, amino acids, fatty acids and citric acids from n-paraffins acid in 1980, producing 2 x 10^5 tonnes of proteins. Japan has also launched a massive programme for the derivation of single cell protein and had already contributed towards establishing a plant in Romania.

Besides, a process for the production of SCP using methanol as a carbon source has been developed by Hoechst AG in West Germany and Imperial Chemical Industries in U.K. Methanol is a preferred substrate because of its solubility in water (unlike methane and n-paraffin). Sulphite waste liquors from paper mills have also been used as a substrate, and fodder yeast has been producced using such wastes in United States. In Finland, a fungus Paecilomyces has also been used to produce SCP when grown on sulphite effluents. France has perfected a technique for producing SCP using Candida tropicalis grown on cassava. The starchy material could be enriched in yeast to the extent of 19 per cent with a good proportion of lysine (7.7%) and sulphur containing amino acids (2-2.7%). The Soviet Union has also been producing SCP on a mass scale amounting to more than 1 million tons per year by using agricultural, forestry and industrial wastes. The raw materials includes sawdust, bisulphite liquors (from paper industry), wood chips and wood pulp, sunflower husks, rice husks, bagasse, molasses and maize cobs. Currently, there is lot of interest in exploiting the vast reserves of deep earth gas as a source of SCP production all over the world.

Of the estimated 2.5 million tons of SCP produced annually throughout the world, mostly comes from cane and beet molasses and about 5×10^3 tons from hydrolyzed wood waste, corn trash and papermill wastes. (Table).

Table

Various Sources of SCP

Agave Juices	Peat
Barley Straw	Plumpere wastes
Cassava	Sunflower seed husks (treated)
Date carbohydrates	Wastes from chemical production of Malefic anhydride
Meat packing wastes	
Mesquite wood	Waste Polyethylene (treated)

Integrated systems can be designed to coup the production of a product or food with SCP production from waste such as saw dust from the lumber industry, which could become a source of cellulase for microorganisms. Forests are the most abundant source of carbohydrates in the form of cellulose (Table). But before it can be used by microorganisms, it must be transformed into the carbohydrate, glucose, by chemical or enzymatic pretreatment. Other major sources of SCP production, currently under intensive investigations, include Petroleum based hydrocarbons and methane or methanol. As an example of the potential significance of SCP, the Soviet Union which is one of the largest producer, expects to produce enough fodder yeast from internally available raw material to be self-sufficient in animal protein food stuffs by 1990.

Table

Various Substrates and Organisms Used for Produce of SCP in Different Countries.

Substrate	***Organism***
Methanol	Methylophilus methylotropus
Sulphite waste liquor	Paecilomyces
Lactoserum (From cheese-making)	Saccharomyces cerevisiae
Coffee wastes	Candida Sp. Trichoderma sp.
Cane molasses	Candida utilis C. tropicalis Rhodoturula gracilis R. rubra.
Cassava	Candida tropicalis saccharomyces fibuligera Candidautilis

Algal Biotechnology for Production of Food

Algal mass cultures represent a successful food production programme bringing together agriculture and industrial fermentation. Investigations have been carried on since the 1940's in several countries, with numerous projects within the laboratory and field studies. Several private enterprises are currently developing conventional heterotrophic fermentation processes for algal products including vitamin C and steroids. Besides, unicellular green alga *Chlorella* has been commercialized as an important health food. Currently, several plants for *Chlorella and Spirulina* (blue-green bacterium) production are operating in Japan, Taiwan, Mexico, Israel, Thailand and U.S.A. The overall production of alga (health food is about 2000 metric tonnes per year, on a dry weight basis). These algae are grown in ponds or in fermenters and then spray dried and sold as powder or pills (Table).

Table

Selected Algal Products Useful in Food Industry

Algal genus	***Products***	***Uses***
Several	Isotopic compounds	Medicine research
Red algal & blue-greens	Phycobiliproteiits	Food coloring Research
Blue-greens & others	Pharmaceuticals	Anticancer antibiotics
Dunaliella	ß-Carotene	Food supplement and food colours
Green algal and some		Chicken feed
Diatoms	Xanthophylls	
Green algal	Vitamin C and E	Vitamins
Chlorella & Spirulina	Health foods	Supplementary food
Porphyridium	Polysaccharides	Viscosifiers gums
Chrysophytes	Bivalva feeds	Seed raising aquaculture
Chlamydomonas	Soil inoculum	Conditioner
Chlorella & blue-greens	Amino acids	Proline, arginine, Aspartic acid
Greens & Diatoms	Vegetable and Marine oils	Foods, feeds and supplement

The development of the Chlorella health food industry in the early 1960s was based largely on plants of a circular pond design with large mixing arms. These were expensive and exhibited poor hydraulics. They could not be scaled up beyond approximately 100 m^2 (0.1 ha). By the 1970's, most new algae production units used the raceway pond design as developed by Oswald at the University of California, Berkeley. These were mixed by paddlewheels like other systems for Dunaliella and Spirulina production. Pond sizes are typically about 0.4-0.5 ha for most commercial systems. Although the costs and efficiencies of alternative mixing devices—airlift, screw, or centrifugal pumps, may be comparable in some applications, paddlewheels appear to have a significant capital and operating advantages in most cases.

Alga(productivities in outdoor cultures obviously varies but in favourable climates, peak production of 35-40 gm-2 per day can be achieved and, over a 300-day growing season, an average productivity of 25 gm^{-2} per day or 75 Tonnes ha^{-1} per year is feasible. This, however, does require the use of both productive algal species and favourable conditions. For all outdoor algal cultures for biotechnological production of food, the following conditions must be met with: (i) The use of highly selective conditions which favour dominance by one species, (ii) The cultivation of algae that spontaneously appear and dominate in ponds, (iii) The use of elaborate methods for preventing contamination.

The production of food from microalgae is still in a stage of extensive experimental trial.

Cyanobacterial Sources of Food

Castillo in 1521 described biscuits named tecuitlatl available in Mexico that were composed of dried mats of Spirulina maxima. In fact, spirulina used to grow in the alkaline waters of lake Texcoco near Mexico and Spirulina mats were recovered from the lake and dried to produce blue-green biscuits. The exceptionally high content of Spirulina protein (more than 60 per cent) was discovered late in 1964 by a Belgian Botanist J. Leonard. This Spirulina has been found to grow in highly alkaline lakes (with pH 11) and it is in a state of monoculture under these conditions. (Table).

Table

Relative Proportion of Various Molecules present in a Cell of Spirulina

Component	*Percent Composition*
Protein	65
Carbohydrates	19
Bile pigments and chlorophyll	6
Lipids	4
Fibre	3
Ash	3

Lot of work has been done on Spirulina as a potential source of protein in several countries including Italy, India, Japan, Mexico and Belgium.

Microbial production of Flavours

Biotechnology has already had a market impact in the sweetener industry. The availability of the enzymes glucose isomerase, invertase and amylose has made the production of high fructose corn sweeteners (HFCS) very profitable. The united States is awaiting the moment for the production of caloric sweeteners. The price advantage of HFCS is expected to result in its continued growth, particularly in the beverage industry. In fact Coca Cola Co. of America announced a few years ago that fructose will soon constitute as much as 50 per cent of the sweetener used in its brand beverage.

Although microbes do not synthesize sucrose (and it is unlikely that such a microbe will ever be discovered), the microbial production of low-caloric sweeteners is a distinct possibility. A lot of work on the sweetener aspartame, monellin and thaumatin is already in progress. Aspartame is synthesized chemically from the amino acids aspartic acid and phenylalanine, which can itself be produced by fermentation. Currently, investigations have been indicated to decide the possibility of using microbes to couple the two amino acids. Besides, the chemical production of aspartame is quite expensive and biotechnological production may be much cheaper. Monellin and thaumatin are natural proteins derived from West African plants. Both the substances are intensely sweet, up to 105 times sweeter than the table sugar. Their microbial production may be competetive with their extraction from plants. Since the physical and biological properties of thaumatin are known, it could also be produced through genetic engineering. Such an approach would not only increase the supply, but would offer new molecules for investigating the physiology of taste. Other flavours and fragrances are relatively less prominent from living cells (Table). Recent work on the formation of flavour and aroma chemicals known as lactones and terpenoids by micro-organisms has been reported. Lactones occur as flavour contributing components in many fermentation products and are formed by microbial reactions. Different pathways have been reported for their microbial formation. Gammabutyrolacetone, which is formed during yeast fermentation is found in sherry, wine and beer. As early as 1930, an organism was isolated from orange leaves that had a peach-like odour and was thought to be Sporobolomyces roseus. The lactones, 4-decanolide and cis6-dodecen-4-olide were found to be responsible for that.

Table

Industrial Chemicals, Amino Acids and Flavour Enhancing Nucleotides

Organism	*Product*
Saccharomyces cerevisiae	Ethanol (From Glucose)
Kluyveromyces tragilis	Ethanol (From Lactose)
Clostridium acetobutylicum	Acetone and Butanol
Aspergillus niger	Citric Acid
Xanthomonas campestris	Polysaccharides
Corynebacterium glutamicum	L-Lysine
Corynebacterium glutamicum	5'-Inosinic Acid And 5'-Guanylic Acid
Candida utilis	Microbial Protein from Paper Pulp Waste
Saccharomycopsis lipolytica	Microbial Protein from Petroleum Alkanes
Methylophilus methylotrophus	Microbial Protein from Growth On Methane or Methanol

Other Products and Generalizing Food Biotechnology

The food processing industry also uses polysaccharides (polymeric sugars) to alter or control the physical properties of foods (Table). Many are incorporated into foods as thickeners, gelling agents and agents to control ice crystal formation in frozen foods. They are used in instant foods, salad dressings, sauces, whips, toppings, processed cheeses and dairy products. Since many of the polymeric sugars now used in food processing are derived from plant sources, microbial polysaccharides have had limited use. However, very few polysaccharides have reached the level of commercial applications. The one in large scale production is xanthan gum. The history of the development of xanthan gum indicates that the commercially significant organisms resulted from an extensive screening programme for gum producers stored

Table

Polysaccharides and Pharmaceuticals

Organism	*Product*
Leuconostoc mesenterodes	Dextran
Xanthomonas campestris	Xanthan Gum
PeniciIJum chrysogenum	Penicillins
Cephalosporium acremonium	Cephalosporins

Organism	***Product***
Streptomyces	Amphotericin B. Kanamycins, Neomycins, Streptomycin, Tetracyclines and Others
Bacillus brevis	Gramicidin S
Bacillus subtilis	Bactitracin
Bacillus polymyxa	Polymyxin B
Rhizopus nigricans	Steroid Transformation
Arthrobacter simplex	Steroid Transformation
Hycobacterium	Steroid Transformation
Hybridomas	Immunoglobulins and Monoclonal antibodies
Mammalian cell Lines	Interferon
Escherichia coli (via recombinant-DNA technology)	Insulin, Human Growth Hor mone, Somatostatin, Interferon

in the Northern Utilization Research and Development Division of USDA's large microbial culture collection. Xanthan gum produced by Xanthomonas campestris NRRL B-1459 was found to have characteristics that rendered it very promising as a commercial product. Many polysaccharides could theoretically be produced for use in food processing. However, this would depend on understanding both the biochemical pathways for synthesis of a given polysaccharide and the systems that control microbial production. Detailed knowledge of the enzymes that are used to modify polysaccharides to more useful forms would also be required.

The role of certain enzymes (Table) in increasing commercial possibilities in the food processing industry is illustrated by the enzyme *pullulanase*. This enzyme degrades pullulan, a polysaccharide, to the maltose or high-maltose syrups that give jams and jellies improved colour and brilliance. They reduce off-color development produced by heat in candies and prevent sandiness in ice-cream by inhibiting sugar crystallization. Maltose has several unique and favourable characteristics. It is the least waterabsorbent of the maltose sugars and, although not as sweet as glucose, it has a more acceptable taste. It is also fermentable, nonviscous and easily soluble. It does not readily crystallize and gives desirable browning reactions. Pullulanase can also break down another carbohydrates amylopectin to produce high amylose starches. These starches are used in industry as quickselling structurally stable gels, as binders for strong transparent films and as coatings. Acetate derivatives are added to textile finishes, sizing, adhesives and binders. In food, amylose starches thicken and give texture to gumdrop candidies and sauces, reduce fat and grease

Table

Vitamins and Enzymes

Organism	***Product***
Eremothecium ashbyi	Riboflavin
Pseudomonas dentrificans	Vitamin Biz
Propionibacterium	Vitamin Biz
Aspergillus oryzae	Amylases
Aspergillus niger	Glucamylase
Trichoderma reesii	Cellulase
Saccharomyces cerevisiae	Invertase
Kluyveromyces fragilis	Lactase
Saccharomycopsis lipolytica	Lipase
Aspergillus	Pectinases and Proteans
Bacillus	Proteases
Endothia parasitica	Microbial Rennet

in fried foods and stabilize the protein, nutrients, colours and flavours in reinstituted products like meat analogues. However, the food processing industry is permitted to use only enzymes that are obtained from sources approved for food use. Since the chief source of pullulanase is a pathogenic bacterium, Klebsiella aerogenes, no significant efforts have been made to apply genetics to improve its production or quality. It may be possible to transfer the pulluiunase trait from K. aerogenes to a microorganism approved for food use.

The application of food biotechnology and genetic engineering will affect the food processing industry in a piecemeal fashion (Table) Isolated successes can be expected for certain food additives such as aspartame and fructose and for improvement of SCP production.

Table

Foods and Beverge from Microorganisms

Organism	***Product***
Saccharomyces cerevisiae	Baker's Yeast, wine, Ale, Sake
Saccharomyces carlsbergenis	Lager Beer
Saccharomyces rouxii	Soy Sauce

Organism	*Product*
Candida milleri	Sour French Bread
Lactobacillus sanfrancisco	Sour French Bread
Streptococcus thermophilus	Yogurt
Lactobacillus bulgaricus	Yogurt
Propionibacterium shermonii	Swiss Cheese
Gluconobacter suboxidans	Vinegar
Peniciflium roquetortii	Blue-Veined Cheeses
Penicillium camembertii	Camembert and Brie Cheeses
Aspergilfus oryzae	Sake (Rice-Starch Hydrolysis)
Rhizopus	Tempeh
Mucor	Sulu (Soybean Curd)
Monascus purpurea	Ang-Kak (Red Rice)

ELEVEN

Uses of Enzymes

Enzymes initiate and accelerate a large number of biochemical reactions in living cells. They process reactions which are otherwise not possible under normal conditions found in the cell. For instance, although, hydrolysis of starch in a test tube requires strong acidic medium and high temperature (boiling), in the alimentary canal it is hydrolysed and digested under normal conditions of acidity and temperature. This is made possible by starch hydrolysing enzymes available in the stomach. In fact, almost all biochemical reactions require one or more enzymes for their completion. Nonenzymatic conversions, though are known, but are very few.

Enzymes are proteinaceous in nature; they can be extracted from living tissues, purified and even crystallized. Under controlled conditions of isolation, they retain their original level of activity and in some cases even exhibit an increased activity. Thus, a purified enzyme can be used to carry on a specific biochemical reaction outside the cell. This property of enzymes has been employed for laboratory experiments as well as for commercial production of several important biochemical compounds, drugs and industrial products. In view of the above, enzyme research has become an important area of biotechnology. Different aspects of the techniques utilized in enzyme research and their commercial application will be described in this chapter. Enzymes can also be variously modified through enzyme modelling and gene manipulations for a variety of purposes, which will be discussed in the next chapter.

Medical and clinical uses

Drugs for digestive disorders— Some digestive enzymes such as **'amylases'**, 'proteases' and 'lipases' are often used as components of drugs to cure digestive disorders.

Deworming medicines— The plant proteases like 'papain' (from papaya) and 'ficin' (from fig) are used as deworming medicines for humans and domestic animals. These enzymes are not inhibited by the secretions of the worms and cause digestion of the worms.

Enzymes to stop bleeding— The enzyme 'thrombin' obtained from beef plasma is used typically to stop bleeding during operations and after tooth extraction. The enzyme converts fibrinogen to fibrin and small peptides, which are then insolubilised causing blood to clot.

Enzymes as surface disinfectants— 'Trypsin' has been used in cleaning away thick, purulent (full of pus) masses in surface infections and in emphysema (enlargement of vesicles of lungs) associated with pleural pneumonia. The area cleaned by its disgestive action heals much more repidly than untreated area. Two other enzymes, 'ficin' and 'streptodornase' are also used sometimes as wound debridement.

Diagnosis of diabetes—'Glucose oxidase' in combination with 'peroxidase' is used for colorimetric determination of glucose in blood. This test is generally conducted to determine the level of blood sugar thus leading to diagnosis of diabetes.

Industrial uses

Textile industry—'Amylases' isolated from bacteria, fungi, pancreas and malt are used in textile industry as softening agents for starched clothes. (Starch is often added to cotton fibres as a stiffening agent, before weaving the fibre into cloth. Since, a starched cloth does not take good colour, the cloth is to be destarched before dyeing it. This is done with an amylase preparation, which hydrolyses starch).

Leather industry—Proteolytic enzymes from certain bacteria and fungi are used in the manufacture of leather. These enzymes digest the collagen or connective tissue holding the hairs to the hide in the skin and thus cause dehairing of the skin. These enzymes are also used for softening or plumping of dehaired skin, a process popularly called bating.

Manufacture of detergents—Proteolytic enzyme 'alcalase' is added in many detergents and helps in removing proteinaceous stains from the cloth.

Manufacture of organic compounds—Enzymes from certain bacteria are also used for commercial production of acetone, butanol, lactic acid, citric acid and many other metabolites by the fermentation of carbohydrate rich substrates. Several enzyme preparations under different trade names are commercially available for such purposes.

Paper industry: cellulase free xylanases—Making paper from wood requires separation of the wood fibres from each other (pulping) and then reforming them into a sheet. In trees, the wood fibres are glued together due to lignin, which needs to be degraded to get chemical pulp (alternatively mechanical pulp can be obtained by tearing the wood fibres apart). From the chemical pulp, the residual lignin is removed by bleaching to get a brighter pulp. Since chlorine is conventionally used in bleaching, effluents from bleaching often contain toxic chlorinated compounds, which is an environmental hazard. Therefore, new techniques are being developed for pulp bleaching, where endoxylanases (cellulase free) are used (lignin-xylan complexes are found both in hardwood and softwood and can be degraded by cellulase free xylanases; presence of cellulase is avoided, since it may destroy the fibres).

Uses in food and food industry

Meat and beer processing— The proteases 'papain' (from papaya) or 'bromelain' (from pineapple) are used to tenderise meat by hydrolysing peptide bonds. It is also used to stabilise chill proof beer.

Manufacture of cheese—'Renin' obtained from the calf stomach is used in the manufacture of cheese, since it converts calcium-casein of the milk to calcium-paracaseinate, which is curd like in appearance. The curd is solidified and processed as cheese after inoculation with an appropriate mixture of micro-organisms.

Another enzyme **'catalase'** is also used for making cheese, since it breaks down H_2O_2 produced during cold pasteurization of milk during cheese making process. **'Lipase'** is added during the processing of cheese for flavour production.

Beverage industry—Enzymes from yeast are used for alcoholic fermentation in beverage industry, since they convert sugars to alcohol and CO_2.

Juice and wine processing—'Pectinases' are often added in canned fruit juices and in wine, since they hydrolyse the pectin making the juice or wine clear.

Chocolate and candies— 'Invertase' is used in the manufacture of chocolate covered berries and other such candies.

Soft drinks industry— 'Glucose isomerase' is used for the production of fructose and high fructose syrups from hydrolysed maize starch to be used in soft drinks.

Ice-cream industry— 'Lactase' is used for the prevention of lactose crystals in icecream.

Uses in research laboratories

Isolated enzymes are often used in the laboratory to study various biochemical reactions. They are also used as standards to detect their presence in unknown samples. Enzyme crystals are often used to study the physicochemical organization and function of proteins.

Isolation and Purification of Enzymes

Enzymes are unstable molecules with a definite physico-chemical organization. Even a slight change in this organization reduces the activity of enzyme and sometimes the enzyme is totally inactivated. Therefore, the enzymes have to be isolated under controlled conditions of pH, ionic strength and temperature. Since they are proteinaceous in nature, standard extraction and purification procedures for enzymes are the same as those used for proteins except that the activity of the enzyme is assayed at each of the following four steps of extraction and purification.

Extraction of enzymes

Fresh tissue is crushed into a paste with an extraction medium (often a buffer) in a mortar and pestle, or in a tissue homogenizes, or in a blender or by ultrasonic vibrations (sonication). The molarity and pH of the buffer is suitably adjusted (which may vary for different enzymes) to achieve maximum solubility and activity of the enzyme. EDTA (ethylene diamine tetra acetic acid) is often included in the extraction medium to remove heavy metals (which otherwise inhibit enzyme activity), and for disrupting the membranes of cells and cell organelles. Detergents such as Triton-X are also used sometimes to solubilise the membranes. Many enzyme proteins contain disulfide (S-S) bonds due to the presence of systeine residues, which are easily broken during enzyme extraction leading to loss of enzyme activity. To overcome this problem are added, thiols such as-mercaptoethanol whose sulfhydryl (-SH) group is able to maintain the S-S linkage in enzymes.

If the extract is not homogeneous, the homogenate (extract) is filtered to remove cell derbis, fibres etc., otherwise filtration may be avoided. All operations of extraction and purification are generally carried out in cold (0-4°C), since most of the enzymes get inactivated at higher temperatures.

Preparation of crude enzymes

Centrifugation— The enzyme extract is centrifuged to remove cell debris, cell organelles and sometimes other molecular aggregates, leading to partial purification of enzymes. It also helps in characterization of an enzyme, since, depending upon its mass and shape the enzyme will move through a solution at a definite speed and occupy a chaıacteristic position in the centrifuge tube.

For most cytosolic enzymes, centrifugation at about 30,000g for 30 minutes is good enough to obtain a fair amount of activity in the supernatant. However, if the enzyme is located in a specific cell organelle, an extract rich in that organelle is prepared through 'preparative centrifugation'. (Centrifugation for different durations at different velocities allows the cell organelles to sediment according to their sizes as shown in) All centrifugation operations are conducted in cold (0-4°C).

Precipitation— Enzymes and other proteins are highly charged molecules, and can be precipitated with appropriate charge neutralizing chemicals. Once, their charges are broken, they form aggregates and settle down as precipitate. When an acid or base is added, the enzyme protein

Table

Cell fractions by different centrifugation protocol

Centrifugation speed (x g)	*Time (min)*	*Cell fraction to settle*
1,000	5	Eukaryotic cells
4,000	10	Chloroplasts, cells debris, cell nuclei
15,000	20	Mitochondria, bacteria
30,000	30	Lysosomes, bacterial cell debris
100,000	180-600	Ribosomes

can be brought to its isoelectric pH. At this pH, there is no net charge on enzyme molecules and electrostatic repulsion between them is low so that they tend to aggregate. Therefore, adjusting the pH to the isoelectric point of a protein causes its precipitation.

Acids and bases, however, often inactivate the enzyme, so that their use for precipitation is not recommended in most cases. Instead ammonium sulphate and other salts are used for precipitation in a process called 'salting out'. Salts can change the structure of the solvent, which can lead to large changes in protein conformation by altering the electrostatic interactions between charged groups on the protein surface. The salt also competes with the protein for solvent molecules and thereby lowers its solvation. In large scale enzyme precipitation, use of many other neutral salts is preferred over ammonium sulphate, which is corrosive and releases NH_3 at higher temperatures.

Some organic solvents like acetone, methanol and ethanol are also used for enzyme precipitation, since water miscible solvents decrease the solubility of proteins, leading to precipitation. They are cooled upto 40-60°C before their use, and precipitation is carried out at 0°C, because precipitation at room temperature causes denaturation of the enzyme in most cases. Organic solvents are added drop by drop to avoid local concentration.

Water soluble non-ionic polymers such as polyethylene glycol, alginate, pectate, carboxymethyl cellulose, polymeta-acrylic and polymeia-acrylic acids, etc., also cause enzyme precipitation. Polyethyleneimine is also widely used as protein precipitant at large scale. They primarily act through the removal of solvent sphere of the enzyme protein.

Purification of enzymes

Enzyme purification involves three steps, i.e., dialysis, centrifugation and electrophoresis. These three techniques and their variants are briefly described in the following text:

Dialysis— Dialysis is the process that is used to remove small molecules from enzyme. For this, enzyme precipitate obtained in previous step is dissolved in a small

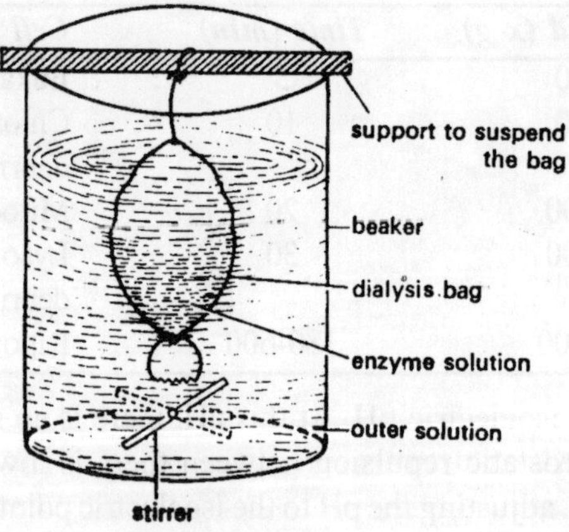

Fig. An arrangement for dialysis to purify an enzyme.

quantity of buffer solution in which the enzyme was originally extracted. The solution is taken in a dialysis bag (may be a cellophane tube) and after sealing securely, the bag is suspended in either distilled water or a buffer of known molarity and ionic composition. Some other salts or chemicals may have to be added sometimes in the outer solution to prevent the loss of enzyme activity during dialysis. The dialysis is carried out for a few hours with regular change of the outer solution or distilled water. At large scale enzyme purification 'dialafitration' instead of dialysis is used.

The enzyme solution is filtered for small molecules through a membrane generally mounted on a fibrous support by pressure dirven operations.

Chromatography— Chromatographic separation of proteins is the most common method of enzyme pruification. Following four types of chromatography are available for this purpose: (i) adsorption or column chromatography; (ii) ion exchange chromatography; (iii) gel filtration chromatography and (iv) affinity chromatography.

(a) Adsorption chromatography— In adsorption chromatography, the protein or enzyme solution suspected to contain other proteinaceous impurities is passed through a column of inert material packed in a glass or steel tube. Most commonly used column materials include finely divided solids such as charcoal, silica, alumina, calcium phosphate, hydroxyapatite, etc. The effluent solution is continuously collected in small fractions of 1.0 to 2.0 ml. The protein in each fraction is estimated by measuring the adsorption at 280 nm using a UV spectrophotometer. The enzyme is also assayed in each fraction. Various spleen enzymes such as basic RNAase, acidic RNAase, acidic DNAse, phosphodiesterase, phosphomonoesterase, etc., are often separated from each other using adsorption chromatography. For large scale chromatographic separation of enzymes, the process is accelerated by using motors and other mechanical devices for packing the column, for loading the enzyme on the column and for eluting the enzyme.

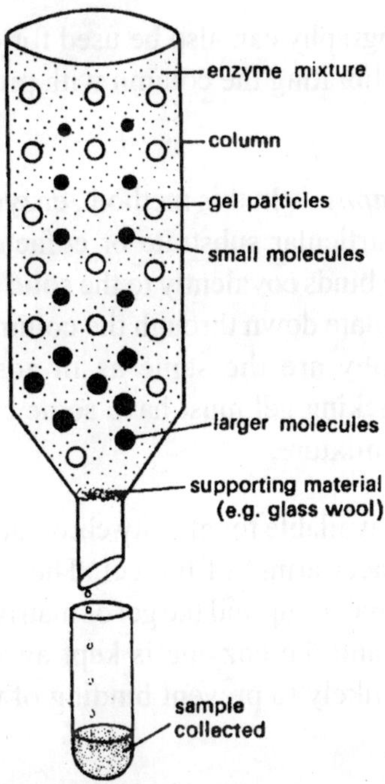

Fig. An arrangement for gel filtration chromatography to purify an enzyme.

(b) Ion exchange chromatography— In ion exchange chromatography, generally a cellulosic ion exchange is taken in the column. The proteins are separated according to their charges. The resolution is quite high and the technique can facilitate large scale protein purification. This has been successfully employed for insulin purification, plasma fractionation and for purification of many other enzymes.

(c) Gel filtration chromatography— In this chromatography, various proteins are separated on the basis of differences in their molecular sizes. This type of chromotography is also known as molecular exclusion chromatography or molecular sieve chromatography.

The basic arrangement for gel filtration chromatography is similar to that for adsorption chromatography. A column made up of glass or steel is taken and packed with a gel. The most commonly used gel is 'sephadex' which is a cross linked dextran produced by certain strains of bacteria. Several types of sephadex, namely G-10, G-30, G-50, G-100, G-150, G-200, etc. are available, which differ according to their pore sizes.

When a mixture of enzymes or proteins is poured on top of the column, different proteins move downwards according to their molecular sizes and come out from the column in order of decreasing sizes; the larger molecules are eluted first. The elution volume is logarithmically proportional to the molecular size.

Gel filtration chromatography can also be used for determining the molecular weight of the protein by celibrating the column with proteins of known molecular weights.

(d) Amity chromatography— In this method , enzymes are purified according to their specificity for a particular substrate or cofactor. One component of the mixture containing enzyme binds covalently to the solid support of the column, and the other components percolate down through the column. The basic requirements for affinity chromatography are the same as in adsorption or gel filtration chromatography but the packing gel must have some component which can bind with one component of the mixture.

Many commercial gels available for affinity chromatography contain functional groups attached to the 'spacer arms' of the gel. The "spacer arm" is a chemical linkage between the functional group and the gel or matrix proper, so that the binding between functional group and the enzyme is kept away from the gel. Thus, the steric hindrance will be unlikely to prevent binding of the specific enzyme to the column.

Electrophoresis— Electrophoresis is a technique in which molecules (enzymes, proteins, amino acids, nucleotides and nucleic acids) are separated by differences in their net charge in the presence of an externally applied electric field. This technique is routinely used in enzyme purification and isozyme separation in the laboratories, although it has found only limited application at large scale, since the technique is time consuming and is a bit expensive.

Various types of instrumental approaches have been used to separate and purify charged molecules using electrophoresis. However, the most common method for purifying enzymes is through electrophoresis on polyacrylamide gel. Polyacrylamide is a polymer of acrylamide and methylene bisacarylamide and when prepared as a gel it is transparent, thermostable, non-ionic and extremely regular in structure. The gel may be taken either in the form of a column or a slab, although the latter is preferred over the former. The protein mixture is loaded in the gel and the components are separated under a direct current of constant voltage. The migration rate of the various components of the mixture is dependent upon their charge and molecular weight.

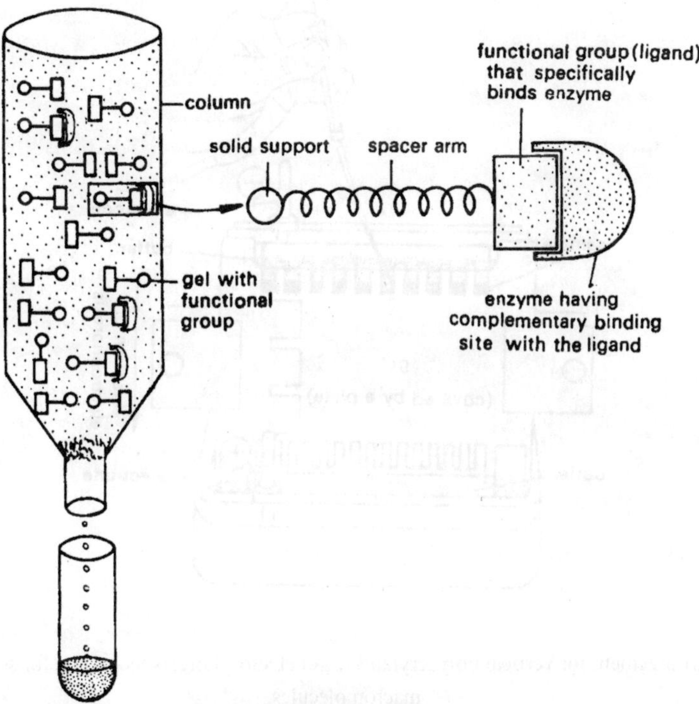

Fig. An arrangement for vertical polyacrylamide gel electrophoresis technique for separation of macromolecules.

A variation of the above polycrylamide get-electrophoresis is the sodium dodecyl sulfate-polyacrylamide gel electrophoresis (SDS-PAGE), which is used to determine the molecular weight of proteins. In this methods, the separation is caused by the seiving action of the gel. The proteins migrate through the gel depending on their shapes and mass to charge ratio. Gel electrophoresis is also used to separate various isozymes of a given enzyme. Isozymes perform the same catalytic function but differ in their regulatory and some kinetic aspects.

Final processing of enzymes

Most of the commercially available enzyme preparations, purified as above, are concentrated and sterile filtered after purification. This is done to reduce both, the volume and the microbial contamination of the sample. Often, before storage and transport, the sample is freeze dried with additives such as sugar substrates and dextrans.

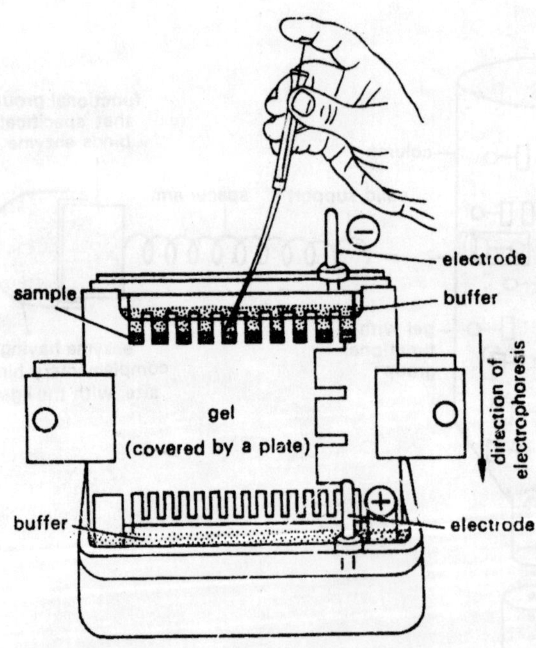

An arrangement for vertical polyacrylamide gel electrophoresis technique for separation of macromolecules.

Immobilization of Enzymes (or Whole Cells)

In recent years, the technique of enzyme or whole cell immobilization has revolutionized the prospects of enzyme application in industry. Immobilization is defined as 'the imprisonment of a biocatalyst in a distinct phase that allows exchange with, but is separated from, bulk phase in which substrate, effector or inhibitor molecules are dispersed and monitored'. In other words, an immobilized enzyme is physically entrapped or covalently bonded by chemical means to an inert and usually insoluble matrix, where it can act upon its natural substrate. The matrix is usually a high molecular weight polymer such as polyacrylamide, cellulose, starch, glass, beads, etc.

The common use of crude enzyme preparation in the production of wine, cheese or in tanning to obtain the desired product is known for long time. Their use as catalysts was limited, however, due to their limited availability, instability and the consequent high cost. This limitation has been largely overcome by the immobilization of enzyme on a support, a phenomenon reported for the first time by J.M. Nelson and E.G. Griffin (1916). They reported the adsorption (immobilization) of invertase on charcoal/alumina without loss of activity. However, the technique of enzyme immobilization could be established only after a lapse of about 40 years, when between 1954 and 1961 many researchers developed relevant procedures and the equipments. Some of the advantages of using immobilized enzyme over free enzyme are as follows : (i) Because of its binding with a matrix, the immobilized enzyme has better stability in many cases. (ii) Efficiency of immobilized enzyme is better. (iii) The enzyme can be recovered at the end of the reaction and can be used repeatedly. Further, the processed product is not contaminated with the enzyme. (iv) Some manipulations of enzyme catalysed reactions are better in immobilized form. For example, the reaction can be slopped rapidly by removing the enzyme from the reaction solution.

The immobilization of enzyme is performed under mild and controlled conditions, so that the enzymes retain their tertiary and quaternary structures, which are necessary for their activity. Immobilized enzymes are preferred over immobilized cells or tissues, since they have a high specificity and yield pure products.

Immobilization Techniques

Various methods developed for immobilizing enzymes, that are described in this section, can be classified into the following five general categories : (i) adsorption method, (ii) covalent binding method, (iii) cross binding method (iv) entrapping method and (v) entrapping by microencapsulation.

Adsorption method— The enzymes may be immobilized by adsorption to several types of materials (adsorbents with charged or neutral surfaces), that are used for the separation of proteins by adsorption chromatography. Since the activity of enzyme may be significantly reduced or lost during adsorption and subsequent release, the adsorbents should be chosen such that enzymes are bound firmly with minimum inactivation. Following are some of the adsorbents often used for immobilization: alumina, amberlite CG-50, bentonite, calcium phosphate gels , carbon, carboxymethyl cellulose, carboxymethyl sephadex, collagen, DEAF-cellulose, DEAE-sephadex, glass, silica gel, titanic (ceramics). The pH and ionic conditions of enzyme and adsorbent solution should be carefully controlled during immobilization. If adsorption involves predominantly ion interaction with very little hydrogen bonding, then a simple shift in pH or ionic strength could exchange the protein ion for another ion and thus adsorption would occur.

In a typical adsorption method for immobilization, adsorbent is packed in a water jacketed column, which is washed with a preconditioning solution. For instance, if we take titanic as an adsorbent, it is washed with 0.5 $NaHCO_3$ and internal surfaces of the adsorbent are rendered air free. Enzyme solution is buffered fairly close to isoelectric point of the enzyme with a low ionic strength (below 0.01 μM), and is then circulated through the column at a desired temperature for several hours. After the enzyme solution is drained from column, the column is washed with water, then with 0.5M NaCl and finally again with water. The immobilized enzyme column is now ready for the delivery of substrate solution and evaluation of performance.

Covalent binding method— An enzyme can be covalently bound to support materials by different methods. The enzyme forms a covalent link with active groups of support material either (i) through the reactive groups on side chains of its amino acids like lysine, arginine, histidine, tyrosine, systeine, serine, aspartic acid and glutamic acid or (ii) with terminal amino and carboxyl groups of the polypeptide chains. Some examples of covalent linkage between the support and the enzyme are as follows: (The reactants required to activate the support are indicated in corresponding figures).

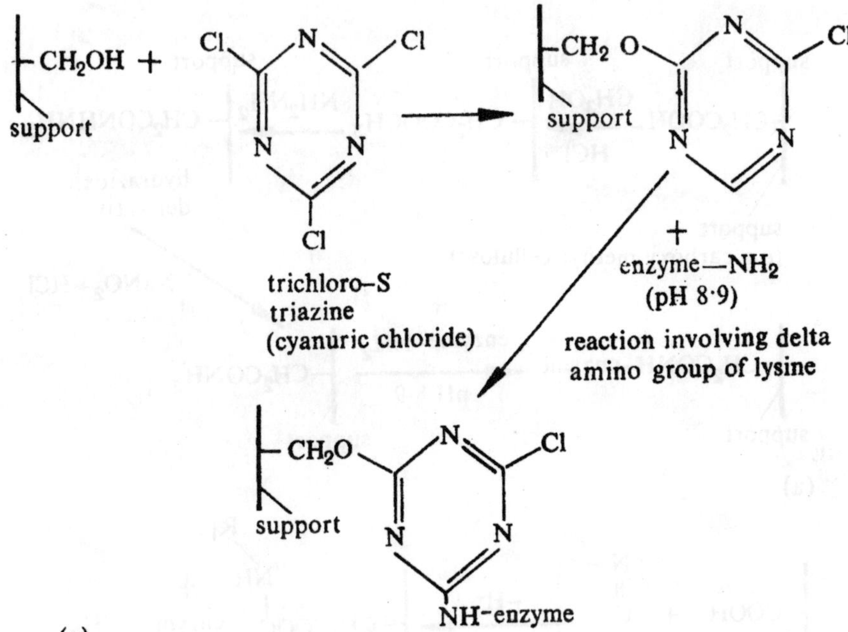

support—CH_2OH + trichloro-S triazine (cyanuric chloride) → support—CH_2O—(dichloro-triazinyl)

+ enzyme—NH_2 (pH 8·9)

reaction involving delta amino group of lysine

→ support—CH_2O—(chloro-triazinyl)—NH-enzyme

(a)

support(CH—OH)(CH—OH) + BrCN (cyanogen bromide) → support(CH—O—CN)(CH—OH) → support(CH—O)(CH—O)>C=NH (iminocarbonic acid)

+ enzyme-NH_2 (pH 8·9)

→ support(CH—O—CONH-enzyme)(CHOH)

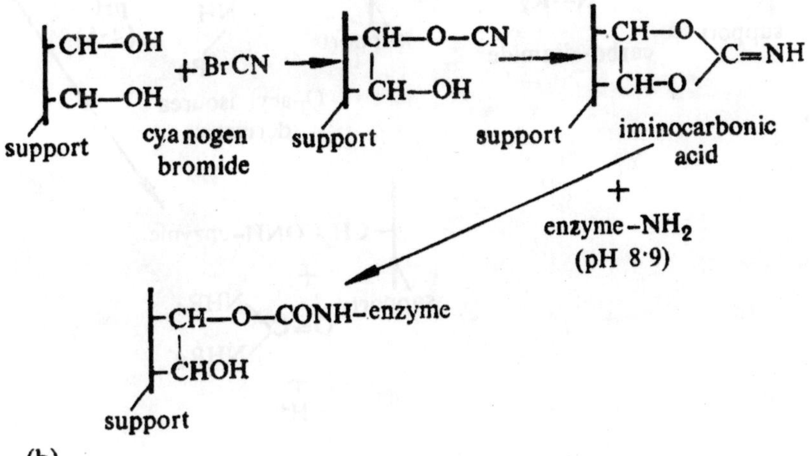

(b)

Immobilization of enzymes, using supports having-011 groups, that are activated by covalent bond formation with (a) triazines or with (b) cyanogen bromide.

(a) Support with -OH group. Such supports can be activated for covalent linking by treating with either triazines or cyanogen bromide. The reaction with enzyme protein in each case involves the -NH2 group of the lysine.

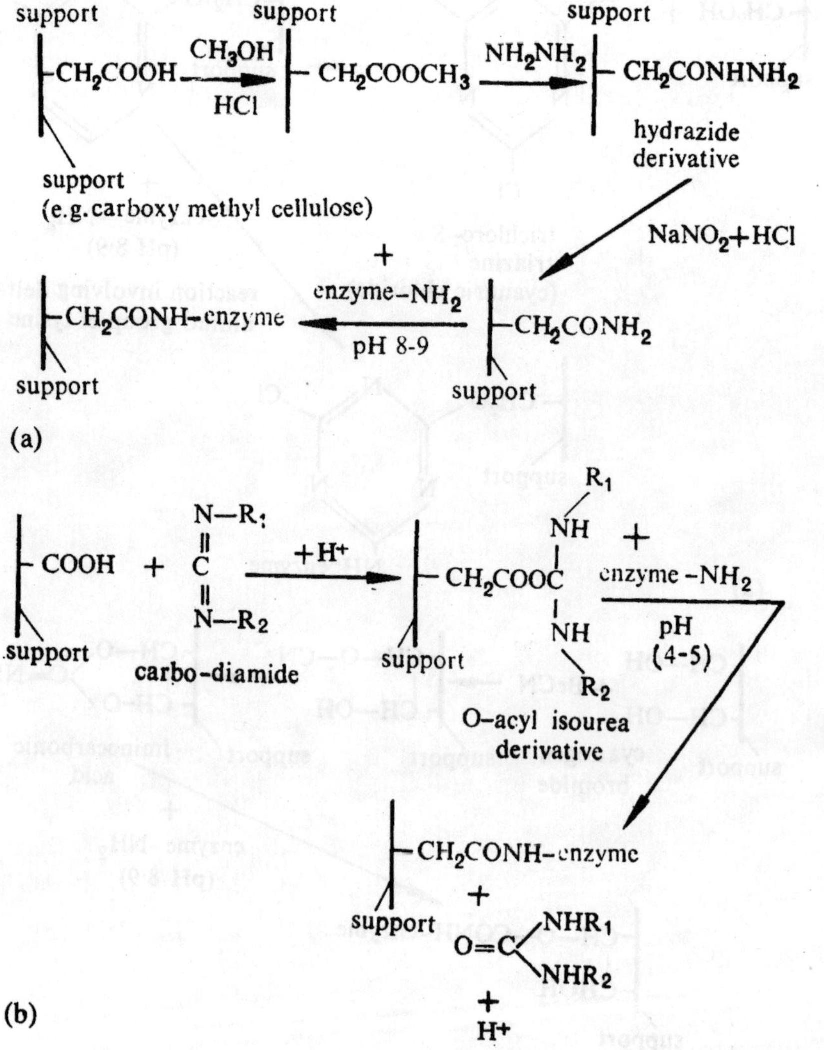

support
support
support
$-CH_2COOH \xrightarrow[HCl]{CH_3OH} -CH_2COOCH_3 \xrightarrow{NH_2NH_2} -CH_2CONHNH_2$
hydrazide derivative
support
(e.g. carboxy methyl cellulose)
$NaNO_2 + HCl$
+
enzyme$-NH_2$
$-CH_2CONH-$enzyme $\xleftarrow{pH\ 8\text{-}9}$ $-CH_2CONH_2$
support
support
(a)

$-COOH + C(=N-R_1)(=N-R_2) \xrightarrow{+H^+} -CH_2COOC(NH-R_1)(NH-R_2)$ + enzyme $-NH_2$ (pH (4-5))
support carbo-diamide support
O-acyl isourea derivative
$-CH_2CONH-$enzyme
+
support
$O=C(NHR_1)(NHR_2)$
+
H^+
(b)

Immobilization of enzymes, using supports having -COOH, with hydrazine (NH2 - NH2) or with -NH= groups through covalent linkage involving (a) azide derivative (b) acyl isourea.

(b) Support with -COOH group— Such supports like carboxymethyl cellulose can be activated via azide derivative or acyl isourea formation. The reaction involves -NH, group of lysine. However, other amino acids such as tyrosine, cysteine and serine are also implicated.

(c) Support with -NH2 group—Support containing amino group can be converted to a diazonium chloride by treating with NaNO2 + HCL. The enzyme protein links with this derivative forming an azo- linkage involving the tyrosine residue of the enzyme protein. Glutaraldehyde is also used sometime to activate the support containing-NH group. The reaction involves a Schiff's base formation between amino group of support and amino group of one of the amino acids of protein. The number of bonds between support and the enzyme molecule during convalent linkage is variable. There are, for example, 17 bonds per papain molecule and 9 for subtilopetidase immobilized on semiadehyde starch derivative.

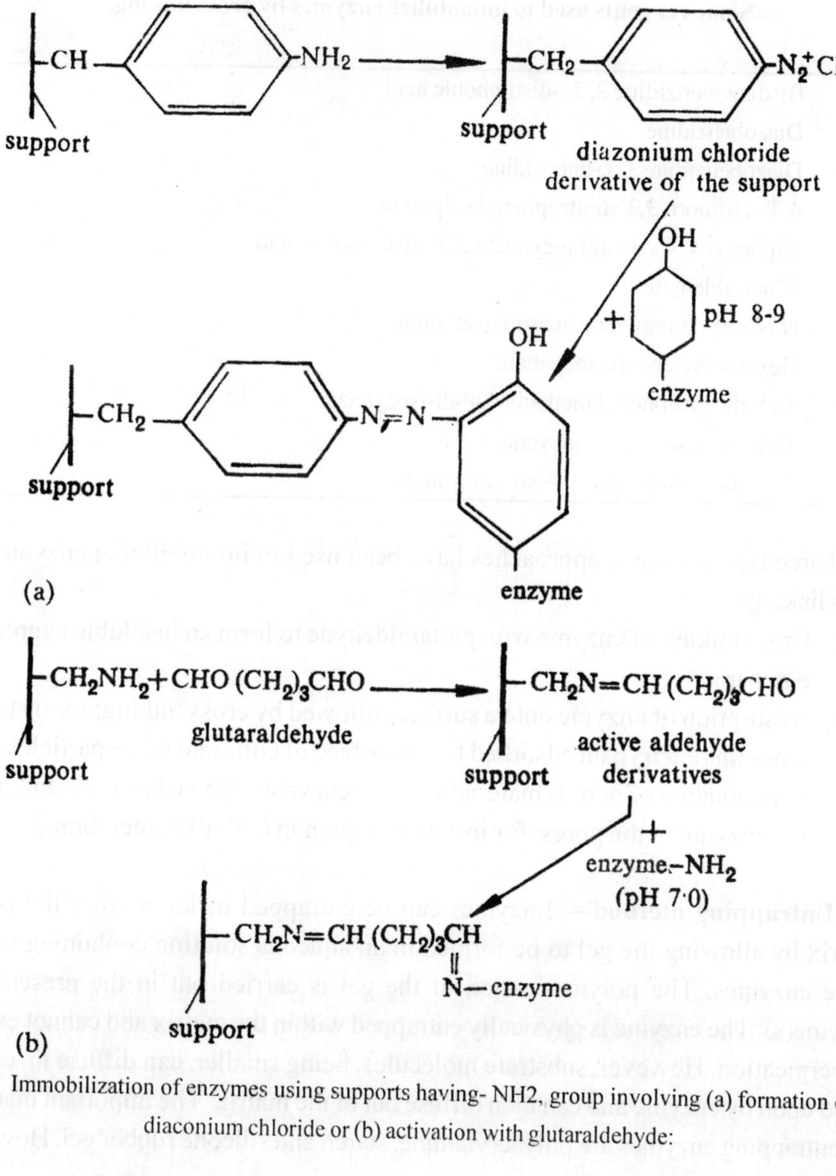

support —CH—C₆H₄—NH₂ ⟶ support —CH₂—C₆H₄—$N_2^+Cl^-$ (diazonium chloride derivative of the support)

\+ OH—C₆H₄—enzyme (pH 8-9) ⟶ support —CH₂—C₆H₄—N=N—C₆H₃(OH)—enzyme

(a)

support —CH_2NH_2 + CHO $(CH_2)_3$CHO (glutaraldehyde) ⟶ support —$CH_2N{=}CH(CH_2)_3CHO$ (active aldehyde derivatives)

\+ enzyme–NH_2 (pH 7·0) ⟶ support —$CH_2N{=}CH(CH_2)_3CH{=}N$—enzyme

(b)

Immobilization of enzymes using supports having- NH2, group involving (a) formation of diaconium chloride or (b) activation with glutaraldehyde:

Papain bound to porous glass is linked through three azo links per enzyme molecule.

Cross binding method— The enzyme can be immobilized by cross linking to a multifunctional reagent without involving a solid support. Alternatively, a chemical bridge (analogous to the spacer arm in affinity chromatography) of some other molecule can be estabilished between the enzyme and a chemical support. The reagent or the bridge chemical reacts with the enzyme and forms a linkage. Some of the commonly used reagents for cross linking enzymes are listed below.

Some reagents used to immobilize enzymes by cross linking.

1.	Bisdiazobenzidine 2, 2, -disulphonic acid
2.	Diazobenzidine
3.	Diazobenzidine 3,3,-dianisidinc
4.	4,4' -Difluoro 3,3' dinitrophenyl sulphone
5.	Diphenyl 4. 4'- diisothiocyanate 2,2'-disulphonic acid
6.	Glutaraldehyde
7.	N,N'- hexamethylene bisisodiacetamide
8.	Hexamethylene di isocyanate
4.	3-methoxydiphenyl methane 4,4'-di-isocyanate
10.	Toluene 2, 4-di-isothiocyanate
11.	Toluene 2-isocyanate 4-isothiocyanate

Three types of basic approaches have been used in immobilizing enzymes by cross linking

(1) Cross linking of enzyme with glutaraldehyde to form an insoluble aggregate; e.g. papain.
(2) Adsorption of enzyme onto a surface followed by cross linking; for instance, cross linking trypsin adsorbed to the surface of colloidal silica particles.
(3) Impregnation of porous material with the enzyme followed by cross linking of the enzyme in the pores; for instance, papain in collodion membrane.

Entrapping method— Enzymes can be entrapped inside a cross linked gel matrix by allowing the gel to be formed in an aqueous solution containing one or more enzymes. The polymerization of the gel is carried out in the presence of enzyme(s). The enzyme is physically entrapped within the matrix and cannot escape by permeation. However, substrate molecules, being smaller, can diffuse in, can be acted upon by enzyme and can then diffuse out of the matrix. The important matrices for entrapping enzymes are polyacrylamide, starch and silicone rubber gel. However, immobilization of enzymes by such an entrapment method is often associated with following three types of difficulties:

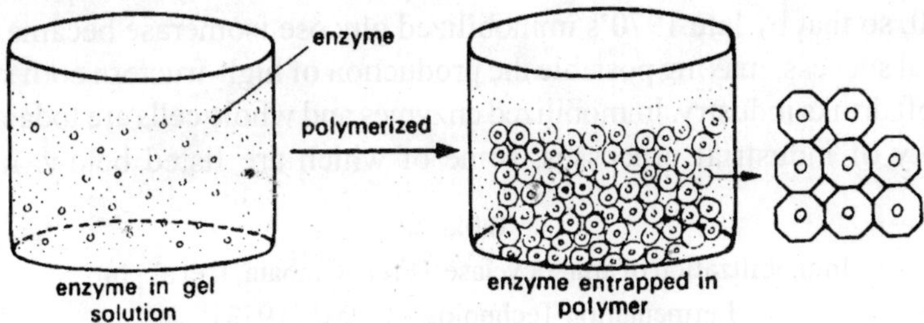

Entrapping of enzyme inside a polymerized gel.

(1) There is continuous leakage of enzyme because of wide pore size distribution in the gel.
(2) There is reduced substrate accessibility to the enzyme.
(3) There is some loss of enzyme activity due to free radicals produced during polymerization of the gel.

Entrapping by microencapsulation— Microencapsulation is another approach in enzyme immobilization by entrapping method. In this approach enzyme can be immobilized within microcapsules prepared from organic polymers, so that the enzyme can not escape, although low molecular weight substrates and products can enter and leave the capsule by diffusion through the membrane. There are two general methods for preparing the membranous capsules for enzyme entrapping:

(1) The membranes are prepared by the process of phase separation, which is similar to homogenization of water in oil. One phase is not miscible with the other but forms a droplet or coacervate with the other phase, when mixed. The enzyme is entrapped within this droplet or coacervate.
(2) A water insoluble membrane is prepared by chemical polymerization and enzyme is entrapped during this polymerization.

Spheres made up of semipermeable nylon or collodion membranes are often used for microencapsulation of an enzyme. Enzymes are also entrapped inside fibres of cellulose triacetate. These fibres may be packed in columns or woven into fabric.

The choice of the method to be employed for immobilization would certainly vary according to the enzyme involved and its application. Sufficient variation has been reported in the level of retention of enzyme activity on immobilization to different supports.

Applications of immobilized enzyme (or whole cell) systems

The use of immobilized enzymes in industry attracted increased emphasis during 1970-1980, so that by late 1970's immobilized glucose isomerase became a major commercial success, making possible the production of high-fructose corn syrup, so vital for soft drinks industry. Immobilized enzymes and whole cells are today utilized in a variety of industrial processes, some of which are listed below. It should

Table

Immobilization of aminocyclase. (From Chibata, I. et al., In "Fermentation Technology Today", 1972).

	Support	*Method*	*Approximate enzyme activity immobilized (per cent)*
1.	Polyacrylamide	Entrapment	53
2.	Nylon	Encapsulation	36
3.	DEAF cellulose	Ionic binding	55
4.	DEAF sephadex A-50	Ionic binding	56
5.	CM sephadex C-SO	Ionic binding	0
6.	Lodoacetyl cellulose	Covalent binding	39
7.	CNBr activated sephadex	Covalent binding	1.0
8.	AE-cellulose	Cross linked with glutaraldehyde	0.6

Table

Some applications of immobilized enzymes and whole cells

Enzyme	*Process*
A. Immobilized enzyme systems	
1. L-Amino acid oxidase	D-amino acid production
2. Flavoprotein oxidase	N-oxidation of drugs containing hydrazine groups
3. 0—'Steroid hydrogenase	Production of prednisolone
4. Ribonuclease	Nucleotide production from RNA
5. Amylase	Corn syrup production
6. Aminocyclase	Resolution of Di. amino acids
7. Glucoamylase	Production of glucose from corn syrup
8. Invertase	Conversion of sucrose into glucose and fructose
B. Immobilized whole cell System	
9. Glucose isomerase	Conversion of glucose to fructose
10. Penicillin amidase	Production of 6-aminopenicillinic acid for semi-synthetic penicillins
11. L-Arginine deaminase	Production of L-citrulline
12. Fumarase	Production of malic acid
13. L-Histidine ammonialyase	Production of urocanic acid
14. Aspartase	Production of L-aspartic acid

however, be appreciated that the immobilized enzymes make only a fraction of the total cost. In developing an immobilized system, once the cost benefit ratio is favourable, immobilized system may be developed and imporved for commercial purposes. For this purpose, the following four areas should receive adequate attention of the decision makers for the use of an immobilized enzyme system.

Enzyme carrier system— The choice of an immobilized enzyme system depends on the following two factors:

(a) Enzyme purification and stability— The manufacturers have to make a choice about the purity of enzyme—the less purified enzyme requires a larger bulk of enzyme while highly purified is more expensive. The higher purity may also avoid undesirable side reactions due to (i) absence of other enzymes and due to the (ii) low durations, for which the substrate has to be in contact with the enzyme.

Since long operating life or maximum yield of product per unit weight of immobilized enzyme is critical for its commercial success, stability of enzyme system is important. Highest total productivity is achieved at lowest temperature, although contamination and capital costs may be higher at low temperature. The stability of the support material as well as the activity of enzyme are also influenced by pH, so that pH profile of enzyme system should also be known.

(b) Nature and composition of carrier system—The charge on the carrier also influences pH. For instance, if carrier is negatively charged, H+ ions will accumulate at its boundary leading to drop in pH and if positively charged, there will be increase in pH. Such changes in pH will reduce stability of the system and may be overcome by incorporation of the ions of interest into the carrier. For example, magnesia incorporated into porous alumina support (developed by Corning) for glucose isomerase increased the observed activity of enzyme by more than 50%.

Reactor types—After the enzyme and the carrier system have been chosen, decision needs to be taken on reactor design, which can be any one of the following four types:

(a) Batch reactor— It is the simplest type of reactor. The immobilized enzyme is placed in a container with the reactants, and the reaction is allowed to proceed until the desired level of conversion is reached. Some stirring or agitation of reaction mixture is also required. Many modifications of these reactors have been designed to simplify recovery and reuse of the enzyme composite.

(b) Continuously stirred reactors— This type of reactors employ a stirred tank, to which reactants are continuously added and products continuously withdrawn.

(c) Fluidized-bed reactor— In these reactors, enzyme system is fluidized by the upward flowing of substrate solution. This helps in eliminating and plugging of enzyme system, although small duration of contact may be insufficient for the desired conversion. This can be overcome by decreasing velocity of solution by different methods. Glucose isomerase and lactase have been shown to double their efficiency due to fluidization when compared with faced beds.

(d) Fixed bed reactors— These are most widely used for large scale commercial operations. Different companies use these reactors for enzymes like (i) aminocylase, (ii) glucose isomerase and (iii) lactase. These reactors will keep on dominating the large scale commercial application due to their (i) high efficiency and (ii) ease and simplicity of operation. There are many types of fixed bed reactors including those with packed bed of particular material to which enzyme is coupled.

Operating strategy— Operating strategy has to be prepared for the use of immobilized enzymes to have maximum production at minimum cost. Since the enzyme activity falls down, the production rate also falls in course of time. This is minimized (i) by using multiple reactor system, in which start-up and reloading time for each reactor is (as done in case of glucose isomerase) staggered with other reactors in series or parallel and (ii) by raising temperature to compensate for loss in activity (as has been done for lactase).

The operating costs have to be kept in mind, while deciding about the operating strategies and will include cost of the following : (i) immobilized enzymes (including carrier system, immobilization, equipment, capital, etc.); (ii) labour (including fringe benefits, supervision and other overhead expenses); (iii) depreciation charges of the equipment used, which may be upto 20% of the capital annually and (iv) raw material. Provision will have to be made for losses at each processing step.

Regulatory considerations— For using immobilized enzymes, particularly for producing food or pharmaceuticals, it is necessary to get approval of appropriate authority (e.g. Food and Drug Administration of FDA in USA) for the use of a particular fixing agent or a carrier system. These fixing agents have to be declared safe by these authorities. Only two fixing agents for glucose isomerase are approved and these are (i) diethylaminoethyl cellulose and (ii) glutaraldehyde. Many of the fixing agents reported in literature are not approved for use in food industry.

Enzyme Engineering

Enzymes can also be specifically engineered to create forms not available in nature. These will then be used in some important commercial applications.

TWELVE

Utilization in Medicines

Medicinal Applications

Biotechnology has applications in the field of medicine. While dealing with diseases, applications of biotechnology include prevention, diagnosis and cure of diseases. Through human genetics, it has found use in genetic counselling, antinatal diagnosis, and gene therapy. In forensic medicine, it has already been used for identification of individuals, who could be criminals (murderers or rapists). These applications will be briefly discussed in this chapter under three headings : (i) animal and human health care, (ii) genetic counselling and (iii) forensic medicine.

Animal and Human Health Care

Biotechnology may have applications in animal and human health care either indirectly or directly. Its indirect significant role involves manufacture of nutritional food through single cell proteins and in detecting food contaminants through molecular probes, etc. (discussed in Chapter). Its direct and more important use, on the other hand, involves efforts to overcome the menace due to a variety of diseases which ran be discussed at four levels of manipulation: (i) providing immunity against a disease through development of a vaccine, (ii) diagnosis of a disease at early stage of its onset, (iii) treatment of the disease and (iv) fertility control. These three aspects of dealing with animal and human diseases will be briefly discussed in this section.

Development of Vaccines for Immunity

Several methods for the production of vaccines were discussed in Chapter. As one of the strategies, vaccines can be prepared from animal material by isolating antigens. In conventional methods, separation of these antigens sometimes has the risk of contamination, and therefore, may become health hazard. Therefore, biotechnology for development of vaccines may be used at least in three different ways, which include : (i) separation of a pure antigen, using a specific monoclonal antibody, (ii) synthesis of an anti with the help of a cloned gene, and (iii) synthesis of peptides to be used as vaccines.

Immunopurification of antigens using monoclonal antibodies—Immunopurification involves separation of a specific antigen from a mixture of very similar antigens. This purified antigen can the be used for developing vaccine against a pathogen. The purification can be very effectively achieved using monoclonal antibodies, which are very specific in their reaction against an antigen. Individual interferons which are proteins having a property of inhibiting viral infection and cell proliferation have also been purified using monoclonal antibodies. After such purification, interferons were used for clinical trials, before they were released recently for commercial use.

Synthesis of antigens through cloned genes—Hundreds of genes in eukaryotes have been cloned either from genomic DNA or from cDNA These cloned genes included a number of genes for specific antigens, an in some cases have been used for the synthesis of antigens leading to the preparation of vaccines. Following two examples can be used to illustrate the use of cloned genes for vaccine preparation: (i) Cloning of Hepatitis, B virus (HBV) genome—The HBV genome has been cloned in the plasmid pBR322 and propagated in E. coli. From this clone, antigen could be produced in good quantity, which reacted with hepatitis B core antibody (HBAb). This has, therefore, been used to produce hepatitis B vaccine, which was later approved for mass vaccination in several countries. (ii) Cloning of human malarial gene—Despite the great menace and threat to human health due to malarial parasite Plasmodium Falciparum, no anti-malaria x vaccine, could be developed so far. Recently with cloning of a gene coding for surface protein of the sporozoite of P. falciparum, there is a hope for developing a vaccine. In human host, malarial parasite passes through several antigenically distinct phases, namely; (i) sporozoite : the form in which the parasite is injected with mosquito bite; sporozoites enter the liver and multiply and develop into (ii) merozoites, which in turn invade and multiply in red blood cells; small fraction of these merozoites in red blood cells form (iii) gametocytes, which may be picked up by a mosquito to start another cycle. Therefore, vaccines can be developed to control any of these phases and will be accordingly called (i) antisporozoite vaccine, which will prevent malaria in the vaccinated individual and also block its spread; (ii) antimerozoite vaccine which will protect or ameliorate the patient but will not check the spread of the disease and (iii) antigametocyte vaccine, which would prevent the spread without helping the patient. Of these, antisporozoite vaccine is in sight due to cloning of gene meant for eireumsporozoite (CS) protein. This gene was obtained directly from DNA of erythrocytic form of parasite, rather than as cDNA from mRNA. This cloned gene may, in course of time, lead to the synthesis of vaccine by synthesizing CS protein by cloned gene.

Synthetic peptides as vaccines—Vaccines can also be prepared through short synthetic peptide chains, which have, therefore, become a subject of considerable research activity. In order to synthesize peptides to be used as vaccines, structure and function of proteins involved should be studied. Since, it is the three dimensional structure (TDS) and not the amino acid sequence, which is responsible for immunogenic response, it may be necessary to find out the protein region involved in immunogenic response. For instance, in Foot and Mouth Disease Virus (FMDV), it is the amino acid 114-160, of virus polypeptide, which can produce antibodies neutralizing FMDV, and thus, provide protection. Neutralization of FMDV was also possible through the region of 201-213 amino acids of the same protein. It has, thus, been shown that small synthetic peptides representing these regions of proteins can show immunogenic response and can, therefore, be used for development of a vaccine.

An alternative approach to find out the immunogenic region of protein is through the study of gene coding the protein. Recently, it has been shown that a cloned gene of an immunogenic protein of a pathogen (feline leukaemia virus) can be cut into fragments by DNAase I and these fragments can be cloned in lambda phage, where they may express. Phage colonies (plaques) having different cloned fragments are screened with a specific monoclonal antibody that neutralizes the pathogen. The fragments, which react with antibody must be synthesizing the immunogenic peptide fragment. This cloned DNA fragment can then be sequenced. It was possible, in this manner to identify a 14 amino acid immunogen of the envelope protein of feline leukaemia virus (FLV). The corresponding synthetic peptide was also found to compete with the virus for antibody. When injected in guinea pigs, such synthetic peptides also elicited a partial immunogenic response. Therefore, there is a great promise for the use of such synthetic peptides to be used as vaccines. Actually, a vaccine for malaria in the form of a synthetic peptide has already been prepared and is being tested for its suitability. This is the first example of a vaccine developed in the form of a synthetic peptide.

Recently, it has been shown that immunognic region of protein of a pathogen can also be identified, by eluting it from purified MHC molecules (MHC = major histocompatibility complex). Different MC allelic variants are available in cells for binding of different proteins and they can be purified using specific T cells. Peptides can be eluted from these purified MHC molecules and sequence of such peptides can be determined and used for manufacturing synthetic peptides to be used as vaccines.

Using the above three approaches, vaccines against several pathogens including the following pathogens have either been produced in recent years or are expected to be produced in the near future. (i) Rabies virus, (ii) Foot and Mouth Disease Virus

(FMDV), (iii) Salmonella Typhimurium, causing typhoid, (iv) Virbrio cholerae, causing cholera, (v) Hepatitis B virus, causing hepatitis (hepatitis B-vaccine produced through recombinant DI Technique has now been approved for mass vaccination in several countries), (vi) Plasmodium Falciparurm, causing malaria, (vii) Feline Leukaemia Virus (FLV) causing cancer, and (viii) Toenia Solium causing cysticerosis.

Diagnosis of Diseases

Diagnosis of diseases caused by protozoa and helminths—In order to supplement the conventional methods available for diagnosis of different diseases, biotechnology has provided monoclonal antibodies and DNA probes as two very effective and most sensitive tools (for techniques monoclonal antibodies and DNA probes, consult Chapters and respectively). Monoclonal antibodies and DNA probes are being prepared and made available for the diagnosis of a variety of diseases. Monoclonal antiboidies can be used through serological tests, which will take only minutes, while the conventional methods may sometimes take weeks, since they may require culturing of bacteria or viruses as in case of herpes virus or other viruses. Similarly, DNA probes, which are even more sensitive than monoclonal antibodies, will, however, take hours (not minutes as monoclonal antibodies) instead of weeks. Readymade DNA probes 4, herpes virus and other human, animal and plant viruses are being prepared. DNA probe kits are also available to help preparation of DNA probe and a market of DNA probes for several hundred million dollars (>$500,000,000) per year is now available. This market will expand and grow in future.

The diagnostic methods at the DNA level have the advantage over other methods, since in these methods genes of the parasite are examined and not the expressed product which changes in different stages of the life cycle of the parasite and also due to the environment. Rapid progress in recent years has been made in the development of nucleic acid based assays for the diagnosis and epidemiological surveillance of human parasites. Probes are now available for a number of human parasites from the group Protozoa and Helminths (both Platyhelminths and Nematyheclminths), which are listed in Table.

Table

Protozoa and helminths where DNA probes and diagnostic tests are available

Parasite	*Probe used*	*Hybridization technique used*
I. Protozoa		
1. Leishmania (kala-azar)	Kinetoplast DNA (kDNA, a repetitive DNA sequence)	Dot blots. Southern blots, in situ hybridisation
2. Trypanosoma (sleeping sickness)	Kinetoplast DNA (kDNA); a repetitive DNA and total parasite DNA	Southern blot (restriction fragment length polymorphism = RFLP)

Parasite	*Probe used*	*Hybridization technique used*
3. Plasmodium (malarian parasite; P. falciparum, P. vivax and P. cyomolgi)	Repeated DNA; synthetic oligonucleotide.	Dot blots; blood samples lysed directly on nitrocellulose filters
II. Helminths		
1. Schistosomes (human blood flukes)	Ribosomal RNA gene	Dot blots; Southern blots
2. Nematodes: Wuchereria and Brugia (filaria); Orlchocerca (river blindness); Trichinella	Repetitive DNA sequences; synthetic oligonucleotides; ribosomal DNA;	Crude DNA, blood samples; dot blots
3. Taenilid cestodes (Taenia solium, etc.)	Ribosomal DNA;	Southern blots

In India also, in (1986) a diagnostic probe for the detection of malaria has been constructed at the Astra Research Centre India (ARCI), which is established as a joint venture between India and AB Astra, Sweden, (inaugurated by Prime Minister, Rajiv Gandhi on Jan. 7, 1987). Methods for the detection of several other diseases are also being developed at this centre.

In the above use of DNA probes, DNA has to be radioactively labeled, which is not very safe in field study. Therefore, techniques for non-radioactive labelling of probes, methods for quick preparation of target DNA and suitable protocols for hybridization are being developed. This, in future,, will allow quick diagnosis of parasites in the field (see also Chapter).

Diagnosis of sexually transmitted diseases (STD) using monoclonal antibodies. (a) Characteristics of human sexually transmitted diseases—The occurrence of sexually transmitted diseases (STD) particularly in the developed countries and in relatively free society in our country has increased in recent years. The most common pathogens responsible for STD's are (i) Neisseria gonorrhoea, (ii) Chlamydia trachununis and (iii) Herpes sintplex virus (HSV). These pathogens also have a role in a wide spectrum of diseases, and therefore, their simple diagnosis tests will be a great help.

(i) Neisseria gonorrhoene causes the following diseases: (i) In men- epididymitis, (ii) In women- urethritis, cervicitis, endometritis and salpingitis and (iii) In both men and women- protiditis and pharyngeal infection. In a number of infected persons, other diseases also appear which include arthritis, dermatitis, endocarditis, meningitis, chorioamnionitis, premature delivery and blindness.

The diagnosis of gonorrhoea is often done through microscopic examination and through culture of thc bacterium on a selective medium taking few days.

(ii) Chlamydia trachomatis also causes infection in urethra, cervix, rectum and conjuctivae, and leads to diseases, some of them common with gonorrhoea. It also causes pneumonia in young infants. The clime similarities of N. gonorrhoea with C. trachomatis are complicated by the co- transmission in many cases, so that the diagnosis of C. trachomatis is very difficult.

(iii) Herpes Simplex Virus (HSV) infections are most common and once caught, it remains life long and is transmitted. It is classified HSV^1 and HSV^2. The diagnosis is mainly done through cell type 1 and type 2 and are distinguished through immunological analyst (this is called HSV typing).

(b) Classical methods for diagnosis—The infectious diseases such as STD are diagnosed usually by any one of the following four classic methods (these methods do not involve biotechnology).

(i) Microscopic examination of tissue specimens and exudates for identification of virus infected cells, bacteria, fungi or other parasites.

(ii) Culture methods with selective growth media permitting multiplication of some microorganisms, which can then be tested for susceptibility to potential therapeutic agents.

(iii) Immunological methods where antigens associated with specific pathogens are identified in tissue or body fluids.

(iv) Measurement of antibody produced in the patient as a result of infection with an organism.

Laboratories often use combination of two or more of the above methods to be sure about the diagnosis, and still the tests are sometime not fool proof, and are time and labour consuming. The use of monoclonal antibody, on the other hand, provides fool proof and very specific tests for diagnosis.

(c) Preparation and selection of antibodies for diagnosis of STDs—The cell hybrids are first prepared as done in hybridoma technology (cons Chapters). Since there is a random loss of chromosomes in c hybrids, screening of cell lines is done through replica plating technique. Hybrid cells are placed in a 96-well microtest plate. Small sample culture fluid from each cell is placed on replica plates, each of which impregnated with a specific antigen. The immune reaction is detected through radio-immunoassay in which ^{125}I-labelled protein A (it binds to protein of human IgG in the immune complex) is added to each cell and then examined by autoradiography. The positive reactions can be traced back to original cells and the cell line then multiplied for production of specific monoclonal antibodies.

Using the above technique, monoclonal antibodies could be prepared that would distinguish between N. gonorrhocae, C. trachomatis and HSV. Antibodies could also be selected which reacted with specific strains of the pathogen. A mixture of three antibodies (4-G5, 2-H1 and 3-C8) identified 99.6% of the 719 isolates of N. gonorrhoeae and did not react with other species of Neisseria. This allowed diagnosis of gonorrhoea with reasonable certainty.

In case of Chlamydia, the infectious form (called elementary body) enters the cell and within 48-72 hours forms a large inclusion body containing several hundred elementary bodies, which on release cause infection in neighbouring cells. When cells were faced in ethanol (18-72 hours after infection) and stained with fluorescein-conjugated antibody, characteristic inclusion bodies were detected as early as 18 hours after infection, by immunofluorescence (IF). Detection was also possible in cultures derived directly from patients, or even on smears on microscope slides prepared from specimens derived from infected tissue of the patient. It has been shown that the direct test takes less than 30 minutes, which is a major advantage over the culture method.

The suitability of monoclonal antibodies in the identification of Herpes Simplex Virus (HSV) has also been demonstrated. A panel of four different monoclonal antibodies allowed distinction between the types HSVt and HSV2. Such tests are described as typing of HSV (typing means to find out the type). Therefore, monoclonal antibodies can be used effectively in diagnosis and typing of HSV directly on primary clinical specimens, derived from oral, genital, mucocutaneous or ocular sites.

Antenatal diagnosis of congenital diseases—When a pregnant woman is known to have a chance of bearing a child with a genetic defect, it may be desirable to diagnose the condition in the fetus. This can be done by taking some cells from the fetus by drawing a few millilitres of amniotic fluid with the help of hypodermic needle. The amniotic fluid has free cells of fetal origin, which can be cultured and tested in various ways e.g. karyotype, enzyme production and restriction fragment length analysis of its DNA. At least 35 diseases which can be identified by this technique are known. If disease is detected through such an antenatal diagnosis, abortion of fetus can then be recommended. However, if abortion is not acceptable to parents, there is no point in carrying out antenatal diagnosis.

It is possible to identify a disease now within two month's of pregnancy, unlike an 18 weeks period required earlier. The number of disease specific DNA probes is also increasing at a fast rate, so that antenatal diagnosis by DNA analysis or 'enzyme linked immunosorbent assay' (ELISA) should be possible for all single gene defects. In recent years, the incidence of the disease thallassaemia, in Cypriot community in Britain has fallen from 30 to 20 per year, due to the use of antenatal diagnosis.

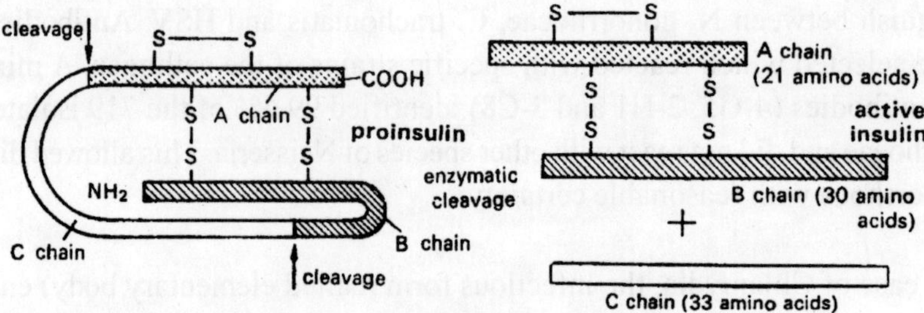

Fig. Structure of insulin and its derivation from a much longer proinsulin molecule.

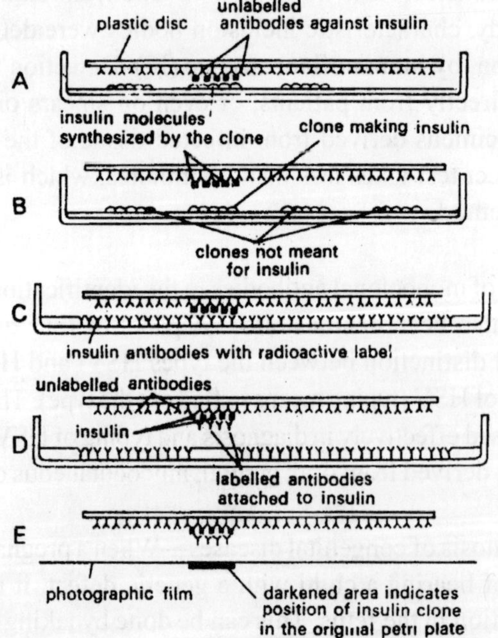

Fig. Isolation of DNA clone encoding insulin.

Treatment of Diseases

Pharmaceutical drugs through biotechnology—Drugs for treatment of diseases can also be manufactured using biotechnology. Treatment of diseases can also be affected through gene therapy. Many of the drugs for treatment are such, which could earlier be obtained only by sacrificing animal life. Two such drugs are insulin for treatment of diabetes and interferons for treatment against some tumour viruses.

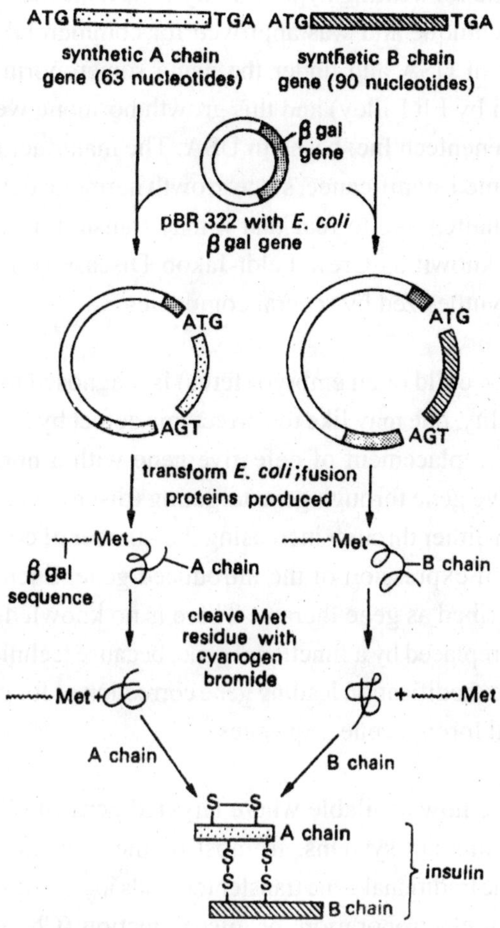

Fig. Strategy used for synthesis of insulin from cloned DNA segment

Such drugs can be manufactured now in bacterial cells in large quantities, if the corresponding genes from human or animals are cloned through plasmid vectors in bacteria, thus making their production relatively very cheap. The gene for insulin cloned in bacteria and could be used for synthesis of insulin (Figs.). Dr. Saran Narang, a scientist of Indian origin, working in Ottawa, Canada was involved in cloning of insulin gene. Synthetic insulin manufactured in this manner is now being sold commercially in North America. About a dozen different interferons are also in different stages of testing and some of them are already being sold commercially. This drug 'interferon' was earlier used to be available at the rate of 10 million U.S. dollars per 50 mg costing a patient, 150 dollars per day. It is estimated that with the availability of synthetic interferons this may reduce to a cost of one dollar per day. A number of other prtoeins like urokinase, factor VII C, human growth hormone (HGH) and many other drugs will soon be available through this technique. In 1985, human growth hormone for treating hypopituitary dwarfism was synthesized using recombinant DNA technique and was approved for commercial marketing under the name prototropin in USA and under the name somatonorm in Britain. Both insulin (manufactured by Eli Lilley) and this growth hormone were manufactured under licence from Genentech Inc. based in USA. The manufacture of prototropin and somatonorm assumed significance, since growth harmone derived from human pituitary glands was halted due to fear that it may transmit a rare and fatal viral neurological disease known as Crew Feldt-Jakob Disease (CJD). More growth hormones are being synthesized by several companies.

Gene therapy—If a child or an embryo (fetus) is diagnosed to carry a defective gene leading to disability, one may like to correct this defect by any of the following three methods: (i) by replacement of defective gene with a normal gene; (ii) by correcting the defective gene through gene targeting (discussed in Chapter) or (iii) by gene augmentation either through increasing the number of copies of the gene or through higher level of expression of the introduced gene. Such a correction of a genetic defect is described as gene therapy. There is no knowledge present, how a defective gene can be replaced by a functional gene, because techniques are available either for targeted gene modification leading gene correction or for gene augmentation by introducing normal foreign gene sequences.

Several reports are now available where targeted gene modification has been demonstrated in mammalian systems. In most of these cases, genes have been introduced by any of the traditional gene transfer methods including calcium phosphate mediated transfection, electroporation, or microinjection (Chapter). This will be followed by site specific mutation as demonstrated for the HGPRT (hypoxanthine-guanine phosphoribosyl transferase) locus and int-2 loci in mouse embryonal stem cells. Therefore, there is a hope that in due course of time, gene correction will be possible by site specific mutagenesis.

More important than the above gene correction method is the gene augmentation method, where normal foreign gene sequences for the defective gene are introduced.

A number of efficient methods for this purpose are already available, where a number of copies of the desired gene area introduced in the celi and are made to express at high level expression and transfer vectors in the form of a number of viruses are now available to achieve this goal. In view of this, maximum progress in the area of gene therapy has been achieved through the above gene augmentation model involving vector mediated DNA delivery system.

Once the gene correction or gene augmentation has been achieved at the cellular level (in the cells obtained from the affected organ depending upon the disease), the modified cells can be implanted into a suitable region either in an organ of the patient or in the embryo. Direct delivery of the DNA (carried by the vector) into the living cells of the body has also been suggested in several cases, so that both in vivo and in vitro introduction of corrected gene has been suggested. Therefore, gene therapy can be used at two different levels : (i) embryo therapy, in which the genetic constitution of embryo at the post-zygotic level is altered, so that the inheritance will also be altered, and (ii) patient therapy, in which cells with healthy gene may be introduced in the affected tissue, so that the healthy gene overcomes the defect without affecting the inheritance of the patient. It is believed that, in future, gene therapy of both types will be possible.

(a) Embryo therapy—This will involve the following steps, which have been tried in case of mouse or rabbit only : (i) in-vitro fertilization of the egg, (ii) insertion of normal gene into embryo at post-zygotic level, either with viruses or directly by microinjection, (iii) integration of inserted gene in host DNA, where it may or may not function. The inserted genes have been found to be inactive generally, but in few animals, where genes have been switched on in a tissue-specific way, their activity is at a very low level. However, it is not yet possible that the therapeutic newly inserted genes function under normal control in the animal in time, space and quantity.

(b) *Patient therapy*—Patient therapy will involve the following steps (i) defective gene should be identified, (ii) normal healthy gene should either be isolated or synthesized, (iii) cells of the tissue, where the normal healthy gene will need to function should be isolated (iv) the normal gene should be placed into a cell, where it can function. The gene will have to be placed into the correct site on the host chromosome so that it may function; or even one may have to delete the defective gene. There are three main problems in this connection. First, that the introduced gene may not function; second, that when corrected cells are reintroduced, these may be outnumbered by the concured resident cells; and third, that there are only few diseases affeciting only a single tissue.

During the last 5-10 years, it has become routine exercise to isolate any gene (see Chapter). The isolated gene may either be directly injected into the cell or be carried by a virus, to which it is linked by recombinant DNA technique. After entering the cell, the gene may become a part of nuclear DNA or remain free in cytoplasm like extrachromosomal DNA. However, in each case, RNA is synthesized only at the rate of few conies per cell in comparison to normal cells where thousands of copies are made.

(c) *Recent experiments on gene therapy*—In recent years, approved human genetic engineering (or gene therapy) experiments have been conducted. These experiments include the following : (i) Neo R/TIL gene marking: In 1989, using Neo R/TIL protocol, the first gene-marked immune cells (tumour-infiltrating lymphocytes =TIL) were successfully transferred into patients with advanced cancer. The TILs are those lymphocytes that are isolated directly from tumour and are then grown to large numbers in tissue culture in the presence of T cell growth factor (interleukin-2 = IL-2). In Neo R/TIL protocol, an aliquot of cells from TIL is taken; a markres gene is transferred to these cells (the marker gene is neomycin resistance gene = Neo R) using retroviral vector; the marked cell and unmarked cells are grown together and then transferred back to the patient. (ii) ADA gene therapy: In 1990, the first trial of actual gene therapy was conducted in USA. A little (4 years old) girl suffering with adenosine deaminase deficiency (ADA), a lethal disorder, was transfected with lymphocytes bearing the ADA gene carried by retroviral vector. In 1991, another girl (9 years old) began treatment in USA and received more than a dozen transfusions by 1992. Both these patients are doing well. In Italy also, a five year old boy was given a mixture of ADA gene corrected T cells and bone marrow cells. In Netherlands also, bone marrow gene therapy protocol for ADA deficiency was approved in 1992. (iii) Cancer gene thrapy: In 1991, two protocols for cancer gene therapy were initiated in USA. Tumour necrosis factor (TNF, an anticancer agent) gene or IL-2 (Interleukin-2) gene is inserted in TIL tumours cell isolated from the patient and grown in culture. These gene-corrected cells are then injected into the body of patient. Other cancer gene therapy protocols have also been approved.

In December 1991, a conference titled ‘Human Gene Therapy’ was held at National Institute of Health in USA and concluded that gene therapy will soon become a potent force in medicine to deal with heart diseases, liver diseases, diabetes, and a variety of cancers. It was hoped that gene therapy will eventually play a role in disease prevention by correcting deficiencies right at brith. The cellular vehicles for gene transfer are also multiplying and will include endothelial cells and myoblasts, besides, lymphocytes.

Cultured stem cells and bone marrow transplantation—As disscussed in Chapter stem cells are undifferentiated cells, which have the potential of differentiating into several kinds of cells. Stem cells can also be of different kinds. For the purpose of bone marrow transplantation, the stem cells used are called 'hematopoietic stem cells'. These stem cells first appear in human embryo in the yolk sac and then migrate to the liver as the fetus develops. In the fetus, the blood cells are created in the liver, but soon after birth they are produced exclusively in bone marrow. These stem cells can be isolated, stored, and are also cultured rarely. They are the main ingredient of bone marrow transplants. If marrow is transplanted to the same individual from whom it was taken for culture, it is described as autotransplantation and if donated to someone else, it is called allogeneic transplantation.

The hematopoietic stem cells produce, in one hour, 3-10 billion platelets, red cells, neutrophils, lymphocytes and even macrophages of the immune system. These cells are used for transplantation in patients, whose immune and blood forming systems have been devastated by leukemia, cancer, chemotherapy, or other unknown causes.

Isolation of stem cells from the mass of cells in the bone marrow has been achieved through the use of monoclonal antibodies which react to a cell marker or antigen called CD34, which is specific for primitive bone marrow stem cells. However, all CD34 positive cells may not be the pluripotent stem cells required for hone marrow transplantation. Therefore, more specialized technique may be used for further screening of CD34 positive cells. Culture systems are also available, which allow multiplication of these cells. These cultures may also be used in future for transplant therapy. Several hormones have also been identified, which stimulate replication of precursor blood cells. These include granulocyte colony stimulating factor (G- CSF), interruptin-6 interleukin-11 and a stem cell factor.

Bone marrow donors are classified by a method called HLA typing, which is used for matching the recipient. However, for autologous transplantation, a patient's own stem cells are collected, frozen and stored outside the body, while chemotherapy or radiation therapy is administered to remove the malignant cells. The bone marrow thus stored is returned directly into a vein (Fig.). However, sometimes bone marrow may be invaded by cover cells, so that healthy cells will have to be separated from malignant stem cells. Monoclonal antibodies linked with toxins can be used for killing the malignant cells. Chemotherapy with extracted bone marrow can also be used for destroying malignant stem cells. It has also been shown that whereas the malignant cells have an advantage over normal cells for growth within the body, they have a relative disadvantage outside the body. In view of this, long term culture systems have been used to remove these cells from bone marrow (Fig.).

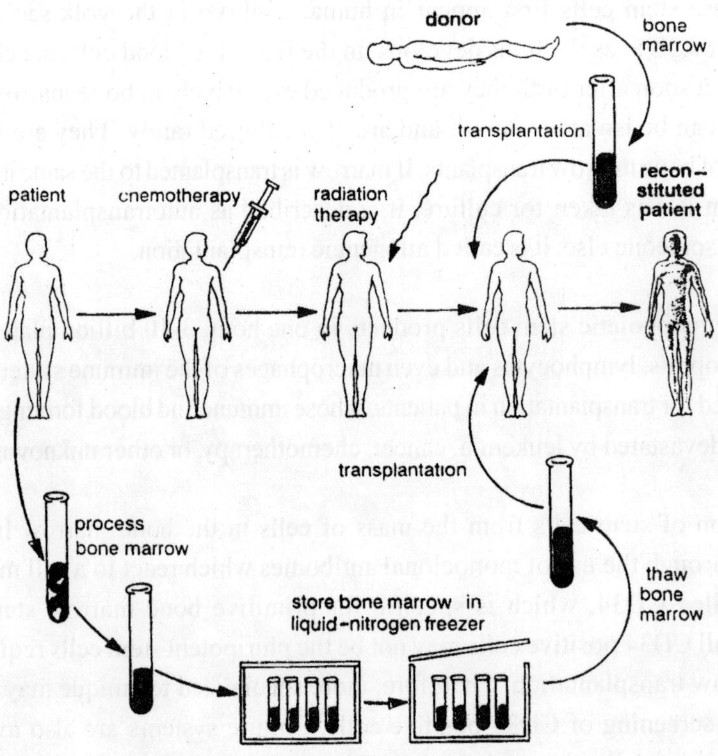

Fig. Bone marrow transplantation; donors and recipients should have compatible tissue types and patient's diseased cells should be killed by chemotherapy or/and radiation therapy

In general, 200×10^6 bone marrow cells are needed for 15 kilogram of body weight. This requires multiple extractions involving 500 — 1000 milliliters of bone marrow for one patient. (One extraction gives 10 – 15 ml of bone marrow). Administering hormones, like CSF and G – CSF, helps since it stimulates blood production. Adding stem cells from peripheral blood to bone marrow may further help engraftment. In future, it may be possible for a healthy person to store his own stem cells permanently and use them whenever required. This will become simpler, if stem cells can be grown and multiplied in culture. Umbilical cord blood (which is normally discarded after birth) may become an important source of stem cells and may be stored for subsequent use during a medical crisis (Fig.).

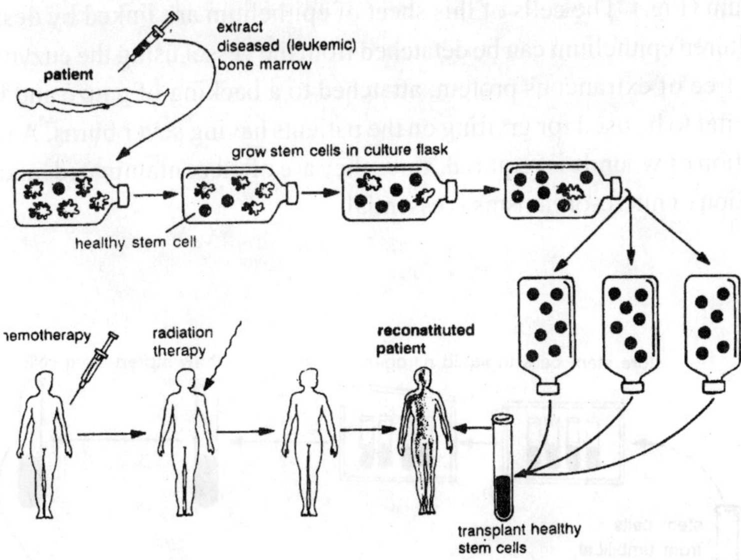

Fig. Autologous bone marrow transplantation; bone marrow cultures seem to encourage healthy stem cells to reproduce more rapidly than leukemic ones, so that over a period one might be able to derive a population of healthy stem cells from diseased marrow; these cultured healthy cells are injected after chemotherapy or/and radiation therapy

Grafting of artificial skin (derived from cultured cells) after severe burns. Skin is perhaps the only organ that can be artificially made from cell culture and used for grafting, when skin is severely damaged due to severe burns. Of the few cell types that can be cultured are keratinocytes, which make 90% of the epidermis of skin (Fig.). They are responsible for giving rise to corneocytes (dead cells) making the external cornified layers of skin, and their proliferation is facilitated by the products of fibroblasts found in the dermis layer of the skin.

Since fibroblasts are useful for culturing keratinocytes, fibroblast cells called 3T3 cells were used (after irradiating with lethal dose to prevent their own growth) to cover the bottom of vessel, before adding epidermal cells for culturing them. Other substances added in culture medium included epidermal growth factors, cholera toxin and a mixture of other growth factors. Only '1-10%' epidermal cells proliferate (after their dissociation with trypsin; see Chapter), others having already started the process of differentiation. These cells form colonies, are separated again and transferred to fresh culture to allow better growth. The process of separating cells from colonies and reculturing them is continued to discourage stratification of cell layers and allow the cell colonies to become confluent forming a sheet of pure epithelium (Fig.). The cells of this sheet of epithelium are linked by desmosomes. This cultured epithelium can be detatched from the vessel using the enzyme dispase, washed free of extraneous protein, attatched to a backing of gauze and brought to the hospital to be used for grafting on the patients having sever burns. A meticulous preparation of wounds is required, since they are often contaminated, and complete elimination of micro-organisms is essential.

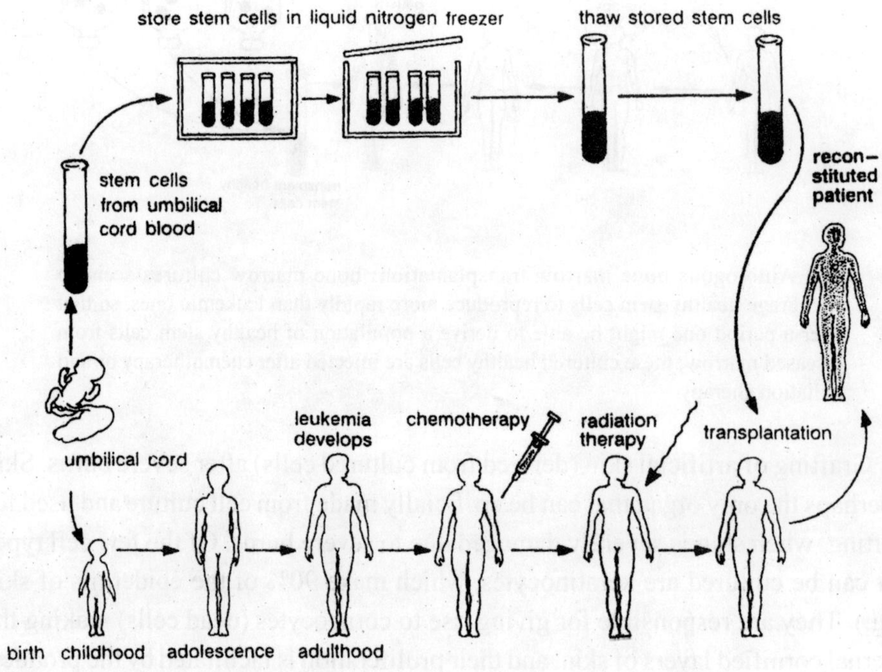

Fig. The use of stored stem cells derived from umbilical cord blood for bone, marrow transplanatation

It is important that cultured kerationcytes used to generate epidermis must come from the unburned portion of the skin of the patient himself; otherwise, these will be soon rejected. Starting from 3 cm^2 skin of the patient, it can be expanded 5,000 fold in 3-4 weeks to supply 1.7m^2 of skin needed for an adult human. The entire body may need 350 grafts, each of 25cm^2 to cover only the front or back surface.

Before the cultured epithelium became available, split thickness grafts were used which involved transfer of 0.3mm thick skin (epidermis + part of dermis) from one part of the body to the other. Such grafting leads to quick recovery and normalization of skin. However, in grafting of cultured skin, if samples of regenerated skin are taken over a period of five years (Fig.), the different elements of normal skin return at different rates, but eventually normal skin, with all essential components, is regenerated.

A variety of diseases have also been treated using cultured keratinocytes. Following are some examples : (i) Scars on skin can be removed using cultured skin. (ii) Cultured oral keratinocytes have been used to regenerate epitheluim of mouth. (iii) Cultured urethral keratinocytes have been used to repair congenital penile defects. (iv) For some middle ear diseases with troublesome discharge also, cultured epidermal keratinocytes are used. (v) Chronic skin ulcers have also been treated with cultured grafts; even allografts (skin from another individual) are successful in these ulcers.

Fig. Engraftment of cultured epithelium followed by regeneration: after one week granular and cornified layers develop; after 3-1 weeps, basement membrane is complete; after one year anchoring fibrics (for attachment of basement membrane to dermis) are fully developed: after 2-5 years, the papillary and reticular dermis regenerates.

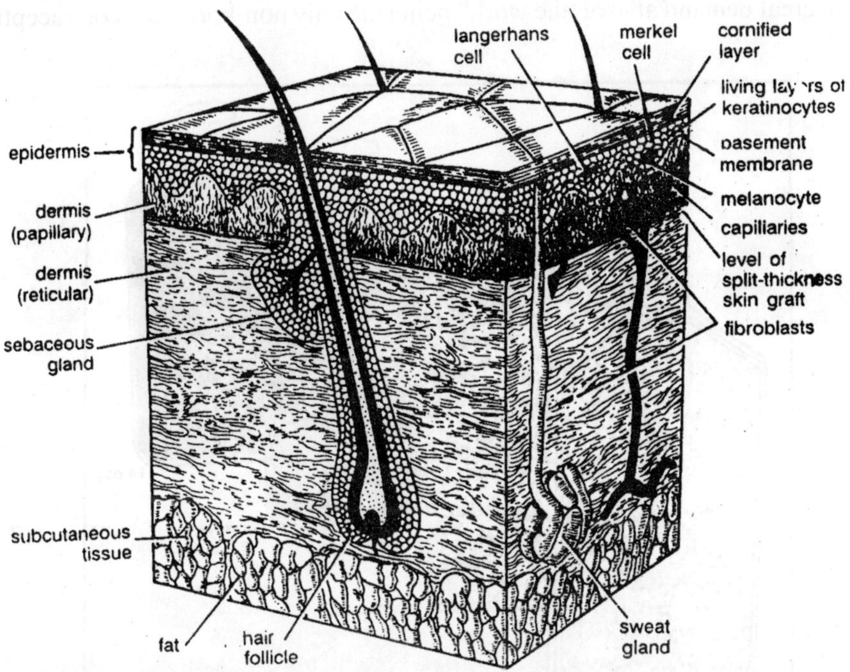

Fig. The structure of skin consisting of epidermis and dermis; outer part of epidermis consists of dead keratinocytes, or corneocytes dermis consists of fibroblast and their products

More than 500 patients throughout the world have already received cultured keratinocytes for treatment of burns, ulcers or other conditions. However, this is a very small number in comparison with the number of patients, who need this help. In future, other cell types may also be successfully cultured and used for following applications : (i) use of cultured endothelial cells for lining of vascular prostheses (artificial limbs etc.). (ii) use of cultured urothelial cells to repair urinary tracts; (iii) use of pancreatic inlet cells to treat diabetes; (iv) use of cultured liver cells to provide for hepatic function; (v) use of cultured myoblasts to treat muscle diseases. A more detailed account of the subject is available in Scientific America (Nov., 1991).

Fertility Control

Efforts have also been made in the past for developing safe, efficacious, cost-effective, long lasting and reversible contraceptives. In this connection, following examples demonstrate the success of Indian scientists in this area.

Once a week oral pill—For the last 20 years, scientists at the Central Drug Research Institute (CDRI), Lucknow, have been working for the development of an oral, once a week, non-hormonal contraceptive pill. This effort resulted in developing a pill called centchroman sold in the commercial name 'Saheli', which was sent to the market in June, 1991. This pill has proved to be free from any side-effect and is now in great demand all over the world, being the only non-hormonal contraceptive pill.

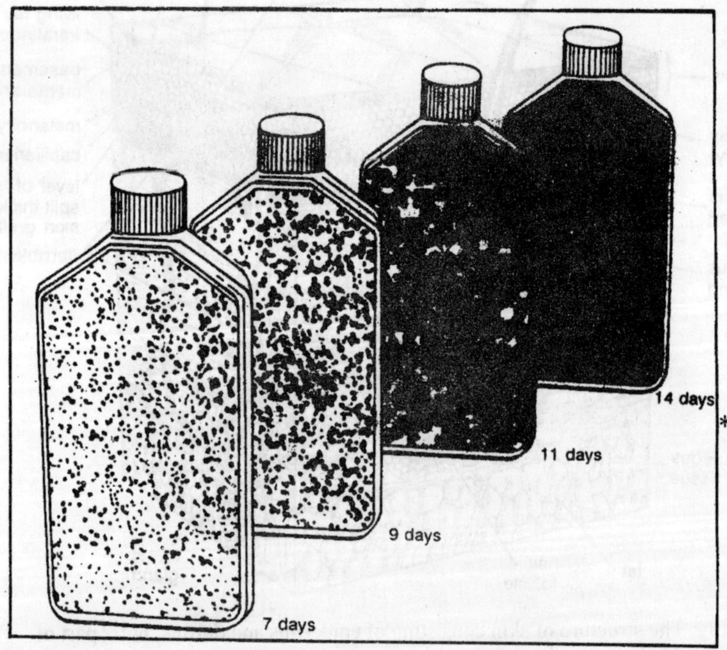

Fig. Proliferation of human epidermal cells, grown in culture for 7, 9, 11 and 14 days; the confluent epithelium is detatched and used as a graft

Vaccines for fertility control—Immunological approaches for development of vaccines have also been tried. One project in India was initiated as a part of family planning programme by Indian Council of Medical Research (ICMR) under the Ministry of Health and Family Welfare, but was transferred to the Department of Biotechnology (DBT) in 1986. Three vaccines are being developed simultaneously: (i) National Institute of Immunology (NII) developed the first ever birth control vaccine based on the hormone hCG (human chorionic gonadotrophin-see Chapter). The vaccine elicited antibodies against both tetanus and pregnancy hormone hCG. Tetanus is a major killer of pregnant women in India in deliveries taking place outside hospital. The vaccine, therefore, has double benefit. Women using this vaccine, continue to ovulate and maintain menstrual cycle. The action of vaccine is reversible, but can be prolonged with booster injections. A clinical trial conducted on 105 women showed that the vaccine was well-tolerated. Phase II trials on 750 women in three cities were in progress during 1991-92. No harmful side-effects were observed in any of the subjects who received the vaccine. International trials have also been conducted in countries like Finland, Sweden, Chile, Brazil, etc., where results of trials conducted in India were confirmed. (ii) At Indian Institute of Science (IISc), Bangalore, scientists found that administration of anti-FSH antibodies to male monkeys renders them sterile without any side effects. Studies on safety, efficacy and standardization are being conducted to undertake phase-I trials. (iii) Another group of scientists at IISc are working on targeting vitamin carrier proteins for contraception. Lowering the concentration of vitamin carrier proteins through the use of antibodies results in rejection of embryo. After the results are established in nonhuman primates, the work would be considered for trials in humans following normal procedures.

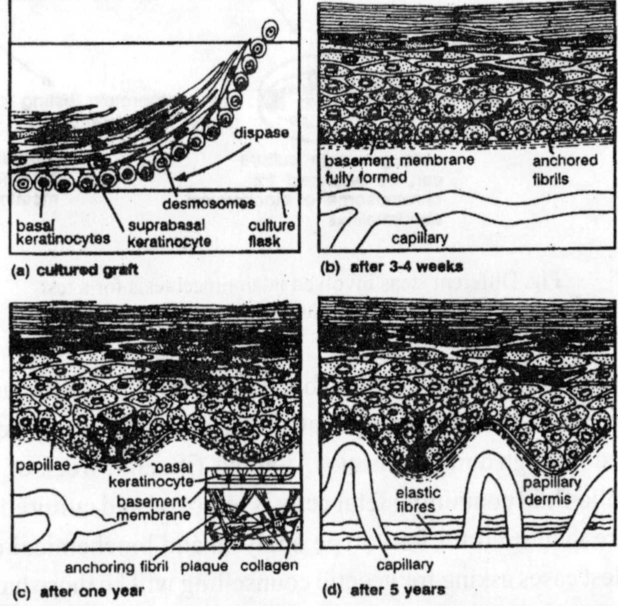

Fig. Engraftment of culture followed by regeneration; after one week granular and cornified layers deveop; after 3-4 weeks, basement membrane is complete; after one year anchoring fibrics (for attatchment of basement membrane to dermis) are fully developed; after 2-5 years, the papillary and reticular dermis regenerates

Genetic Counselling

Chance of Having a Child with Congenital Defect

Genetic counselling for couples who believe that there may be a risk of producing a child with a congenital disease has now become routine aspect of medical practice, particularly in the developed countries. The parents may either voluntarily abstain from having any child or under selective abortions on suspicion or after ascertaining it through antenatal diagnosis as discussed earlier.

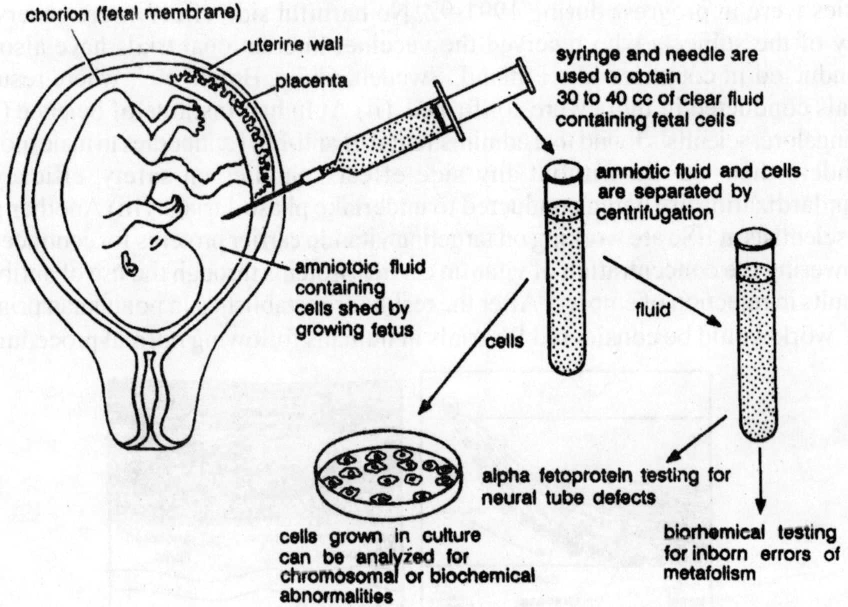

Fig. Different steps involved in amniocentesis for a test of hereditary/developmental defect

A genetic counsellor should first be able to identify the disease and therefore, should be first a clinician and then a geneticist. The hereditary or developmental defects are tested through amniocentesis. At about fifteenth week of pregnancy, a sample of amniotic fluid containing fetal cells is removed and cultured in vitro. The cultured cells are then used for karyotype analysis and biochemical enzyme tests (Fig.) The simplest cases asking for genetic counselling will be those having a family history of disease and the parents may like to know the chances of having a child free of that disease. In the simplest case, a couple may have one abnormal child (carrying disease) and would like to know the chances having a normal child on the next pregnancy. In such a case, if the defect is controlled by single recessive gene and both parents are normal, the chance is three in four of having a normal child, although even a normal child will have a two third chance of being a carrier. The parents are likely to give birth to such a child who may be a carrier, because the chance of his or her spouse also being a carrier will be remote. However, in such cases, even the possibility of having the abnormal grand child can be worked out, if the frequency of heterozygotes in the population available for marriage of the child is known. For instance, in case of fibrocystic disease (cystic fibrosis) of the pancreas, the frequency of heterozygotes in general population (in UK) is 1/22. The chance of the normal child to be a carrier being 1/22, the chance of the grand child to be abnormal would be 2/3 x 1/22 x 1/4 = 1/132 (since 1/4 is the chance that a child born to both heterozygote parents will be defective). A risk of one in 132 may or may not be worth talking depending upon temperament of the parents and the circumstances. More detailed calculation in these simple cases and also in cases of polygenic nature and variable (B.C.-22) penetrance can be done. The readers are advised to consult a book on human genetics for further details.

In some cases, detection of a heterozygote may also be useful and possible. Following are the situations where it is possible. (i) When heterozygote, though phenotypically normal, produces a particular enzyme activity intermediate to those found in two homozygotes, the heterozygote can be identified. In such cases, if biochemical laboratory is available, deficiency like HGPRT (lesch-nyhan syndrome) in heterozygous condition can be detected from a blood sample or some skin cells. (ii) When the mutant produces an altered form of gene product, the heterozygote may produce two different forms of protein that can be separated by electrophoresis to enable the identification of a heterozygote. (iii) If the defect is associated with a chromosome structure, then with the availability of a cytogenetics laboratory, such an abnormality in heterozygous condition can be identified.

Making a Choice of Baby's sex

Married couples have always been interested in knowing the sex and sometimes in making a choice about the sex of their baby. The chromosome techniques have made it possible to know the sex of developing fetus by drawing amniotic fluid and preparing karyotypes from cells derived from the fetus and floating in this fluid (Fig.). There are clinics available now which can advise pregnant ladies about the sex of the developing fetus, so that the ladies can decide early about the abortion of fetus, if it belongs to the unwanted sex.

Recently, techniques have also been developed, which will not require preferential abortion but will allow preferential fertilization by male (carrying Y chromosome) or female (carrying X chromosome) determining sperms. There are techniques available now, which allow separation of sperms carrying Y chromosomes, from the ejaculation of a man (through Ericson's method developed by R. Ericson of U.S.A.) to be used for insemination of ovulating women. This technique (using quinacrine stain) has been used with 80% success in 47 sperm centres in the world including one in Bombay at Khar Road. Ericson has actually established a company named Gametries Ltd. in California, U.S.A. which specializes in separating sperms with Y chromosome and hundreds of male children have been produced with its help.

Techniques have also been developed to separate sperms carrying X chromosome for artificial insemination leading to the birth of female children. This technique involves 'sephadex gel column', in which sperms with Y, being lighter are trapped in gel and those with X being heavier reach the bottom of the column, and can be used for insemination.

The techniques permtting choice of sex of the baby have been condemned by many sociologists, who fear that this may disturb the sex ratio leading to a variety of problems. But some doctors argue that this will help couples in planning their-families, since there are also couples who may like to have a female child. This may also allow selection against sex linked abnormalities in the children.

Forensic Medicine (Identification of Murderers, Rapists, etc.)

The techniques of DNA fingerprinting developed in (1985, 86) by Alec Jeffreys and his colleagues at Leicester University in U.K. and that of autoantibody fingerprinting described by A.K. Francoeur (1988) will completely revolutionize forensic biology. In this field, identity of a person with the help of blood stains, semen (sperms) stains, hair roots, tears, saliva and perspiration will be possible with almost absolute certainty. These two techniques will be briefly discussed in this section.

DNA Fingerprinting Using Minisatellite DNA

In this technique, DNA will be isolated from blood stains, semen stains or hair roots will be subjected to Southern blotting and DNA hybridization with the help of specific DNA probes. The probes correspond to hypervariable minisatellites in DNA, each made up of tendem rapeats of short sequences. A large number of these minisatellites are scattered throughout the human genome which were first detected in an intron of human myoglobin gene. This will reveal polymorphism in DNA, which has a very stable inheritance.

It is speculated that the above technique will allow the identification of criminals in murder cases, rapists in rape cases and of mother and or father in case of doubtful parentage. This technique will allow identification even when the stains on victim's clothes, etc. are several years old and with much more certainty than has hitherto been possible through techniques of blood groups etc., since the number of blood groups available becomes a limitation. The technique of DNA fingerprinting reveals such a great polymorphism that the possibility of two persons having the same pattern of DNA fingerprints is very remote.

Autoantibody Fingerprinting Using Dipsticks

Autoantibodies that react with cellular components occur in high frequency in patients with systemic rheumatic diseases. However, a novel class of autoantibodies that react with cellular components has now been identified in normal humans and other animal species. Unlike the disease associated autoantibodies that are restricted in number, these human autoantibodies increase in number from birth upto the age of two years and then remain constant for decades, if not lifelong. The complement of these autoantibodies present in an individual is unique and for this reason they have been named individual specific (IS) autoantibodies. These IS autoantibodies when physically separated comprise an antibody fingerprint that can serve to identify people just like DNA fingerprints discussed above. For these autoantibody fingerprints, body fluids such as blood, semen, tears, saliva, and perspiration can be used. The advantages of autoantibody fingerprinting over DNA fingerprinting include the following : (i) sensitivity (less than 10 μl blood needed); (ii) rapidity (only few hours needed; no DNA extraction required, which may take time); (iii) simplicity (no equipment required); (iv) cost effectiveness; (v) portability; (vi) autoantibody fingerprints though are similar in the newborn child and the mother can distinguish between genetically identical individuals like identical twins which can not be distinguished through DNA fingerprinting.

The autoantibody fingerprinting protocol involves (a) preparation of antigen and (b) antibody fingerprint assay. Following steps are involved (i) A panel of antigens is prepared from an extract of human cells (e.g. HeLa cells) cultured in vitro: the antigens are first separated by molecular mass with denaturing polyacrylamide gel electrophoresis (SDSPAGE); (ii) the antigens are then transferred electrophoretically to a nylon membrane; (iii) unbound sites on the membrane are blocked with the help of a blocking agent; (iv) the membranes are cut into strips called dipsticks, which are (v) incubated with dilutions of sera or plasma for 1 hour, (vi) washed with buffer, (vii) incubated in detector molecule (1 μCi/assay of ^{125} I protein, which binds to the F_c protein of human 1gG in the immune complex formed) for 1 hour; (viii) rewashed and dried and (ix) finally used for autoradiography on X-ray film or scanned with a gamma scanner or an optical scanner. The results obtained can be utilized for the identification of individuals in humans and animals like dogs, mice, cows, horses, rabbits, etc.

THIRTEEN

Protein Engineering and Drug Designing

Protein Engineering

Prortein engineering and drug designing is a multidisciplinary approach.It involves computer aided molecular modelling (CAMM) based on the study of three dimensional structures of receptor and functional macromolecules. During the last 10 years, following two major developments made it possible to change the structure of macromolecules in a predetermined way leading to their desired functional role. First is the possibility of site-directed mutagenesis (already discussed) permitting changes in genes at known sites leading to modification of function in a predetermined way; and second is the availability of computational and graphical tools which allow, on the screen, the display and exploration of the three dimensional structures of proteins. These two new developments allow us to decide about the three dimensional structure of protein/drug needed for a desired purpose, before using recombinant DNA technology (for proteins), or organic synthesis (for drugs). The different steps generally involved in protein engineering are shown on the next page, and include the following: (i) The protein sample is prepared from an organism and it is characterized with a ligand (e.g. enzyme-substrate, receptor-hormone or antibody-antigen interaction); (ii) The three dimensional (3-D) structure is studied through NMR (nuclear magnetic resonance) and X-ray diffraction patterns of crystals; (iii) 3-D structure is displayed with interactive computer graphics and the available information is used to suggest a novel design suiting the needs; (iv) The desired DNA sequence that is expected to give the novel designed protein is then either synthesized or obtained by site directed mutagenesis of an available gene; (v) The novel gene is introduced into a suitable expression system and the gene product is purified and characterized biochemically; (vi) If biochemical characterization does not satisfy the earlier predicted structure, the cycle may be repeated again, till the desired structure is available.

It should be noted that the above steps of protein engineering share steps used for designing of drugs, insecticides, herbicides, and peptide vaccines as shown in. The only difference is that the drug designing concerns only the ligand, which can be modified more effectively by chemical synthesis, while in protein engineering recombinant DNA technology is used.

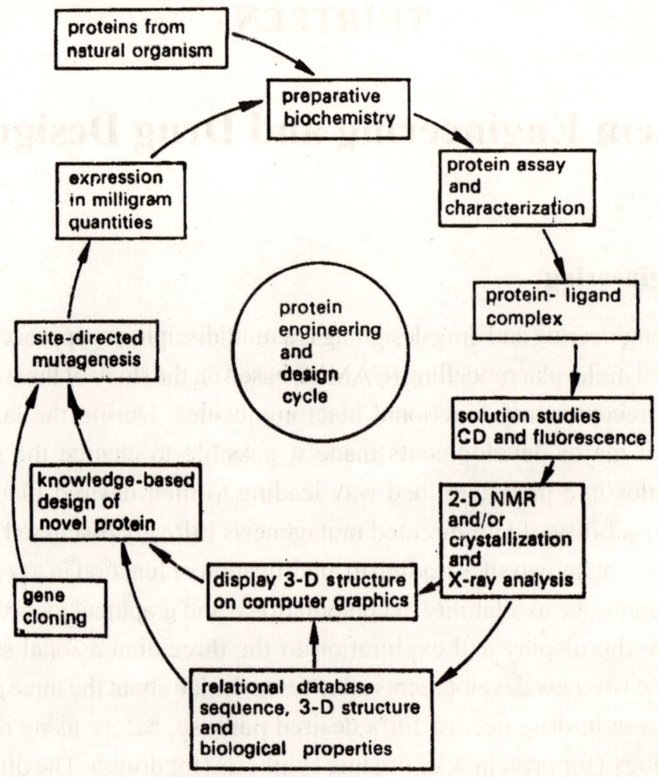

Fig. Different steps involved in protein engineering. (Redrawn from Proc. Trans. Roy Soc. Lond, B. 324. 447-460; 1989).

The three dimensional structures for a few related proteins will normally be available to a protein engineer, while planning production of novel protein or a drug for the desired needs. Therefore, as discussed above, a procedure is needed for generating the three dimensional structure of the macromolecule (protein or drug) of interest. This may be achieved de novo or through modelling from homologous or analogous proteins, if they are available. This is described as knowledge based modelling and designing through the use of computer software including computer graphics, computer simulations and databases (computer software includes the entire set of procedures and programmes to be used on the computer to solve problems along with the operating aids). The computational aspects of knowledge based approaches in the designing of macromolecules (proteins, drugs, vaccines, etc.) are shown and the curious readers will find details elsewhere (Proc. Trans. R. Soc. Lond. B. 324:447-460, 1989).

Protein structure

During the last twenty years, rapid progress has been made in the analysis of protein structure and function. Sequences of amino acids are now available for as many as 8000 proteins, but the three dimensional structures of only about 400 proteins have been resolved through X-ray crystallography. It is realized that the analysis of protein structure and function has now reached only a level which for DNA existed in 1970's. Such a comparison gets support from the fact that only one centre in the world is available for the storage of protein data (Brookhaven Data Bank), while several such centres are available for DNA sequence data. From the three dimensional structure of 400 proteins analysed so far, it is apparent that if two proteins

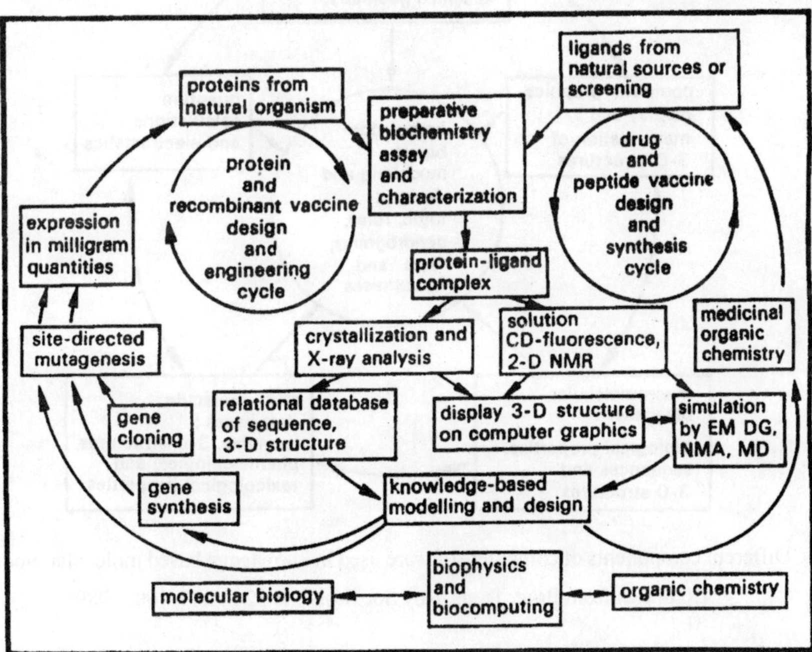

Fig. Cycles involved in protein engineering and drug designing showing common steps in the two cycles. (Redrawn from Proc. Trans. Roy Soc. Lond. B 324 : 447-460; 1989r.)

are similar in their amino acid sequence, they will tend to fold into similar three dimensional structures. The rules that will translate the amino acid sequence of a protein into its three dimensional structure are sometimes referred to as the second half of the genetic code. These rules are not fully known yet, but as these rules become known, it will be possible for us to predict three dimensional structure of a protein from its amino acid sequence. This will then also enable us to identify a model structure for a protein and predict its function. The model can then be used for the synthesis and modification of a gene that will give the desired sequence of amino acids leading to the three dimensional structure for a specified function. An alternative approach of protein engineering may be to modify a protein by a chemical reaction to make it more suitable for a specific function. This area of protein research (modification of protein either by recombinant technology or by chemical reaction) is currently receiving considerable emphasis and is popularly described as protein engineering.

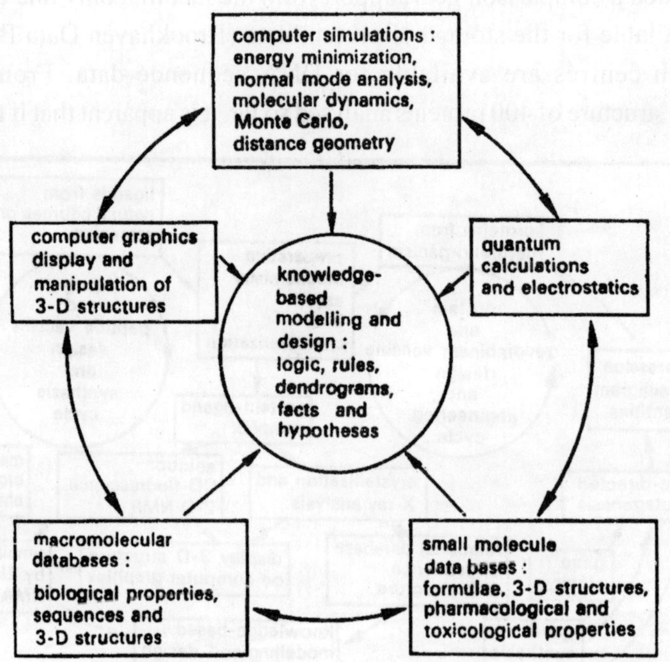

Fig. Different components of computer software used in knowledge based molecular modelling. (Redrawn from Proc, Trans, Roy Soc. Lond. B 324 : 447-460. 1989)

Through recombinant DNA technology, a gene can now be cloned in an expression vector and made to express in bacteria. From such bacterial cells having the desired gene cloned in an expression vector, a protein can be obtained in abundant quantity. Since a gene can be artificially synthesized and modified using recombinant DNA technology discussed in this will allow production of novel proteins through genetic engineering, an area popularly described as protein engineering. The proteins receiving attention in this new and exciting area of protein engineering would include enzymes, synthetic peptides, storage proteins and drugs to be used in medicine, industry and agriculture. The objectives of protein engineering include the following: (i) to create superior enzymes to catalyze production of high value specific chemicals; (ii) to produce enzymes for large scale use in the chemical industry; and (iii) to produce biological compounds (including synthetic peptides, storage proteins and specific drugs), that are superior to natural ones. Some success in this area of protein engineering has already been achieved and much progress is expected to be made in this area in future. The different aspects of this fascinating area of protein engineering will be discussed in this section of the chapter.

Rationale of Protein (enzyme) Engineering

Although thousands of proteins have been characterized in prokaryotes and eukaryotes, only few became commercially important. This is due to the high cost of isolating and purifying enzymes in sufficient quantities. Although the cost aspect has been overcome by producing an enzyme in large quantities in bacteria, for its industrial application, an enzyme (outside the cell) should also have some characteristics in addition to those of enzymes in the cells. These characteristics may include the following: (i) enzyme should be robust with a long life; (ii) enzyme should be able to use the substrate supplied in the industry even if it differs slightly from that in the cell; (iii) enzyme should be able to work under conditions (e.g. extremes of pH, temperature and concentration) of the industry even if they differ from those in the cell.

In view of the above, enzyme should be engineered to meet the altered needs Therefore, efforts have been made to alter the properties of the enzymes. Following is the list of properties that one needs to alter in a predictable manner for protein or enzyme engineering:

(1) Kinetic properties of enzyme-turnover and Michaelis Constant, Km.
(2) Theremostability and the optimum temperature for the enzyme.
(3) Stability and activity of enzyme in nonaqueous solvents.
(4) Substrate and reaction specificity.
(5) Cofactor requirements.
(6) Optimum pH.
(7) Protease resistance.
(8) Allosteric regulation.
(9) Molecular weight and subunit structure.

For a particular class of enzymes, variation in nature may occur for each of the above properties, so that one may like to combine the optimum properties to get the most efficient form of the enzyme. This aspect of protein engineering will be illustrated using the example of glucose isomerases, which convert glucose into other isomers like fructose and are used to make high fructose corn syrup vital for soft drink industry. It exhibits wide variation in its properties.

Sometimes, it may not be possible to get a combination of optimum properties. For instance, an enzyme with highest activity may not be the most stable. Therefore, a compromise in properties may have to be made, if we have to select an enzyme from the available variability or even if we create variability by mutagenesis. However, if structure and function relationship of an enzyme is known, the structural features for desirable function may be combined and protein engineering techniques may then be used to create a novel enzyme exhibiting a combination of all desirable functional properties. Glucose isomerase belongs to a TIM barrel family of enzymes which resemble each other in having a highly characteristic domain called TIM barrel, with active site for catalytic action at one end. This TIM barrel may be found in enzymes that may differ in sequences and may catalyze different reactions. As earlier discussed, since similarities of structure of protein meant similarities in function, TIM barrel presents a challenge to this concept. However, it is curious that some enzymes in this family appear in pairs in their metabolic pathways so that they catalyse two consecutive steps thus showing coupling of their functions. As an example of two enzymes of TIM barrel family, while 'triose phosphate isomerase' is one of the most efficient catalysts, 'glucose isomerase' is relatively very inefficient. Therefore, if 'glucose isomerase' enzyme is redesigned to use the highly efficient domain of TIM barrel family, it will be a remarkable achievement for soft drink industry. Efforts in this direction are being made (see later for methods of protein engineering).

Table

Variation in properties known in the natural glucose isomerases.

	Property	*Range*
1.	Turnover number	63-2151 glucose molecules/enzyme/ minute at 60°C
2.	Km	0.086 to 0.920 molar
3.	Molecular weight	52,000 to 191,000
4.	Temperature optima	50 to 90°C
5.	pH optima	4 to 11

Basic Assumptions for Protein Engineering

While attempting protein engineering, one should recognize the following properties of enzymes: (i) many amino acid substitutions, deletions or additions lead to no change in enzyme activity, so that they are silent mutations; (ii) proteins have a limited number of basic structures and only minor changes are superimposed on them leading to variation; (iii) similar patterns of chain folding and domain structure can arise from different amino acid sequences, which show little or no homology (although same amino acid sequence never gives different folding and domain structures).

The above properties suggest that while many major changes sometimes may lead to no alteration in function, some of the minor changes at specific positions may lead to the desired favourable change. For example, a single amino acid replacement (glycine to aspartic acid) in E. coli asparate transcarbamylase leads to (i) loss of activity and to (ii) an alteration in the binding of catalytic and regulatory subunits. Another example involved the engineering of a single chain biosynthetic antibody binding site (BABS), which is though only one-sixth of the size of the complete antibody, but retains its antigen-binding specificity. This synthetic fragment has heavy and light chain variable regions (VH and VL) connected by a 15 - amino acid linker. A synthetic gene has also been prepared for the fragment, which expressed in E. coil. This fragment binds to digoxin, a cradiac glycoside. Single amino acid replacements in BABS fragment have sometimes led to major changes in its binding affinity.

In view of the above, it is necessary to examine not only the crystal structure but also the active sites therein, so that the gene may be modified or artificially synthesized for protein engineering to meet the desired needs.

Steps Involved in Protein Engineering

A Study of Protein Structure-three Dimensional— A study of three dimensional structure of a protein (including a study of active sites) is the first step in any exercise on protein engineering. For such a study, protein should be available in crystal form. X- ray diffraction methods have also been developed to generate data at a fast rate, to enable us to predict the three dimensional structure of the proteins through model building. Observed differences in structures of different enzymes can be related with the differences in their functions, so that the knowledge from such studies can be used for protein engineering. X-ray diffraction data at different temperatures also allow us to learn about the changes expected in protein structure due to change in temperature suggesting that the protein structure is dynamic. Similarly, NMR (nuclear magnetic resonance) technique will allow us to know the structure of enzymes in solution rather than in crystals, so that we can bypass the time consuming process of crytallization thus speeding up the study of protein structure. The information generated from all these techniques is utilized in protein modelling, which is then utilized for an exercise in protein engineering.

When crystals of a protein are not available for study, but amino acid sequence data is available, the identical sequences in any two proteins are matched and effects of substitution are examined. This is simetimes described as 'sequence baring' and allows us to know which amino acids are involved in catalytic binding at active sites.

Protein Modelling— Utilizing the data generated through X-ray diffraction and NMR studies, models can be constructed with the help of computer graphics. There are computer programmes available (interactive colour graphics programmes) with the help of which a protein structure can be fitted to the electron density map (obtained from X-ray diffraction) by simultaneous display on the screen of computer monitor. Similarly, Van, der Waals surfaces for the protein can be displayed and interaction between several molecules simulated. There are also other interactive molecular graphics which can be used (with the help of programmes) to find out the perturbations (disturbances) in protein structure that will result from specific modifications of amino acid sequences.

We know that to some extent the three dimensional structure of a protein can be predicted from the amino acid sequence, but we still have to depend partly on X-ray diffraction patterns for determining the three, dimensional structure. In future when the three dimensional structure can be accurately predicted from amino acid sequence data, this will lead to long term success in protein engineering. The models of proteins, made on the basis of amino acid alterations, can then be tested for the prediction about structure-function relationships.

Perturbation Theory— A perturbation theory for protein engineering is being formulated, which will enable us to calculate the effects of small changes in amino acid sequence accurately. With the availability of such a theory, there will be no need to perform experiments at each step of protein engineering process to find out the outcome (change in function of suggested modifications). Such a theory will result from the available methods for (i) protein structure refinement and (ii) calculation of theoretical structure of a protein which will be homologous to other proteins of known structure. This perturbation theory can then be tested by using families of monoclonal antibodies which differ only slightly in amino acid sequence 1, once tested for its accurate prediction, it can be used for engineering of a variety of proteins.

Engineering Methods

A variety of methods have been used and proposed for future use it protein engineering. In this connection mutagenesis, selection, and recombinant DNA are being used and will be increasingly utilized in future. These techniques for protein engineering will be briefly discussed in the following sections:

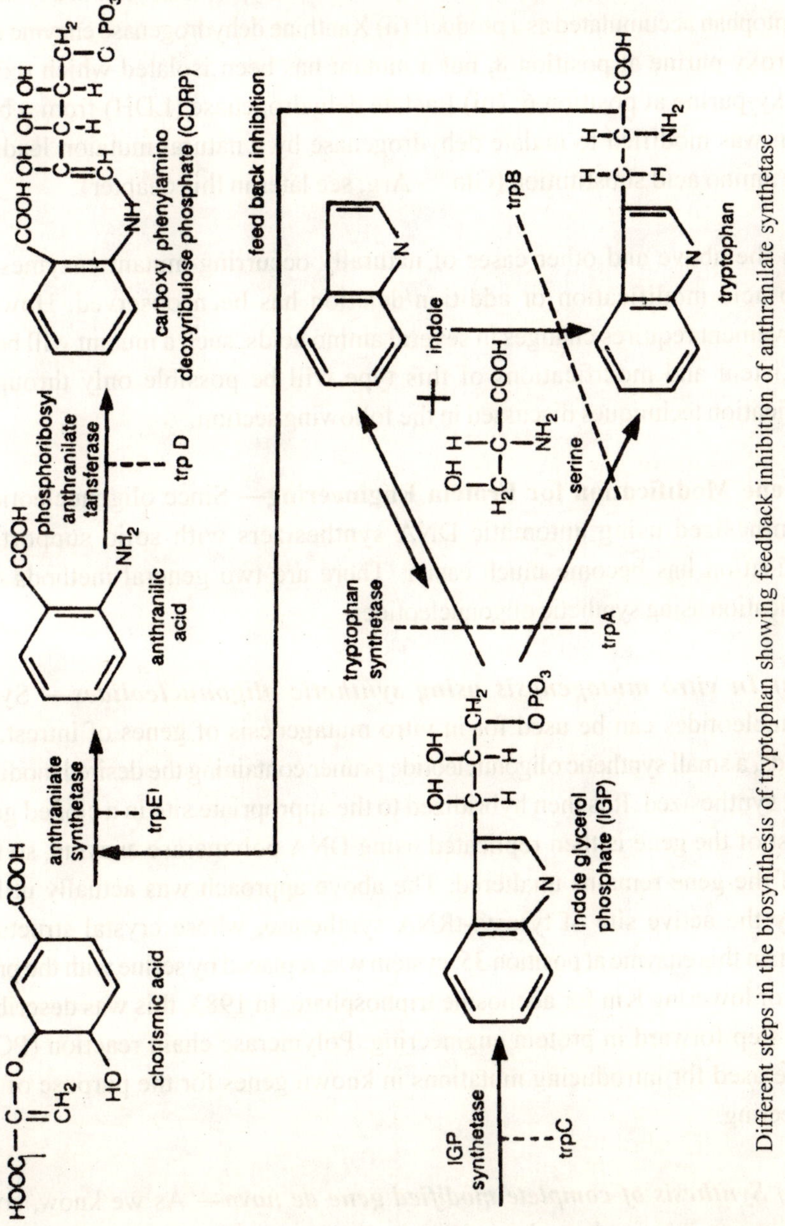

Different steps in the biosynthesis of tryptophan showing feedback inhibition of anthranilate synthetase

Mutagenesis and selection for protein engineering— Mutagenesis and selection can be effectively utilized for improving a specific property of an enzyme. Following are some of the examples of selection of mutant enzymes: (i) E. coli anthranilate synthetase enzyme is normally sensitive to tryptophan inhibition due to feedback inhibition as shown in. An MTR 2 mutation of E. coli was found to possess an altered form of enzyme anthranilate synthetase that is insensitive to tryptophan inhibition. They may help in continuous synthesis of tryptophan without any inhibition by tryptophan accumulated as a product; (ii) Xanthine dehydrogenase enzyme oxidizes 2 hydroxy-purine at position 8, but a mutant has been isolated which oxidizes 2 hydroxy-purine at position 6; (iii) Lactate dehydrogenase (LDH) from a bacterial system was modified to malate dehydrogenase by a natural mutaion leading to a single amino acid substitution (Gln^{102}- Arg; see later in this chapter).

In the above and other cases of naturally occurring mutant enzymes, single amino acid modification or addition/deletion has been observed. However, if improvement requires changes in several amino acids, such a mutant will be rare or nonexistent and modifications of this type will be possible only through gene modification techniques discussed in the following section.

Gene Modification for Protein Engineering— Since oligonucleotides can be synthesized using automatic DNA synthesizers with solid supports, gene modification has become much easier. There are two general methods of gene modification using synthetic oligonucleotides.

(a) In vitro mutagenesis using synthetic oligonucleolides— Synthetic oligonucleotides can be used for in vitro mutagenesis of genes of intrest. In this methods, a small synthetic oligonucleotide primer containing the desired modification is first synthesized. It is then hybridized to the appropriate site in a cloned gene and the rest of the gene is then replicated using DNA polymerase enzyme, so that the rest of the gene remains unaltered. The above approach was actually utilized to modify the active site of tyrosyl-tRNA synthetase, whose crystal structure was known. In this enzyme at position 35, cystein was replaced by serine with the predicted effect of lowering Km for adenosine triphosphate. In 1983, this was described as a major step forward in protein engineering. Polymerase chain reaction (PCR) can also be used for introducing mutations in known genes for the purpose of protein engineering.

(b) Synthesis of complete modified gene de novo— As we know, complete genes in some cases have been chemically synthesized in the form of several oligomers (e.g. genes for insulin, somatostatin and interferon), that were ligated in the correct order to produce a complete gene. The sequence of the synthetic gene can be designed in a modular fashion so that it may have sites for restriction enzymes at convenient positions. By convenient positions, we mean that subsequently the gene can be cleaved at these sites giving fragments, in which modifications may be introduced leading to alteration in function. Since a number of oligomers are involved, modifications can be introduced in different oligomers and these modified oligomers can be used in ligation mixture for synthesis of complete gene, which will have several desired modifications. This method can be used for making extensive changes in the amino acid sequence of the protein. This can be illustrated by using the example of interferon gene, which was artificially synthesized. It was possible to modify portions of the interior of this artificially synthesized gene, so that a variety of artificial interferon genes could be prepared. These genes were found to express in E. coli.

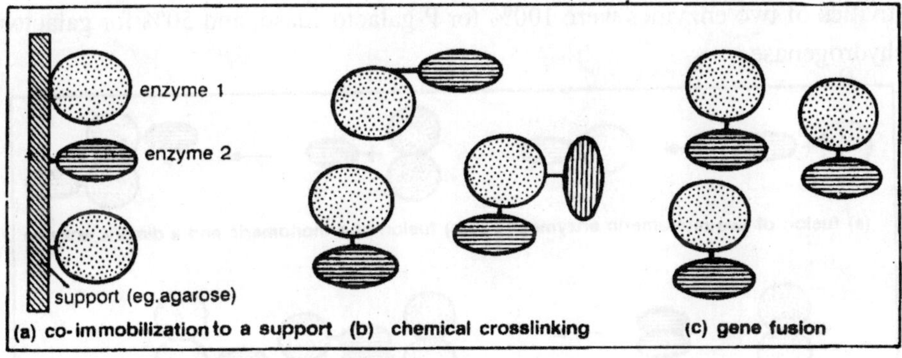

Fig. Three basic methods for generating proximity between two enzymes.

Multienzyme Systems (bi-and polyfunctional enzymes) by Gene Fusion.

Multienzyme systems have been artificially synthesised, which can catalyze sequential reactions in many biotechnological production processes. Although proximity of more than one enzymes can also be achieved by co-immobilization and chemical cross-linking, gene fusion appears to have the highest potential in enzyme technology. The gene fusion technology, for preparation of bi-and polyfunctional enzymes, involves joining of structural genes of two or more enzymes. The translational stop singal at the 3' end of the first gene is removed and ligated in frame to the ATG, start codon of the second gene. Alternatively, short linkers (2-10 aminoacids) are used. The novel chimaeric gene gives a single polypeptide chain carrying active sites of both genes. This fusion may involve (i) two monomeric enzymes; (ii) a monomeric and a dimeric enzyme or (iii) two dimeric enzymes. During the past few years, a variety of artificial bifunctional and polyfunctional enzymes have been prepared by gene fusion in vitro, and their physical and chemical properties analysed. Some of them include the following:

(a) ß-Galactosidase-galactokinase— Sequenial hydrolysis of lactose to glucose is achieved by P- galactosidase (dimeric) and galactokinase (dimeric) enzyme. The gene product of the fused genes (both genes from E. coli) was found to be a tetramer, each polypeptide carrying an active site for each of the two enzymes.

(b) ß-Galactosidase-galactose dehydrogenase— The structural gene for ß-galactosidase (dimeric) from E. coli was fused with the gene for galactose dehydrogenase (tetrameric) from Pseudomonas fluorescens. The hybrid enzyme catalyzes the hydrolysis of lactose to galactose, which is oxidized to galactolactone. The enzyme was composed mainly of hexameric and tetrameric aggregates. The activities of two enzymes were 100% for P-galacto.:dase, and 50% for galactose dehydrogenase.

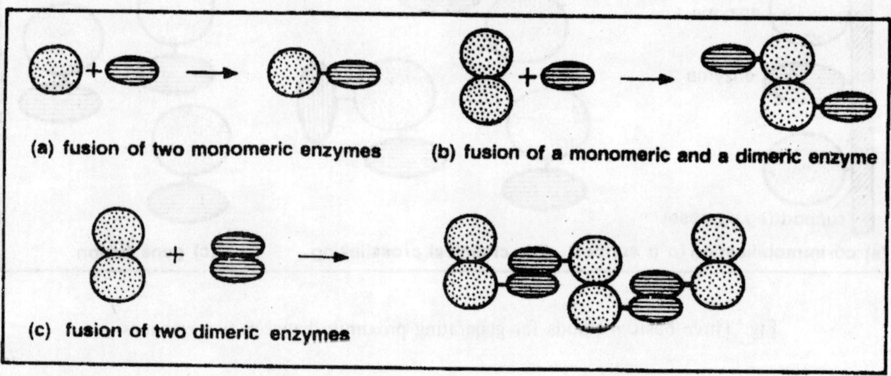

Fused polymer enzyme (multienzyme) system, showing interaction between component enzymes.

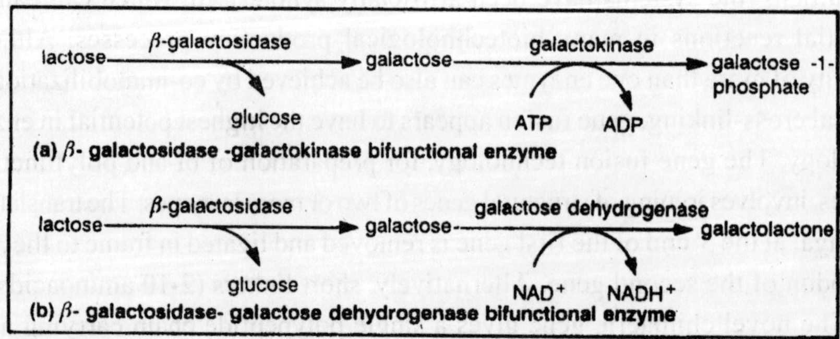

Two reactions catalysed by bifunctional enzyme constructs.

(c) ß-Galactosidase-galactose dehydroginase-galactokinase. A trifuctional enzyme carrying all the three activities on the same polypeptide was prepared using a chimaeric fused gene = cloned in vector pDZK 10. The enzyme was a tetramer or an octamer.

(d) Galactose dehydrogenase-luciferase. Bacterial luciferase is a heterodimer with two subunits (a and ß encoded by LuxA and luxB. Only luxB

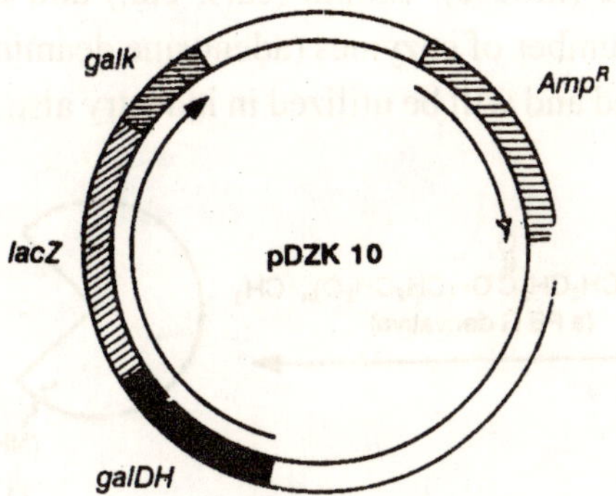

Chimeric plasmid pDZK 10 for an inframe fusion between genes for galactose dehydrogenase. (gal DH), # galactosidase (lac Z) and galactokinase (gal K).

was fused with the gene for galactose dehydrogense. The bifunctional enzyme generated much higher bioluminescent peaks compared with a mixture of native component enzymes.

Advantages of multienzymes— Following are some of the additional advantages of multienzymes: (i) Their purification is simplified; (ii) These are useful in connection with biosensors; (iii) They are also useful in the development of metabolic engineering. It is, therefore, believed that the preparation of such artificial multienzyme systems will be most valuable in the construction of micro- organisms by inserting these fused genes through transformation.

Chemical modification of enzymes— We know that the proteins synthesized under the control of gene sequences in a cell undergo post-translational modification. This leads to stability, structural integrity, altered solubility and viscosity of individual proteins. This may also alter the chemical reactivity. These alterations can be achieved in vitro and may sometimes even create entirely new enzyme, by creating new active sites or modifying the old ones. Some of the examples will be described in this section.

***(a) Enzyme-PEG conjugates*—** An enzyme L- asparaginase (isolated from microbes) has anti-tumour properties, but is toxic with a life time of less then 18hr thus reducing its utility. It has been shown that E. coli L-asparaginase can be modified by polyethylene glycol derivatives to produce PEG-asparaginase conjugates which differ from the native enzyme in following features: (i) it retains only 52% of the catalytic activity of native enzyme; (ii) it becomes resistant to proteolytic degradation; (iii) it does not cause allergy. In view of this, PEG-asparaginase has been used to treat malignant murine (mouse), canine (cats, etc.) and human tumours. PEG conjugates of a large number of enzymes (adenosine deaminase, uricase, catalase, etc.) have been prepared and will be utilized in industry also.

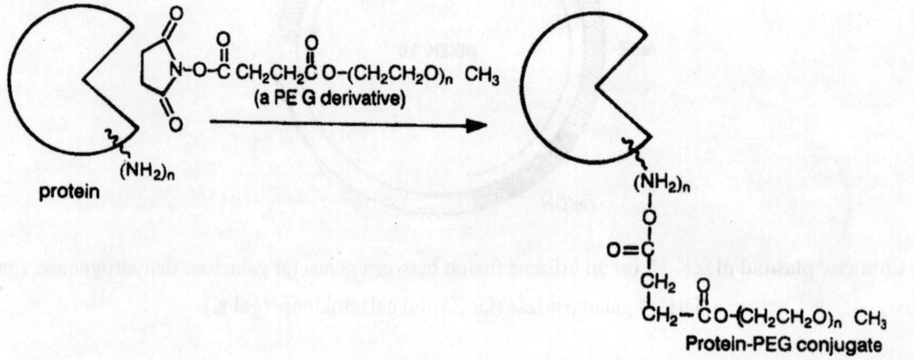

Modification of a protein with polyethylene glycol (PEG) producing a protien-PEG conjugate, which is non-immunogenic and resistant to proteolysis, increasing its therapeutic potential.

***(b) Modification of proteases into peptide ligases*—** Peptide ligation to native enzymes may lead to high specificity and stereoselecitivity, and may suppress side reactions. Therefore, synthesis of any enzyme that may catalyze peptide ligation

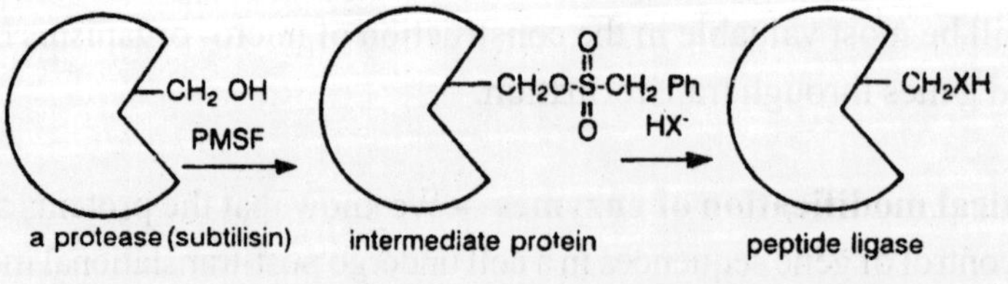

Chemical modification of a protease (subtilisin) by phenylmethyl sulphonyl fluoride (PMSF) to produce a peptide ligase; the alcohol side chain of Ser is replaced by HX (hydrogen sulphide or hydrogen selenide).

will be most welcome. Protease 'subtilisin' has been modified (by converting a serine into cysteine or seleno-cysteine) into thiol-and selenolsubtilisin, the two semi-synthetic enzymes (they are damaged proteases), which can catalyse peptide ligation. Both these damaged proteases are efficient peptide ligases. Similarly, histidine residue can also be modified to yield peptide ligases.

(c) Production of site specific nucleases (e.g. restriction enzymes)— The DNA recognition and binding properties of proteins can be combined using chemical cleavage agents. Cys[178] of E. coli CAP protein has been modified using '5-iodoacetamide-1, 10-phenanthroline' yielding a DNA-cleaving agent that recognized and cleaved DNA at the centre of the recognition site (22 bp) for CAP. This may give restriction enzymes recognizing upto 20 bases instead of 6 or 8 bases

A two step procedure for producing a nuclease from a DNA binding protein.

Coupling of an olignucleotide to a nonspecific phosphodiesterase to generate a site specific nuclease.

and may, therefore, be useful for isolating long DNA fragments needed for sequencing and mapping.

Nucleases may also be produced by fusion of non-specific phosphodiesterases to oligonucleotides of defined sequence. For a nuclease from Staphylococcus modified by this approach, it was shown that oligonucleotide component of fused product pairs with its complementary sequence and the hybrid enzyme hydrolyses single stranded DNA or RNA adjacent to the oligonucleotide binding site. This approach thus can also be used for developing artificial restriction enzymes.

(d) Production of artificial semi-synthetic oxido-reductases (flavo enzyme)— Artificial oxido-reductases can be prepared by covalently attaching redoxactive prosthetic groups to existing sites. Linking of 10-methylisoalloxazine derivatives (as redox-active groups) to specific sites of several proteins has been achieved. The efficiency of these semisynthetic enzymes (e.g. flavopapain) compares favourably with that of naturally occurring flavoenzymes.

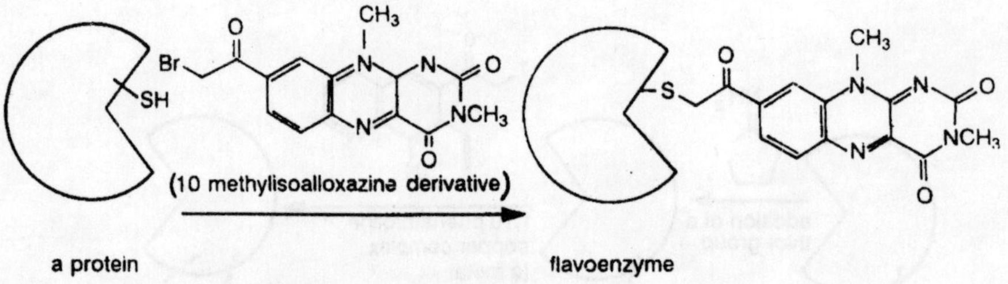

(10 methylisoalloxazine derivative)

a protein flavoenzyme

Alkylation of protein binding sites with a reactive 10 methylisoalloxamnic derivative yielding a semi-synthetic tlavocn-ryme which combines the reactivity of the catalytic cofactor with the specificity of the template protein.

Some achievements of protein engineering

A number of proteins are known now, where efforts have been made to know the effects of site pecific mutagenesis involving substitution of one or more amino acids. Efforts have also been made to study in detail the function of different regions of a protein. Following are some results of such efforts:

ß-laetamase— This enzyme functions in the periplasmic space of bacterial cells. The enzyme hydrolyses and inactivates the beta- lactam ring of penicillin derivatives and helps in transport across the inner membrane. During transport, a polypeptide (signal sequence peptide of 23 amino acids) is cleaved off. Detailed analysis suggested that transport and processing does not depend on this polypeptide of 23 amino acids alone. An active site involving amino acid serine has also been identified, since its replacement by cysteine leads to reduction in the activity of this enzyme.

Dihydrofolate reductase—In this enzyme, replacement of a single amino acid, aspartic acid (ASP) by asparagine (ASN), led to a decrease in specific activity by a thousand fold, suggesting that aspartic acid is very important for the active site. Other similar modifications were also examined.

Insulin— It consists of A and B chains linked by C-peptide of 35 amino acids. It was shown that a sequence of 6 amino acids for C-peptide was adequate for the linking function.

Lactose permease (product of 'y' gene of 'lac' operon)— This enzyme is involved in transport of lactose and a cysteine to glycine substitution showed that this amino acid was not essential for transport. Furhter, out of four histidine residues, two at positions 35 and 39 do not play any essential role in transport, while the mutation in any of the other two histidines at positions 208 and 322, lead to loss of transport function.

T4 lysozyme— A mutation of isoleucine to cystine in this enzyme leading to formation of a disulphide bridge led to thermal stability and a 200 fold increase in enzyme activity even at 67°C.

Human beta interferon— Removal of one of the three cysteine residues led to an improvement in stability of the enzyme.

Repressor— This protein could be engineered to develop a specific site for cro-protein, since the alteration led to development of a cro recognition site.

Acetylcholine receptor— This protein is involved in transport of acetylcholine through the membrane. Specific regions of this protein involved in acetylcholine binding and channel formation have been identified.

Cytochrome C— A phenylalanine residue has been identified to be non-essential for electron transfer but is involved in determining the reduction potential of the protein.

Trypsin— It could be redesigned to have altered substrate specificity.

Subtilisin— Another successful alteration of substrate specificity involved the enzyme subtilisin reported in 1986-87.

Lactate Dehydrogenase— A lactate dehydrogenase (LDH) from Bacillus stearothermophilus was modified separately by each of the three substitutions of amino acids (resulting from mutations; Asp ^{197}→ Asn; Thrm^{246}→Gly; Gln;^{102}→.Arg). The substitution, Gln^{102}→Arg, led to change in specificity from lactate to malate, with high efficiency, comparable to that which the native LDH had for lactate.

Lactic protease—Substrate specificity of lactic protease (in E. coli) has been shown to be dramatically modified by replacing active site methionine by alanine (Met^{192}→Ala; Met^{213}→Ala).

Possible Candidates for Future Protein Engineering

Following are some of the proteins, whose structures are known and for which cloned genes are available, so that they can be used for protein engineering: (i) immunoglobulins; (ii) glucose isomerases; (iii) amylases; (iv) p-hydroxybenzoate hydroxylase; (v) ribulose — 2, 5, bisphosphate carboxylase (ruBP); (vi) repressor proteins, and (g) restriction edonucleases.

Immunotoxins

By creating novel gene fusions using sequences coding for antibodies and toxic peptides, it may be possible to create cytotoxic drugs or 'magic bullets' called immunotoxins. These drugs will attack specific targets in the patient. Some success in this direction has already been achieved in the synthesis of an immunotoxin using a toxic peptide which has properties similar to those of melittin. Immunotoxins are conjugates of cell binding antibody or antigen, covalently bound to a plant or a bacterial toxin, so that when an immunotoxin is supplied to a patient, the antibody or antigen helps in recognition of the target cells to be killed and the toxin component helps in killing these cells, which may be harmful being cancerous. Thus, an immunotoxin will have two parts: (a) a toxin polypeptide or a part of it having toxin activity; this may be called A chain and (b) a cell binding recognition polypeptide (or antibody or a part of it having binding site), which may be called B chain. These immunotoxins represent a new approach to pharmacology.

Ricin, a Plant Toxin used as Immunotoxin

Ricin is a toxin polypeptide (A chain) and is attached to a cell binding polypeptide-a lectin (B chain) which binds to glycoproteins or glycolipids (containing galactose), present on the cell surface. The cells are thus recognized by B chain and the immunotoxin enters the cell by a mechanism not fully understood (perhaps receptor mediated endocytosis). The A chain (ricin) then inhibits protein synthesis by enzymatically inactivating the EF2 binding portion of 60S ribosomal subunit. It has been shown that in this immunotoxin, the B chain can be replaced by any binding moiety such as a hormone, a growth factor or an antibody, so that the binding specificity is changed, but the toxicity effect is retained.

The immunotoxin 'ricin' has actually been used for a study of its effect on mouse tumour cells (B cell tumours). There can be a variety of these tumour cells each having its own characteristic ideotype (specific immunoglobulin or surface antigen), so that an anti-ideotypic antibody can by raised and conjugated with the toxin (ricin or ricin A, which is a chemically synthesized toxin resembling natural ricin) to give an immunotoxin. When tumour cells were incubated with such an immunotoxin, protein synthesis decreased by 70-80%, while the protein synthesis in control cells had no effect. Further, if the tumour cells had a different ideotype (different specific immunoglobulin, for which the antibody present in immunotoxin is not an anti-ideotype), then no effect of the immunotoxin was observed.

Experiments have also been performed on bone marrow cells (treated with immunotoxin) used for reinfusion into a patient treated with high doses of irradiation or chemotherapy leading to complete killing of the patient's own bone marrow. Such infusions did not develop tumours, since the reinfused cells were pretreated with immunotoxin to kill tumour cells, if any present. In vivo experiments were also conducted with mice to study the effect of immunotoxin for cancer therapy. It is believed that these immunotoxins and other types of drug designing will find increasing application in future therapeutic uses for cancer treatments.

Drug Designing

Drug designing is another area of biotechnology research, which is being actively pursued. Several approaches are being used for this purpose and some of them will be briefly described in this section to give an elementary idea about this important and fascinating area of research. However, all these approaches are based on an exercise where we first acquire knowledge about active site that needs to be attacked by the drug and then utilize this information to design a drug that will not only bind to the chosen site but will also possess all other properties required of a drug. Therefore, drug designing will involve at least three steps: (i) Knowledge will have to be acquired about the site structure, upto the level of atomic resolutions, so that the accessible surface, which needs to be attacked, is known; (ii) A ligand will have to be designed that will fit the binding site; for this purpose the steric outline of the ligand should be complementary to the site and, therefore, these complementary features will have to be built in the ligand at appropriate positions. This should result in the maximum interactions between atoms of the ligand and the site, and is mainly achieved with the help of computer-aided methods; (iii) Once the ligand is designed to fit the binding site, ligand needs to be modified to have pharmacological and toxicological properties, while maintaining its affinity for the binding site; (ii) This and the above step will require three dimensional constructions. Computers have provided methods which will allow automated site-directed drug designing.

Drug Designing by Blocking Enzyme Activity

Since most drugs act by modifying or blocking the activity of an enzyme, a deeper understanding of suitably chosen target enzymes should lead to major advances in rational drug designing. Following two examples will illustrate the utility of drug designing as an important area of biotechnology research.

Trimethoprim (TMP)— This clinically important antimicrobial drug provides an important example to demonstrate drug designing. This drug inhibits the enzyme dihydrofolate reductase (dHFR) in bacteria and is frequently used to treat urinary tract infections, but in high concentrations, it attacks even human dHFR, thus becoming harmful (toxic), if used by the patient. In view of this, the structures of different dHFRs have been compared, so that TMP with greater specificity for bacterial enzyme may be designed. Since TMP shows flexibility in its three dimensional structure in association with enzyme dHFR, efforts have been made to synthesize TMP, which will have a rigid three dimensional structure in association with bacterial enzyme dHFR, so that it may not be able to attack human dHFR.

Inhibitor of renin— Inhibitors of the enzyme 'renin' are also being actively modelled. The enzyme catalyses the first in a series of reactions that lead to elevated blood pressure. Models of renin have been prepared on the basis of known structures of similar other proteins like aspartic proteinase and pepsin. This led to the designing of nonpeptide inhibitors that mimic intermediate product in the reaction of renin with its substate, so that native renin will not function. These inhibitors may be useful for treating hypertensions.

Drug Designing through Blocking Hormone Receptors

Drugs have also been produced through a study of receptor molecules that are responsible for the disease, followed by identification of chemicals that will block such receptors. Utilizing this strategy, at least two drugs were developed by Dr James Black (of UK), who shared with Drs. G. Elion and G. Hitchings (of USA), the 1988 Nobel Prize for Physiology and Medicine. These two medicines are briefly described in the following text:

Propranolol— This drug is widely used for treating heart disease and high blood pressure. Heart diseases result due to the presence (in excess) of hormones 'norepinephrine' and 'epinephrine' due to their role in contraction and relaxation of cardiac muscles. These hormones act through two receptors a and P. The drug 'propranolol' is a locker and relaxes the heart muscles. It has been shown that the drug not only cures the heart diseases but reduces the death rate of heart disease survivors by 25%. The drug is also used for treatments of 'angina pectoris' and 'high blood pressure'. Obviously by blocking the receptor 'P', the drug inhibits the activity of hormones which act through a receptor molecule.

Cimetidine— This drug is antiulcer and is used for treatment of ulcers. Ulcers result from acid production due to histamine release in the stomach and therefore attempts were made to treat these ulcers through antihistamines that were earlier used to combat respiratory allergies. These antihistamines proved ineffective for ulcer treatment and James Black attributed this to use of receptors that were different in stomach lining and respiratory tract. He was able to characterize histamine receptor of stomach lining and called it H2 receptor, which is blocked by the drug 'cimetidine', thus leading to healing of ulcers by preventing acid production due to nonavailability of histamine. Therefore, 'cimetidine' marked the beginning of a new era in ulcer treatment, since earlier ulcers, that did not heal, had to be removed surgically.

Drug Designing Through Inhibition of Nucleic Acid Synthesis

Through their detailed studies of antimetabolites or inhibitors (they block key enzymes) of DNA synthesis, G. Elison and G. Hitchings (both of USA) were able to develop drugs for treating 'cancer', 'gout', 'malaria' and viral infections such as 'herpes'. They found that there are drugs like '6-mercaptopurine' and 'thioguanine' which inhibit DNA synthesis leading to inhibition of cell division and therefore are effective in cancer chemotherapy. Another drug developed by them is 'azathioprine' used to fight transplant rejection, or diseases like rhematoid arthritis, in both cases the drug acting through attacking the immune system (immunosuppressive effect).

Elion and Hitchings also studied the differences in nucleic acid synthesis of prokaryotes and human cells, which led to the development of Burgs like 'pyramethamine' for malaria, 'trimethoprim' for bacterial infection and 'acyclovir' for herpes virus infection. 'Azidothymidine (AZT)' which also working on the principle formulated by Elison and Hitchings, has been recently developed for AIDS.

SECTION - 4

LATEST DEVELOPMENTS

FOURTEEN

Spotlights

Biotechnology is primarily the industrial use of biological systems and processes. The modern revolution in biotechnology is based on our new-found abilities to exploit molecular recognition system. As the lines between basic and applied science blur, more and more breakthroughs are being reported in medicine, industry and agriculture. The last ten years have witnessed the birth of an industry based on this discipline which is quite old but newly-organized. The present chapter is aimed at describing some crucial achievements in the area of biotechnology over a wide range of social needs.

Hybrid Antibiotics

Since the original discovery of Penicillin 60 years ago, over 5000 different antibiotics have been isolated from various microbes, primarily from Streptomyces. While this is an impressive number, only a small percentage of these are clinically useful and only about 100 have commercial value. The continuous demand for new antibiotics and the limitations of the methods currently used for their discovery have led several groups to developing alternative approaches for the generation of new antibiotic structures and one approach is to divert the biosynthetic machinery of antibiotic producing organisms. The basis for this approach is that many enzymes do not have an absolute substrate specificity, but can handle a variety of compounds resembling their normal substrates in the same way as the original. A system called *mutasynthesis* has been developed to produce these hybrid or aminoglycoside antibiotics. Here in a multipathway convergent biosynthetic system, the organism is mutated so that the endogenous synthesis of one of the antibiotic precursors is blocked. The missing component is then replaced by analogs added to the cultures resulting in the formation of analogue structures. Similarly, inhibitors can sometimes be used to block precursor synthesis. A prominent example is the use of the antibiotic Cerulenin to block the formation of polyketides, which form the backbone of many antibiotics. This approach has been termed *hybrid biosynthesis* (Fig.).

The aminoglycoside antibiotics (Table) represent a complex interrelated family including at least 150 natural compounds as well as semisynthetic derivatives with many possibilities for making specifically modified or hybrid antibiotics by gene transfer. The structures of some typical aminoglycosides are shown in Fig.

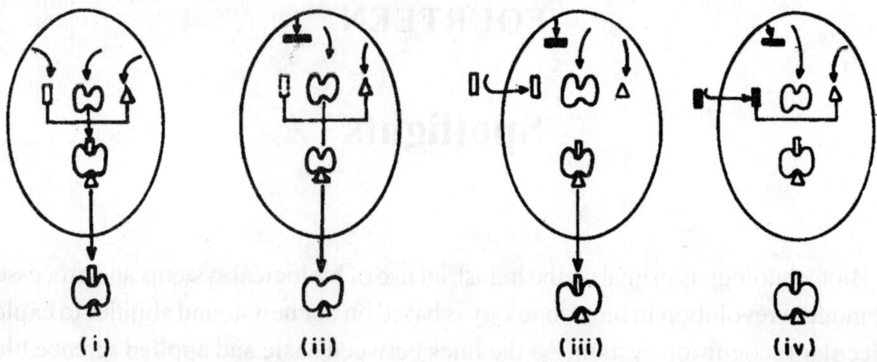

Fig. Mutasynthesis—Development of New Antibiotics

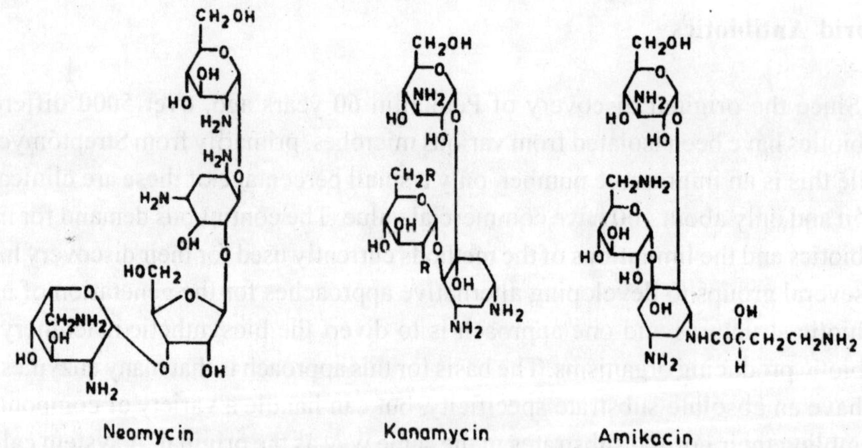

Fig. Molecular Structure of some Aminoglycoside Antibotics

Table

Major Amino glycoside Classes

Aminoglycoside	*Aminocyclitol component*
Streptomycins	Streptidine
Hygromycins	N-methyl-2-deoxystreptamine
Neomycins	2-deoxystreptamine

Aminoglycoside	*Aminocyclitol component*
Kanamycins	2-deoxystreptamine
Gentamicins	2-deoxystreptamine
Fortimicins	fortamine
Validamycins	validamine

A rational approach to the improvement of antibiotic yields require a knowledge of the biosynthetic pathway and identification of the rate-limiting steps that may be susceptible to manipulation. The biosynthetic pathways of antibiotics are extremely complex and are not understood fully. Considering the interrelationships of all biosynthetic steps which determine net flux through a pathway, amplified expression of any one gene could increase yields to some extent. Over-expression of all genes by cloning on multicopy plasmids might provide all maximal production. Although this sounds like a formidable task, it will undoubtedly be facilitated in many cases by the observed genetic clustering of antibiotic biosynthetic genes on the chromosome (e.g. chloramphenicol, actinorhodin, undecyloprodigiosin, oxytetracycline or rifamycin).

It is difficult to isolate a gene coding for a specific antibiotic. With one exception, the initial hopes that genes for antibiotic biosynthesis in actinomycetes might be conveniently separated from chromosomal genes, 'cloned' on, naturally occurring plasmids, have not been realized. Although plasmids are found in about one out of three streptomycetes, only in the case of methylenomycin biosynthesis it has been shown that the coding capacity of an extrachromosomal element contains structural genes necessary for antibiotic production. This early report, as well as the erroneous assumption that elimination or alteration of antibiotic production by various plasmid curing agents represents plasmid loss, have led to many unsubstantiated claims that antibiotic production genes were coded by plasmids.

Several approaches can be used to identify clones that harbor plasmids carrying antibiotic biosynthetic genes. Mutants which are blocked at various steps in aminoglycoside production pathways are available. These blocked mutants can be used as recipients for production genes from shotgun cloning experiments. Shotgun cloning is the isolation of a specific DNA sequence from a large population by cloning random DNA sequences (in this context genomic fragments) and screening for the desired phenotype. The plasmids isolated from transformants in which antibiotic biosynthesis is restored by cloned genes would then be introduced into the production strain for yield improvement. The technique of insertional mutagenesis can be used to obtain both mutants and cloned DNA in a single experiment.

The successful isolation of overproducing strains may depend on the ability of the organism to remain productive in the presence of higher concentrations of potentially toxic products (i.e. aminoglycosides or their precursors). There is circumstantial evidence to support a direct causal relationship between antibiotic resistance levels and production. High level aminoglycoside resistant strains are often overproducers. Alternatively, the loss of resistance is associated with loss of production of neomycin or streptomycin. Thus, it is important to understand the effects of the antibiotic on the producing organism. Aminoglycosides inhibit protein synthesis by binding to ribosomes. Lethal aminoglycoside effects in antibiotic-producing actinomycetes are prevented in at least three ways: (1) modification of the ribosome (2) modification of the aminoglycoside by acetylation or phosphorylation; or (3) exclusion of the antibiotic from the cytoplasm.

The fact that the aminoglycosides are all close structural relatives suggests many possibilities for the production of new molecules by gene cloning. Fig. is a composite outline of aminoglycoside biosynthesis which emphasizes structural similarities and interrelated pathways used by organisms which produce neomycin, ribostamycin, butirosin, paromomycin or gentamicin. Aminoglycosides are synthesized by branched pathways which often generate twenty or more related molecules in the same organism, some of which are produced in small amounts. In principle, genetic manipulations of regulatory genes might be used to focus the metabolic flux through a preferred pathway. This could increase yields, thus simplifying the purification of existing aminoglycosides, or revealing a previously undetected antibiotic. The discriminatory use of a variety of specifically resistant indicator bacteria can be used as a simple screening method to distinguish different types of modified aminoglycosides.

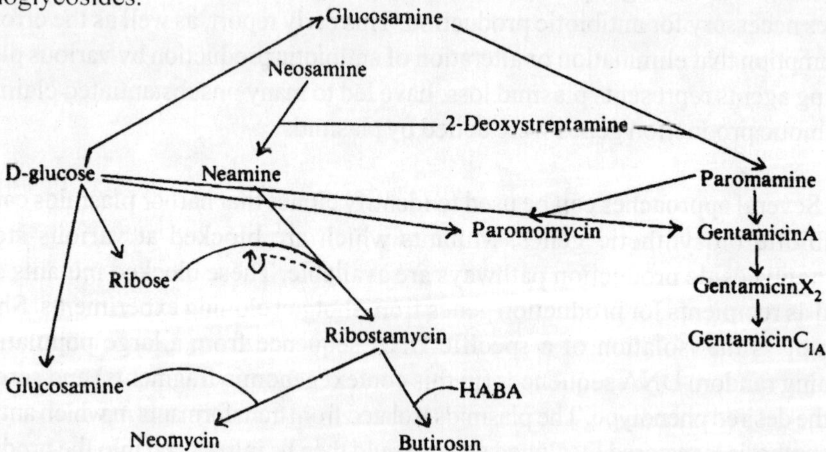

Fig. A General Outline of Aminoglycoside Biosynthetic Pathways. Deoxystreptaminebased Aminocyclitals are produced via the Simplified Pathways shown below. Parts of the above Network are shared by Organisms which Produce Neomycin, Ribosotamycin, Butirosin, Paramcmycin or Gentamicin. Dashed Lines indicate Pathways whose Intermediates are not shown.

Since the enzymes involved in the biosynthesis of aminoglycosides have broad substrate specificities, the transfer of genes between species which produce different aminoglycosides might also be used to generate new antibiotics. This has been suggested by the technique of mutasynthesis in which a mutationally blocked producing organism is able to incorporate a precursor analogue to produce a modified form of antibiotic (see Fig.). If genes which code for the synthesis of the chosen precursors were cloned interspecifically, many existing aminoglycosides as well as new hybrid molecules could be made by one-step fermentations. The precursor may be a natural molecule or chemically synthesized analogue. There are limitations to this approach, such as the cells' ability to take up the analogue and their enzymes acceptance of it. Despite these limitations many interesting and useful hybrids (such as hybrimycins, mutamicins, hydroxygentamicin and others) have been obtained by mutasynthesis.

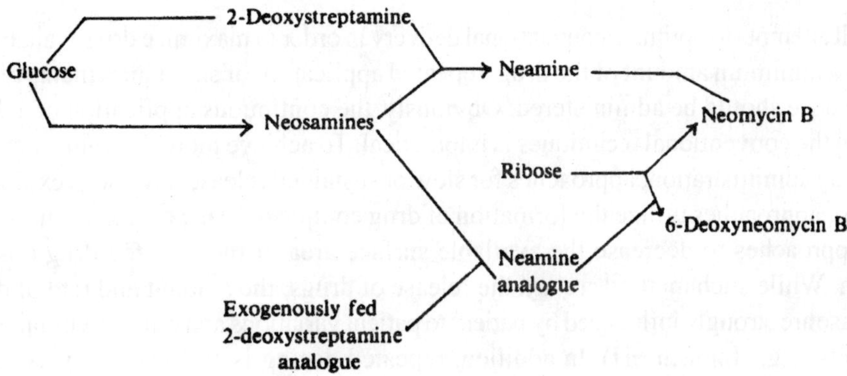

Fig. Production of New Antibiotics by Mutasynthesis

The same approach might also be used for more economic production of other products which are currently produced by chemical modification of existing aminoglycosides. Specific chemical modifications of aminoglycosides are expensive and often difficult because amino groups have to be protected and de-protected in order to give the desired change. The specificity of microbiological conversion is a great advantage. Amikacin is one of the most effective aminoglycosides; production of this molecule chemically from kanamycin makes it one of the most expensive. An aminoglycoside-producing strain of Bacillus circulans can convert kanamycin into amikacin by the addition of hydroxyalninobutyric acid. It may be possible that interspecific transfer of this gene could be used to persuade a kanamycinproducing streptomycete to make this molecule more economically.

In spite of this potential, major problems exist. The field of streptomycete molecular biology is largely unexplored. The transcriptional organization of streptomycete genes has not been studied. A lot more remains to be worked out to improve finally the antibiotic production and quality.

Release of Bioactive agents

In conventional drug delivery modes such as an injection or the taking of a pill, each time a person takes medicine, the drug concentration in the blood rises, peaks and then declines. Since each drug has a therapeutic range above which it is toxic and below which it is ineffective, the plasma drug concentration in any patient at a particular time depends on compliance with the prescribed routine (problems arise when minimum effective levels and maximum safe levels exist and are close together). The drug concentrations below the effective level or above the toxic level represent not only wasted drug but increased risk. As an example, it is generally felt that the complications resulting from insulin dependent diabetes are caused by wide fluctuations in blood-glucose concentrations, even when patients take daily injections of insulin.

It attempts to optimize conventional delivery in order to maximize drug availability with a minimum amount of the drug, repeated application of small increments of the total dose should be administered. Obviously, the continuous application of a drug using the conventional techniques is impractical. To achieve more desirable patterns of drug administration, approaches for slow or sustained release have been explored. These approaches utilize the formation of drug complexes, suspensions, emulsions or approaches to decrease the available surface area of the injected drug dosage form. While such methods retard the release of drugs, the amount and rate of drug release are strongly influenced by patient to patient variations and other environmental effects (e.g. stomach pH). In addition, repeated dosing is still necessary. A more promising method is to protect the drug by incorporating it in a solid polymer from which the drug is released at a controlled rate over a prolonged period of time.

Advantages of controlled release systems include:

(a) The plasma drug concentrations are continuously maintained in therapeutically desired range.
(b) Harmful side effects from systemic administration can be reduced by localized delivery of the drug to a particular body compartment.
(c) Drugs that have short half-lives *in vivo* may be protected from degradation.
(d) Patient compliance may be improved.
(e) Reduced need for follow-up care.
(f) Potential decreased wastage of drug.

Types of controlled-release systems

(i) Diffusion controlled systems

There are two basic types of diffusion controlled systems: The first is a reservoir device in which the active agent forms a core, surrounded by an inert diffusion barrier (Fig.). These systems include membranes, capsules, microcapsules, liposomes and hollow fibers. The second type is a monolithic device in which the active agent is dispersed or dissolved in an inert polymer (Fig.). As in reservoir systems, drug diffusion through the polymer, matrix is the rate-limiting step, and release rates are determined by the choice of polymer and its consequent effect on the diffusion and partition coefficient of the drug to be released.

(ii) Bioerodible controlled systems

The rationale for using bioerodible (or biodegradable) systems is that the bioerodible devices are eventually absorbed by the body, and thus, need not be removed surgically. Polymer bioerosion can be defined as the conversion of a material which is insoluble in water into one that is water-soluble. There are three types:

Type 1 erosion encompasses water-soluble polymers that have been insolubilized by hydrolytically unstable crosslinks. Type II erosion includes polymers that are initially water-insoluble and are solubilized by hydrolysis, ionization or protonation of a pendant group. Type III erosion includes polymers that are converted to small water soluble molecules by cleavage of their backbones.

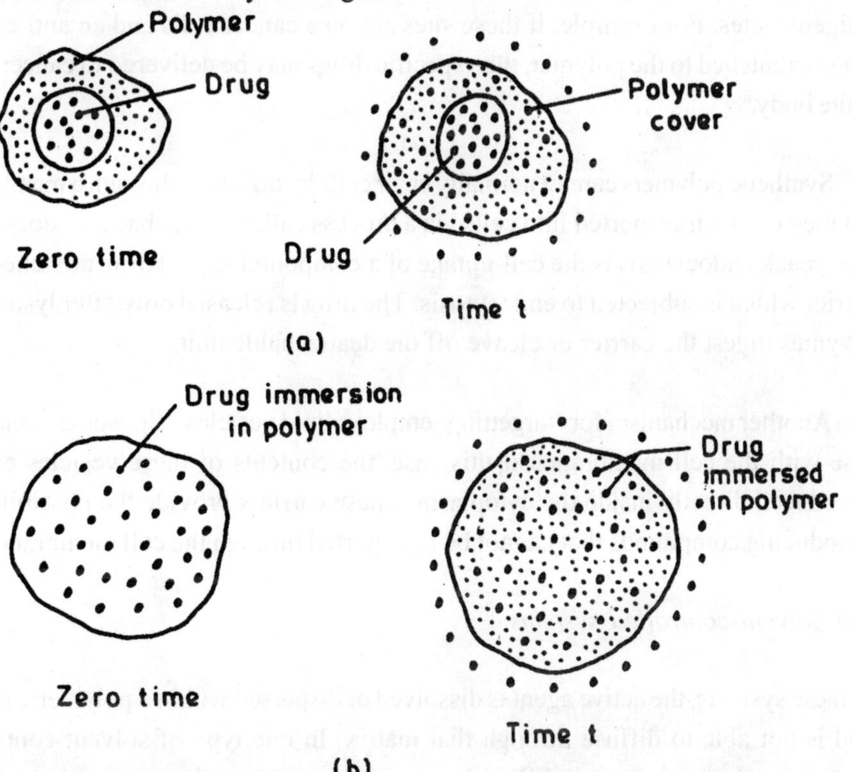

Fig. (a): Cross section of reservoir system (b): Cross section of matrix system.

In a bioerodible system, the drug is ideally distributed uniformly throughout a polymer in the same way as in monolithic systems. As the polymer surrounding the drug is eroded, the drug escapes. This ideal situation where surface erosion is the only process contributing to drug release has rarely been observed in practice. Most polymers that have been utilized for controlled-release have been selected because earlier studies have shown them to be biocompatible (e.g., as in bioerodible sutures). These polymers are generally hydrophilic, and therefore, bulk erosion may occur in addition to surface erosion. Furthermore, diffusion may occur in addition to erosion; this makes it difficult to predict the release process. The use of hydrophobic polyorthoesters and polyanhydrides represent current attempts to develop surface erosion controlled systems.

(iii) Polymeric drug carrier systems (targetting)

Attaching drugs to polymeric carriers affords the possibility of targeting them to desired sites in the body. Thus, if the targeting moiety is a particular antibody attached to the polymer along with the drug, the polymeric particle will be attracted to specific antigenic sites. For example, if these sites are on a cancer cells and an anti-cancer drug is attatched to the polymer, then specific drugs may be delivered to cancer cells in the body.

Synthetic polymers cannot normally enter cells by diffusion through a membrane, but they can be transported into cells via a process called 'piggyback' endocytosis. Piggyback endocytosis is the cell-uptake of a compound, e.g., a drug attatched to a carrier which is subjected to endocytosis. The drug is released only after lysosomal enzymes digest the carrier or cleave off the deattachable unit.

Another mechanism for 'targetting' employs lipid vesicles, 'liposomes', that can fuse with the cell membrane. In this case, the contents of these vesicles can be introduced directly inside the cytoplasm. These carriers provide the possibility of introducing compounds that cannot be transported through the cell membrane.

(iv) Solvent-controlled systems

In these systems, the active agent is dissolved or dispersed within a polymeric matrix and is not able to diffuse through that matrix. In one type of solvent controlled system, as the environmental fluid (e.g. water) penetrates the matrix, the polymer swells and its glass transition temperature is lowered below the environmental (host) temperature. Therefore, the swollen polymer is in a rubbery state and allows the drug contained in it to diffuse through the polymer.

(v) Pump Systems

Two types of artificial devices are being developed. The first is generally referred to as a closed-loop system and consists of a sensor, a minicomputer to interpret the sensor's findings, and a pump to deliver the active agent (drug) in response to the actual requirements of the patient. The second device, commonly known as the open-loop system does not include the sensor component. Thus, the drug is delivered in a predetermined pattern which can be adjusted according to the patient's needs. Both externally worn and implantable pumps have been developed. In both cases, the driving force is a pressure difference, which results in the flow of a drug solution through a small orifice. This pressure difference can be generated by pressurizing a drug reservoir, by osmotic action or by direct mechanical actuation.

Drug delivery at a constant controlled rate is often preferable to the conventional methods of drug administration. However, a better method may be drug delivery in response to the physiological needs of the body. An example is insulin administration where constant rate delivery is not sufficient. Several research groups are developing systems which would more closely resemble the normal homeostatic process in which the amount of drug released can be affected according to physiological needs.

Recently, glycosylated insulin bound to the lectin concanavalin A (COA) have been synthesized. Con A is immobilized on Sepharose beads. The glycosylated insulin, which is biologically active, is displaced from the Con A by glucose in response and proportional to the amount of glucose present, which competes for the same binding sites. This system offers the potential advantage of releasing insulin in amounts controlled by the body's glucose concentration. The development of a polymeric membrane to control delivery of insulin at rates dependent upon the external glucose concentration has also been explored.

Recent studies now suggest the feasibility of delivering macromolecules as large as 2 x 10 mol. wt by embedding them in a hydrophobic nondegradable polymer, such as ethylene-vinylacetate (EVAc). The system can then be implanted in a convenient site from which the macromolecule is slowly released. This approach has proved successful in the treatment of diabetic rats using insulin for periods of 140 days. This system has served as a basis for the development of a system which can deliver macromolecules of increased rates on demand. In this case, the system consists of small magnetic beads embedded in the polymer along with the drug. Release rates can be enhanced when desired by an oscillating external magnetic field. Parameters that affect the extent of modulation include magnetic field strength, magnet orientation and frequency of the applied field. Up to 30-fold increases in release rates can be achieved when the magnetic field is turned on.

The pharmaceutical industry has gone through several evolutionary stages starting with the introduction of antibiotics in the thirties and forties to the development of drugs produced by genetic engineering techniques today. There is little question, however, that the development of controlled release technology is essential for the use of many drugs and will have an increasing impact as biotechnological processes begin producing the drugs of the future. The impact of controlled release technology on biotechnology can best be appreciated by realizing that macromolecular drugs such as polypeptides (e.g. interferon, growth hormone), human derived antibodies and monoclonal antibodies will probably be difficult to administer orally because they will be degraded by enzymes and have difficulty being absorbed. They will also not be likely to be administered effectively transdermally because of their slow diffusion through the skin. Thus, in most cases they will have to be given parenterally and this will be extremely complicated since, in their native form, most of these polypeptides have half lives of twenty minutes or less. Thus, it will become absolutely essential to have controlled drug delivery systems capable of releasing these drugs over long periods of time and yet protecting them from degradation. For biotechnology to have a significant impact on the pharmaceutical industry, it is critical that the development of new drugs be coupled with methodologies for delivering them appropriately.

In addition, biotechnological developments should facilitate the creation of novel drug delivery systems. The development of approaches for drug targetting may depend upon highly specific monoclonal antibodies. In addition, enzymes may become valuable in developing delivery systems that operate in direct response to physiological needs. While controlled release technology is already a multi-hundred million rupees industry, it seems clear that its promise for the future is far greater and that it offers the potential for improved therapy for many pharmaceuticals currently being developed.

Cell Immobilization and Production of Biomolecules

Suspension cultures of microbes and plant cells have been in use for a long period for understanding basic and applied problems. These cells are in a state of mobilization in the culture medium and can occupy any one of the three dimensional configuration. However, in recent years, cells are "forced" by several techniques, including altering the entire metabolism to be in a static or immobilised condition. This is just like blocking the movement of an individual and then studying in detail the changes occurring in that immobilized state. The use of such immobilized microbial and plant cell components as biocatalysts are a very important area of biotechnology. The obvious advantages of immobilized cells, as compared to batch and continues culture are numerous:

1. Increased rate of biochemical reactions primarily due to increased density (mass per unit volume).
2. Increased overall cell metabolism as well as transport of substances across the membranes.
3. No dilution and washout of cells.
4. Increased possibility of choosing a particular cell configuration.
5. Direct control of several catalytic processes.
6. Quick and easy recovery of cellular products.

Cells can be immobilized by several methods including (i) adsorption, (ii) entrapment in gels or polymers (iii) covalent coupling and (iv) cross linking to insoluble matrices. A large number of matrices are now available for cell immobilization including agar, carrageenan, diatomaceous earth, foam polymers (polyvinyl) and calcium alginate gels. The usual procedure for preparing immobilized cells is to grow the cells in a culture vessel, harvest them, preferably in the log phase, and then immobilize to a high cell density. This is the preferred technique for entrapment of cells in agar, calcium alginate and other gels. Cells are also immobilized in reticulate foam pieces by inoculating a culture vessel containing the growth medium and the carrier matrix and then allowing the cells to colonize the foam during growth. Immobilized cells including several algal and cyanobacteria are increasingly used in studies on production of a wide variety of products such as ammonia, hydrogen, polysaccharides, glycerol lipids, amino acids, antibiotics and several enzymes.

Judging by both the numbers of patents and current applications, Japan is ahead of all other countries in enzyme technology. In 1969, Tanabe Seiyaku Co. were the first to find a satisfactory industrial application for an immobilized enzyme: this was amino acylase, used for continuous production of L-amino acids from acyl-DL-amino acid. Since then, they have also developed industrial applications of immobilized microbial cells, applying them in the continuous production of L-aspartic acid from ammonium fumarate using immobilized E. coli with an enhanced aspartase activity. Immobilized enzymes may be used to replace soluble enzymes in virtually all enzymatic analyses. Because the enzymes are in the solid phase they must be readily separated from the reaction mixture and reused, while sample contamination by enzymes is eliminated.

In 1973, Tanabe Seiyaka Co. Ltd. introduced a continuous process using immobilized microbial cells to convert sodium fumarate into L-malic acid. Since then, the process has been modified in several major and minor respects to create a system which is twenty times more efficient than the original one and is capable of producing about one tonne of malic acid per day from a 1000 litre medium.

The potential of cell immobilization is great, including the production of fuel (such as Hz) and fertilizers (such as ammonia). However, further basic research is needed to develop more sophisticated systems in the pharmaceutical, chemical, agricultural and food industries.

Bioreactors and Production of L-amino acids

Amino acids constitute a multibillion rupee market. Traditional uses have been as food and feed additives, and in nutritional and therapeutic preparations. As the technology for manufacturing amino acids improves, the price of certain amino acids will drop sufficiently to make these interesting as candidate feedstocks for biologically active products. For example, glycine is already used to make the herbicide glyphosate; threonine is the backbone of the antibiotic azthreonam; phenylalanine and aspartic acid are constituents of the sweetener aspartame. The importance of amino acids as metabolites in all forms of life makes them likely starting points for the development of herbicides, pesticides, antibiotics, pharmaceutical peptides and flavors. The similarity of structure among amino acids also suggests methods of using inexpensive amino acids to make more expensive ones, such as alanine from aspartic acid, serine from glycine, and tryptophan from serine.

Recently, Japanese and American workers have combined recombinant DNA techniques and enzyme technology to develop bioreactor methods for enzymic production of L-amino acids. Enzyme and whole-cell bioreactor technology competes well with chemical methods in the amino acid field because the optically active Lstereoisomers, produced by sterospecific enzyme reactions, are generally needed. Table compares the bioreactor approach with fermentation, another method

Table
Bioreactor Vs Fermentation

Advantages	***Disadvantages***
Higher product concentration	Greater feedstock cost
Less purification required	Cofactor requirements
Higher productivity	
Less capital intensive	

frequently used for L-amino acid manufacture. When a suitable enzyme and inexpensive feed stock combination can be found, and requirements for enzyme cofactors can be economically met, then the bioreactor is preferable. Several examples of bioreactor processes are now available for the production of L-aspartic acid, L-phenylalanine, L-serine, and L-tryptophan.

L-Aspartic acid via aspartase

Aspartase is an intracellular microbial lyase that catalyses the stereospecific amination of fumaric acid to L-aspartic acid. Some form of immobilized aspartase has been used for industrial production of L-aspartic acid in Japan since 1973. During this time, immobilized microbial whole-cell systems have replaced free-cell systems, resulting in an overall cost reduction of 40% and providing simple operations. Many other reports have described the use of immobilized aspartase or whole cells for L-aspartic acid production. For manufacturing purposes, immobilized whole cells are preferred over immobilized aspartase because of the added costs of enzyme isolation.

Recently, workers from Genex Corporation in United States have developed a bioreactor for the continuous production of L-aspartic acid that uses whole cells (containing aspartase) immobilized on vermiculite. A strain of E. coli expressing high aspartase activity is grown in an aerated fermentor at 37°C for 12 h and the cells recovered by centrifugation. The cell paste is immobilized at high density (500 g wet cell paste per liter of bioreactor volume) using the vermiculite carrier system and proprietary entrapment technology. The resulting immobilized whole-cell support particles are loaded into a column bioreactor. A 1.8 M solution of ammonium fumarate containing 1 mM magnesium sulphate, is passed through the column continuously at a rate of 2.1 liters of solution per liter of reactor volume per hour. The fumarate was converted to L-aspartic acid (237 g/1) with almost 100% yield. The productivity of the column under these 'highconversion' conditions was 0.5 kg of L-aspartic acid per liter of total bed volume per hour (3.7 mol 1^{-1} h^{-1}).

Productivities at 37°C as high as 0.85 kg per liter of bed volume per hour have been achieved with 71% conversion of the 1.8 M fumaric acid substrate. The L-aspartic acid was easily recovered in high yield using an acid precipitation/filtration scheme.

L-Phenylalamine via PAL

Over 20 years ago, it was discovered that phenylalanine ammonia-lyase (PAL) converts L-phenylalanine (L-Phe) to ammonia and trans-cinnamic acid (CA). It has been recognized for several years that a process for manufacturing L-Phe from CA and ammonia might be developed by applying PAL.

Recently, Genex workers have been able to convert approximately 90% of CA PAL in an input stream to L-Phe with a single pass through a PAL bioreactor. L-Phe titers higher than 60 g/1 in bioreactor product streams have also been achieved. Genetics, fermentation technology and immobilization have all contributed to this work.

To achieve the aim, a strain of the yeast Rhodotorula rubra was genetically manipulated to obtain a strain that produces high titers of intracellular PAL when grown aerobically on an inexpensive fermentation medium. Special operational procedures were developed for the R. rubra fermentation to achieve a stable peak PAL titer. On the laboratory scale, whole cells of R. rubra, containing the PAL enzyme, were recovered by centrifugation and immobilized using proprietary entrapment technology with vermiculate as the carrier. The immobilized biocatalyst was installed in a bioreactor, and a solution containing 7.85 M total ammonia and 0.37 M CA, was fed to the bioreactor. The product stream from the bioreactor contained 0.35 ML-Phe (59 g/1). The crude L-Phe in the bioreactor product stream was recovered by a simple crystallization procedure. A modified version of this process is now being run on large scale to produce purified L-Phe in dry powder form.

L-serine via SHMT

Serine hydroxymethyltransferase (SHMT) acts on a wide range of substrates. One catalytic function of SHMT is to reversibly convert glycine and formaldehyde to L-serine in the presence of a cofactor, tetrahydrofolate (THF), as shown in the following scheme:

$$\text{THF + HCHO} \rightleftharpoons \text{N}^5\text{,N}^{10}\text{ methylene-THF} \qquad (1)$$

$$\text{N}^5\text{, N}^{10}\text{ methylene-THF + glycine} \underset{}{\overset{\text{SHMT}}{\rightleftharpoons}} \text{L-serine + THF} \qquad (2)$$

The overall reaction between glycine and formaldehyde, with SHMT acting as biocatalyst, has been utilized for production L-serine bioreactor.

Genetic engineering for SHMT biocatalyst

To provide an inexpensive SHMT biocatalyst for the serine process, a production strain with the following characteristics were needed:

(i) The host organism should be inexpensive to grow to a high density.

(ii) Expression of the gene should be at a high level to provide desirable SHMT specific activity.

(iii) The strain should be stable, particularly with respect to plasmid loss.

This was achieved by introducing into Klebsiella aerogenes a multicopy plasmid containing the *E. coli* SHMT gene and stabilized by nutritional selection. The multiple copies of the SHMT gene resulted in the desired high enzyme specific activity.

Klebsiella can use starch, which is the least expensive carbon source. In various parts of the world, molasses might be less expensive. Klebsiella will use molasses too. The best results have been obtained by feeding the carbohydrate during fermentation and maintaining a low dissolved oxygen concentration. These steps slow down the growth rate which is important for obtaining high enzyme specific activity. A titer of approximately 30 000 units/ 1 was achieved in a fermentation with a cell density of about 30 g/ 1 dry weight. Fermentation of this type produces SHMT biocatalyst very economically for serine bioreactors which use permeabilized whole cells or whole cell lysates. The average productivity was about 10 g of L-serine per liter of bioreactor volume per hour with at least 80% molar conversion of glycine to L-serine.

L-Tryptophan via tryptophanase

The essential amino acid tryptophan can be produced by chemical synthesis, fermentation from sugar and methods of enzymatic synthesis using tryptophanase or tryptophan synthase. Numerous microbial fermentation methods have been described for the biosynthesis of tryptophan, including those employing strains modified by genetic engineering. Genetic engineering has increased productivity but has not solved the primary problem of low final tryptophan titer (maximum 15-17 g/1) in the fermentation broth.

Genex workers have used E. coli tryptophanase (Tase) for bioreactor production of tryptophan. The enzyme normally catalyses the degradation of tryptophan to ammonia, pyruvate and indole. Indole (at low concentration) inhibits the degradation reaction. By increasing the concentration of ammonia and pyruvate, tryptophan can be synthesized. Tase also efficiently converts indole and L-serine into L-tryptophan and this reaction is the basis for the L-tryptophan bioreactor.

In the process, crude L-serine from the SHMT fed-batch tank reactor system was used directly for tryptophan production by the addition of Tase biocatalyst (from disrupted E. coli) and indole. Sequential SHMT and Tase bioreactions yielded 200 g of tryptophan per liter with 95% milar yield based on indole as substrate and 93% yield based on serine (or 84%) yield based on glycine). The insoluble L-tryptophan was removed and the unused glycine and L-serine recycled to another tryptophan production bioreactor without any detrimental effects. The bioreactor processes for producing other amino acids are likely to be commercialised in the years to come. Several L-amino acids are currently manufactured by using enzymes or whole-cells contained within bioreactors. Production of L-aspartic acid in Japan is one of the earliest examples of this technology.

High Temperature Resistant Proteases

Proteases could be the most important group of industrial enzymes and certainly form a major portion of world-wide enzyme sales (Table). High temperature resistant proteases, with their high specific activities and their superior chemical and

Table

Selected Proteases and Their Source

Protease type	*Source/organism*
Thermolysin	Bacillus thermoproteolytices (B. stearothermophilus)
Thermitase	Thermoactinomyces vulgaris
S Protease	Streptomyces rectus
Aqualysin 11	Thermus aquaticus YTI

physical stability characteristics, would seem to be good candidates for current and furture biotechnological applications. Historically, proteases have played a very important role in the processing of natural products. For many centuries, protease containing cultures, extracts and waste products have been used diversely in the tanning of hides and the production of condiments. Today, proteases are probably the most important of the classes of industrial enzymes, with world-wide sales representing about 60% of the total enzyme market. Three enzymes (bovine and microbial rennins, used in cheesemaking and Bacillus megaterium neutral protease, used in the detergent industry) are responsible for nearly two-thirds of the protease market.

These three are, of course, industrial enzymes which are available in bulk and are relatively cheap. They are quite distinct from the speciality enzymes, which are usually highly purified, very expensive, and used on a much smaller scale (for example, in research work). The desirable properties for a commercial protease will depend primarily on whether the enzyme is intended for the industrial or specialist market. Within these categories, the requirements are extremely varied, depending entirely on the nature of the application. For example, a protease suitable for the detergent industry would ideally possess both high activity and stability in moderately alkaline conditions (pH 10-1 1), should be unaffected by oxidizing and chelating agents and should be available in bulk at low cost. However, as a tool in protein sequencing, the critical requirement for a protease is that it has a narrow and well-defined peptide bond specificity. The cost is of relatively little importance.

New proteases are constantly being isolated. Although almost all types of proteases have some potential application, one of the reasons that most are not even considered for commercial use is that any possible advantages are heavily outweighed by the necessary development costs. Even some which seem to have substantial advantages (e.g. the alkophilic proteases) have not yet found great commercial acceptance. It seems reasonable to predict that any new protease would be more likely to find application as a speciality enzyme than as an industrial enzyme.

In the following discussion, we will outline some of the properties of thermophilie proteases and attempt to identify areas of biotechnology where their novel characteristics could impart some process advantage.

Proteases from thermophilic organisms and Production

Many proteases have been isolated. The most detailed studies have been on thermolysin, a zinc metalloprotease isolated from Bacillus thermoproteolyticus, a strain of B. stearothermophilus. The distinction between thermophiles (optimum growth temperature, 55-65°C) and extreme thermophiles (optimum growth temperature above 65°C) is becoming less distinct with the isolation of new, high-temperature strains of B. stearothermophilus. To date, few proteases from extreme thermophiles have either been reported or studied in detail. Characterization studies are limited to the extracellular protease from Bacillus caldolyticus and four proteases from different strain of the genus Thermus.

Protease-producing thermophiles are routinely grown on undefined complex media. In the laboratory, yeast extract and commercially available protein digests are normally used, while large scale commercial producers substitute cheaper nutrient sources such as corn-steep liquor, fish meal and soyabean extract which usually have high C:N ratios. While the thermophilic bacilli are rather nonspecific in their nutrient requirements, Thermus species grow poorly in media containing high concentrations of carbohydrate. Although there are numerous examples of the induction and repression of the synthesis of proteases by mesophilic microorganisms, no other data on similar effects in extremely thermophilic bacteria are currently available.

Recently, the structural gene for a B. stearothermophilus neutral protease has been cloned into a plasmid. When the recombinant plasmid was expressed in B. stearothermophilus, protease production increased 15-fold. There have been no reports of attempts to increase protease production in cultures of extreme thermophilies by genetic manipulation.

Structure, Function and Uses

The extracellular proteases of all the thermophilic bacteria so far reported are serine or neutral metallo-proteases. Cysteine proteases appear to be restricted to the thermophilic fungi. The catalytic mechanism of a protease is an important consideration in commercial application. Cysteine proteases are susceptible to inhibition by many reagents, particularly metal ions, and lose their activity by oxidation. Optimal activity is often only obtained in the presence of reducing agents and such additions are often not industrially feasible. Neutral metalloproteases are rapidly inactivated by the removal of the catalytic metal ion and thus cannot be used in application where chelating agents are present (e.g. in some detergents). The serine proteases, however, are not readily inhibited, although many are stabilized by metal ions, and are used in most established biotechnological applications.

The stabilization of proteins against thermal denaturation is generally conferred by small changes in the amino acid sequence. The stability of thermophilic proteases is not restricted to temperature but also includes resistance to denaturing agents, detergents and organic solvents. Protein stability can be increased by various chemical modification procedures. Reports of the increase in thermostability of thermophilic proteases by immobilization and metal ion substitution have been published. Other methods of protein stabilization such as by covalent cross linking are well-established.

The ability of thermophilic proteases to retain a significant level of activity after extended periods at high temperatures in aqueous solution (e.g. 4-5 h at 85°C) must impart certain biotechnological advantages. The difficulty lies in identifying which existing enzymic processes could be satisfactorily performed at higher temperatures; most are currently carried out at less than 60°C. A multitude of industrial protein recovery or solubilization processes are likely candidates.

Both the study and biotechnological applications of thermophilic proteases have just stated. Relatively, few proteases from extreme thermophiles have been isolated, let alone characterized in detail, and applications are largely hypothetical. Their future lies in the development of unique applications to match the unique properties of these enzymes. The discovery of proteases produced by the extremely thermophilic, archaebacteria could pave the way to the realistic use of proteolysis at temperatures above 100°C. A large number of thermophilic cyanobacteria are known and it may be worthwhile to use the intrinsic stability of their enzymes including proteases for various biotechnological operations in the near future.

Table

Some Current Commercial and Industrial Applications of Mesophilic Bacterial Proteases

Industry	*Example of application*
Washing/cleaning	Detergent preparations
Tanning	Dehairing/bating
Protein recovery/hydrolysis	Soy protein
	Meat and fish hydrolysates
	Gelatin hydrolysis
	Tenderization
Organic synthesis	Aspartame synthesis
Beverage	Clarifying wine and beer
Milling/baking	Gluten hydrolysis
Medicine	Digestive aids, treatment of burns, ulcers, etc.
Photography	Recovery of silver from emulsions

Hepatitis Vaccine Production

Presumably every gene cloning company maintains an active interest in yeast as a possible host for the expression of cloned genes. Recent results with hepatitis B surface antigen (HBsAg) expression have shown that in at least one case a yeast system has distinct advantages over E. coli. It allows the production of adequate amounts of HBsAg in an immunologically active form.

Hepatitis B virus infections affect large numbers of people world wide. In USA alone, it is estimated that there are 100,000 to 200,000 cases annually of which 5-10% become chronic carriers, and there have been estimates of 170 million carriers in the world. Especially at risk are those who work with blood or blood products, or those who need multiple transfusions or blood products manufactured from multiple donations (e.g. factor VIII required by haemophiliacs) and there is an obvious need for a suitable vaccine. HBV replicates only in human and chimpanzee livers, hence the great attraction of producing a suitable surface protein antigen by recombinant methods. HBsAg is a glycosylated polypeptide of molecular weight 27,000 (23,000 in its non-glycosylated form) which can aggregete into non-infectious, but immunologically active 22 nm particles in the serum of human hepatitis carriers. These particles lack the core antigen and the nucleic acid of the intact virus and are much more immunogenic than the HBsAg monomer.

There are now a substantial number of reports of the cloning and expression of the HBsAg gene in a range of organisms including *E. coli, Saccharomyces cerevisiae* and transformed mammalian cell lines. In bacteria, the yields were low, and there were some possible objections to the use of transformed mammalian cells for producing vaccines suitable for human use. However, in 1982, it was reported that cloning of HBsAg in Saccharomyces cerevisiae using the 5-flanking region of the yeast alcohol dehydrogenase I (ADHI) gene led to the formation of the non-glycosated HBsAg monomers and that some of these assembled into particles similar to the 22 nm particles secreted by human cells. Since then, several other groups have reported success with yeast using expression vectors with the promoters associated with the yeast genes for 3-phosphoglycerate kinase. All of these groups reported that these 22 run-like particles were immunogenic in laboratory animals.

It has recently been reported that the successful vaccination of chimpanzees against HBV using yeast-derived particles in what may prove to be the first examples of vaccines produced from recombinant cells that are effective against a human viral infection. The yeast-derived HBsAg particles can be obtained in a highly purified form by immune affinity chromatography and HBsAg vaccine from yeast now seems to be a very attractive proposition compared to the vaccine produced from human plasma. It has the potential to provide a lower cost product that is safer to produce and use, and clearly there is now a case for clinical trials to be performed on human volunteers.

Thaumatin-the intense Sweetener

Thaumatin is a sweet-tasting protein extracted from the arils of the fruit of *Thaumatococcus daniellii.* This plant is indigenous to Africa. Its fruits have been used for centuries to improve the taste of the locally produced palm wine, and as a sweetener in cooking. Aqueous solutions of thaumatin have an intense sweetness, about 5,500 times that of sucrose when compared with a 0.6% sucrose solution, and about 1,600 times in relation to a 10% sucrose solution. For Western and Oriental markets, it has been used as an intense sweetener and as a flavour potentiator capable of improving or extending the flavour characteristics of various food products. Being soluble in aqueous ethanol, it can be added directly to flavour essences or oil. In Japan, it is used in a range of food products but in other countries, it is only used in chewing gum and pharmaceuticals. Taste-enhancing effects of thaumatin were observed just below the threshold concentration (5 x 10%) of thaumatin. Talin (the trade name for thaumatin produced by Tata & Lyle (PLC)) not only enhances flavours associated with sweet products such as peppermint but is also capable of improving savoury flavours. It reduces the harshness of tobacco and improves its 'smoothening' effect when incorporated in filter tips. A flavour-enhancing effect in cultured milk has also been reported.

Biochemistry of thaumatin

At least five different forms of thaumatin, called I, II, III and b and c, can be isolated from the arils by ion-exchange chromatography. Their molecular weight is about 22,000 and they are all basic proteins, with isoelectric points at a pH of nearly 12. Thaumatins I and II predominate in the fruit although their quantities may vary from batch to batch; both proteins have a single polypeptide chain of 207 amino acid residues and their amino acid sequences are almost identical.

Beside being an attractive molecule for studying structural aspects of sweetness perception, the intense sweetness and flavour enhancing properties of thaumatin make it a possible sugar substitute and a competitor with other low-calorie sweeteners like saccharine and aspartame. Biotechnological (as opposed to natural) production of thaumatin is attractive because T. daniellii does not bear fruit outside its natural environment. For these reasons, the genetic information encoding thaumatin was cloned.

Cloning thaumatin-encoding genes

First, the mRNA-derived cDNA molecule encoding thaumatin II was obtained and then five naturally-occurring genes were isolated from a gene bank containing leaf DNA of T. daniellii.

Although thaumatin-encoding mRNA species are abundant in the polyadenylated mRNA population isolated from arils (the five forms of thaumatin constitute about 50% of the aril's mass), identification of the cloned cDNA sequence was hampered by the large number of clones containing copied ribosomal RNA. Thaumatin-encoding mRNA species were, therefore, purified by extraction from a polyacrylamide gel, converted into cDNA and double-stranded DNA and then cloned.

Polyacrylamide gel electrophoresis showed that the protein was synthesized as an intact molecule. However, yield is quite poor with about 500 molecules per cell in E. coli. Because of this low yield, trials have been made for expression of the gene in yeast cells. Both *Saccharomyces cerevisiae* and *Kluyveromyces lactis* have been utilized for gene cloning. It has been discovered recently that yeast can process the plant protein precisely and completely indicating that the plant signal sequence is recognized by yeast and is essential for expression.

Thaumatin is a potential low-caloric substitute for sugar and attractive compound for studying the structural aspects of sweetness perception. Since thaumatin has other properties such as flavour enhancement, it could also be used in products other than low-calorie foods.

Production of L-Malic acid by Immobilized cells

Nearly fifteen years ago, Tanabe Seiyaku Co Ltd. introduced a continuous process using immobilized microbial cells to convert sodium fumarate into L-malic acid. Since then, the process has been modified in several major and minor respects to create a system which is 21 times more efficient than the original one and is capable of producing about one tonne of malic acid per day from a 1,000 litre column, with 70076 of the maximum theoretical yield. These improvements were achieved, particularly by changing to the use of Brevibacterium flavum and a novel immobilization matrix, x-carrageenan.

L-Malic acid-uses and Production

L-Malic acid is employed in the food industry as an acidulant in fruit and vegetable juices, carbonated soft drinks, jams, candies, etc. In the pharmaceutical industry L-malate salts of basic amino acids are used in amino acid infusions. Precise figures for the current world production of L-malic acid are hard to come by, but it may be as much as 500 tonnes per year.

Although L-malic acid can be prepared by isolating it from natural fruit juices, or by separating it from the recemic mixture of D and L forms of malic acid formed by chemical synthesis, or by fermentation, the only industrial process employed for its production is enzymic. L-malic acid is formed from fumaric acid by the action of the enzyme fumarase, shown in reaction:

$$\begin{array}{c} COOH \\ | \\ CH \\ | \\ HC \\ | \\ COOH \end{array} + H_2O \xrightleftharpoons{\text{fumarase}} \begin{array}{c} COOH \\ | \\ CH_2 \\ | \\ HCOH \\ | \\ COOH \end{array}$$

fumaric acid **L-malic acid**

L-Malic acid could be produced by a batch process employing microbial cells which contain fumarase, but a continuous reaction using immobilized enzymes should be more economical. Recently, it has been reported that fumarase isolated from microbial cells can be efficiently immobilized on cellulose triacetate, making it possible to develop an economically attractive method for producing L-malic acid. However, no industrial process utilizing immobilized fumarase has been brought into operation.

Continuous production using microbial cells immobilized with polyacrylamide

Rather than immobilizing the enzyme itself, Japanese workers in 1974, employed an industrial process in which whole cells of Brevibacterium ammonia genes with high fumarase activity were immobilized in gel matrices of polyacrylamide. A disadvantage of this system is that by-products are formed in considerable amounts; unconverted fumaric acid and succinic acid accumulate in the reaction mixture. Although fumaric acid can be easily precipitated by acidifying the reaction mixture with hydrochloric acid, it is very difficult to separate succinic acid from L-malic acid. The key to a successful process is, therefore, to prevent the formation of succinic acid. It is now produced on an industrial scale by the enzymatic conversion of fumaric acid using immobilized cells of Brevibacterium flavum. Recent improvements to this system, especially the use of x-carrageenan supports, have resulted in a continuous process capable of yielding 30 tonnes of L-malic acid per month.

Japan has been operating this improved immobilized B. flavum cell column for the industrial production of L-malic acid since 1980, with about 70% of the theoretical yield. The unconsumed fumarate is recycled. Thirty tonnes of L-malic acid can be produced in one month using a 1,000 litre column fed at a low rate of 450 litre/h of 1 M sodium fumarate.

This L-malic acid production system is the third industrial application of immobilized cells at Tanabe Seiyaku, the others yielding L-aspartic acid and L-alanine, and in 1969 the industrial production of three amino acids, L-valine, L-phenylalanine and L-methionine, was instituted using immobilized amino acylase. These processes and similar immobilized systems employed by other companies to produce high-fructose syrups and 6-amino penicillanic acid amply demonstrate that this area of biotechnology has a very bright future.

Biotechnology and Biodegradation

Several countries including India are confronting a hazardous-waste disposal problem which may appropriately be considered a national problem, whose scope demands creative solutions. Biotechnology can provide a useful tool to develop new and safer approaches for hazardous-waste disposal and cleanup. However, such developments have been slowed by regulatory uncertainties and the lack of appropriate incentives to promote alternatives to landfill disposal of hazardous waste. Hence, the government needs to clarify the current regulatory picture for the products of biotechnology and to expand incentives for the development of new technologies.

Biodegradation is the use of microorganisms or their metabolites to treat toxic chemicals and render them harmless. There is a growing belief that once developed and proven, biodegradation is potentially less expensive than any other approach to neutralizing toxic wastes. Such systems involve a low capital investment, have a low energy consumption, and are often self-sustaining operations.

In biodegradation, a microbe feeds on toxic waste or, in the presence of a particular waste, secretes an enzyme to degrade the waste. As early as 1914, researchers were isolating bacterial strains that efficiently degrade noxious sewage and reduce the volume of organic matter. Today, microorganisms that degrade some of the most toxic of chemicals are being 'isolated from the soil of hazardous-waste sites. These organisms can be selectively bred in the laboratory and returned to the environment as more efficient scavengers.

Genetic engineering can enhance microorganisms, enabling them to degrade a whole range of man-made chemicals. The potential range of toxic substances that can be degraded by microorganisms is limited now only by our ability to isolate and manipulate the genes for enzymes that break these compounds down.

Biodegradation techniques are versatile and can be used at several different stages. Some applications of biodegradation include removal of contaminants from raw materials before processing, treatment of pipeline wastes before discharge from the factory, decontamination of soils and surface and ground water, and the cleanup of dump sites. These operations may be conducted in a number of ways. (i) Microorganisms or their active products can be released directly into the contaminated environment. (ii) Microorganisms already present in the environment can be enhanced by the addition of suitable nutrients (bioaugmentation). (iii) Microorganisms can be used in contained or semicoptained reactors.

For biodegradation to be more fully utilized, however, several scientific problems will have to be overcome. The first step toward practical applications of biodegradation systems for hazardous-waste treatment is identification of microorganisms capable of degrading the waste. To do this, increasing our understanding of catabolic pathways used by microorganisms to break down the waste will be important. For both new and existing microorganisms, it is important to learn more about how they survive in the environment. Once the microorganisms have been identified or created, process technologies need development. Process technologies make possible the efficient and effective use of microorganisms in highly variable, realworld conditions.

The prospects for constructing or finding catabolic pathways in nature to detoxify hazardous chemicals are encouraging. There are many sources of catabolic genes, including in vitro genetic engineering, laboratory selection techniques, naturally occurring organisms, and environments yet untapped (extreme, anaerobic, and marine). Developments in this area will depend on an increased understanding of microbial physiology, the basic biochemical and genetic bases for microbial detoxifications, and an understanding of microbial resistance to heavy metals. The microbial degradation processes also must be made to work fast enough to be of practical use. This will require sufficient information on the kinetics of bacterial and cyanobacterial growth and biodegradation.

In addition to having the correct catabolic potential, microorganisms used for biodegradation must possess the full array of ecologically important traits needed for success in the environment; that is, the useful organism must have a competitive advantage over the indigenous organisms so that it survives long enough to accomplish its task. To achieve this, genetic enhancement of ecologically important traits may be required.

There are other significant scientific problems to be overcome in developing suitable microorganisms for field use. These include:

— *The complex mixture of most waste.* Any one bacterial strain will generally degrade only one particular chemical or closely related groups of chemicals. Often, however, waste contains a variety of toxic chemicals in differing concentrations. Publications contain little work on using microorganisms to treat complex waste mixtures.

— *The use of carbon by the bacteria.* In many cases, waste contains carbon sources that microorganisms may choose as nutrient sources instead of the targeted toxic chemical.

— *The high toxicity of the waste.* No microorganism has yet been grown on a defined substrate at concentrations at which pollutants are frequently of practical concern (10 ppbpb).

— *The survival of the organism.* In open systems, the effectiveness of the technology is influenced by environmental conditions such as the temperature, type of soil, the presence of other naturally occurring microorganisms, and the presence of water, oxygen, and acidity.

— *The possibility that the organism will mutate.* Since many of the waste pool constituents are mutagens, the organism could mutate to become ineffective.

To date, some 40 bacterial strains and species have been isolated worldwide in several laboratories that can degrade many of the compounds of the greatest hazard to the environment and to health, although, in many cases, degradation is not fast enough to be of practical interest. The compounds include almost all those of major environmental concern. (Microorganisms have been isolated to degrade organophosphate pesticides, such as parathion, malathon, carbofuran, alachlor, DDT, carbon tetrachloride, and other halogenated hydrocarbons; 2, 4-D, 2, 4, 5-T, and dioxin; PCB's benzene, toluene, biphenyls, and naphthalene; heavy metals, such as mercury, cadmium, uranium and selenium, as well as their organic derivatives).

Although these results are promising, there is a major gap between laboratory studies and applied technology. To assure success in a growing market, laboratory research efforts must be closely linked to process engineering.

Before a microorganism can be tested outside a contained environment, an evaluation of what happens after it has performed its task is needed. Predictive ecological research is essential to ensure that biodegradation of waste does not unintentionally create new and potentially serious environmental problems.

Molecular biological approaches to hazardous-waste control are scientifically promising; however, commercial application of these techniques is contingent on substantial additional research into basic science, process technology, and predictive ecology (environmental fate and monitoring). Since most research is basic, and consequently, long term and inherently economically risky, it is not undertaken by many companies and clearly not the kind of research that could be undertaken by small biotechnology companies.

Another research project at the Agricultural University at Wageningen has resulted in a bacterium that breaks down the carcinogenic compound vinyl chloride at a high enough rate to be used for detoxification of industrial wastes.

Vinyl chloride is liberated in the production of PVC, a raw material of all kinds of plastics. The total estimated expulsion of vinyl chloride in the Netherlands is more than a million kilogrames per year. Since vinyl chloride is carcinogenic, Dutch PVC producers have been looking for a method to degrade the compound at acceptable costs. The only reliable method found until now (combustion) has proved to be too expensive, taking into account the production costs of foreign PVC producers. Most companies expel vinyl chloride through a high chimney in the hope that the compound will be diluted far enough before it reaches the ground. Alternatively, vinyl chloride is dissolved in water and drained into a river. Since, there is no no-effect-dose for vinyl chloride, the pollution of the environment with the compound is a serious threat.

The Wageningen researchers, De Bont and Tramper, have now found a bacterium that will degrade vinyl chloride. They isolated the bacterium (likely a Mycobacterium) from the soil near the PVC producing company, Polysar in Arnhem, The Netherlands, where 100 kilogrammes of vinyl chloride are expelled per day. De Bont and Tramper and their students developed ideas for a purification installation consisting of the bacteria immobilised in alginate. The industrial waste gas is conducted through the installation before being expelled into the atmosphere. The calculated costs of an installation degrading 99% of a presented amount of 100 kilogrammes of vinyl chloride per day amounts to an investment of Df 1.500,000 (About Rs. 2,80,000) plus an-annual expenditure of Df 1.150,000 (ICs 800,000), which is acceptable for a company producing PVC at a rate liberating 100 kilogrammes of vinyl chloride per day.

Human Plasminogen activators

The blockage of blood vessels by clots is the commonest cause of death in the developed world. Blood clots in the vessels which supply the heart (coronary heart disease), brain (stroke) or lungs (pulmonary embolism) account for more than half of all deaths in the West.

The processes involved in the formation and subsequent destruction of a blood clot (thrombus) are complex and finely balanced. An array of enzymes and other substances ensure that clots normally form only when and where they are required to prevent blood loss, and that they are broken down once their task is completed. The basis of a blood clot is a network of fibres of the protein, fibrin. This network can be destroyed by the action of plasmin, a fibrinolytic enzyme, which is formed in the blood from an inactive precursor, plasminogen. The role of plasminogen activators (PAs) is to remove part of the plasminogen molecule to yield plasmin (Fig.).

Sources of plasminogen activators

PAs are found in blood, tears, saliva, urine, semen and cerebrospinal and other body fluids. They are also found in many tissues and cells, including cancer cells. In addition, Streptomyces bacteria manufacture an enzyme, streptokinase, which is also able to activate plasminogen.

The best known human PA is urokinase, which is synthesized in the cells which line the kidney tubules; urokinase is secreted into the urine from which it can be

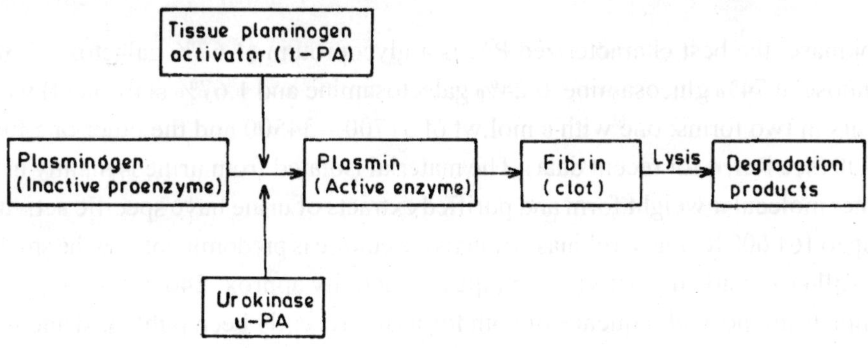

Fig. Activation of plasminogen and the break down of blood clots.

isolated. PAs which are similar to urokinase and some perhaps identical with it, are found in several other tissues and cells including the cervix and overian cancers. Enzymes of the urokinase type are termed u-PAs.

The other distinct group of enzymes, the tissue-specific PAs (t-PAs) are made by many different types of cell, including several types of cancer. Table 1 summarizes the human cells which have been reported to produce PAs. Although t-PAs are generally not well-characterized and are available only in very small quantities, they are at the focus of much medical interest since they have several advantages over the two PAs currently in clinical use and streptokinase.

Both urokinase and streptokinase are relatively non-specific in their action; they cleave plasminogen molecules wherever they encounter them, giving serious risk of haemorrhages. Furthermore, streptokinase, being a bacterial enzyme, can often produce a severe immunological reaction in the patient. Finally, urokinase is very expensive; a complete course of treatment could cost as much as Rs. 100,000 using urokinase extracted from up to 5,000 litres of urine. By contrast, t-PAs are far more specific in their action, binding strongly to fibrin clots and activating plasminogen mainly in the area where it can attack the offending thrombus. Clearly, if t-PAs can be produced on a large scale at an economic cost they should find an important place in medicine. In the meantime, despite its cost, urokinase will continue to attract attention. The major market, at present, is in Japan where it is estimated that Rs. 22,000 million worth is used each year to attack blood clots, and as an adjunct to cancer chemotherapy.

Physico-chemical properties and Production of PAs by Cell Culture

Urokinase, the best characterized PA, is a glycoprotein (1.67% galactose, 1.38% mannose, 0.74% glucosamine, 0.24% galactosamine and 1.67% sialic acid) which occurs in two forms; one with a mol.wt of 31700—34500 and the other of 54000-54700 (see Ref. 4 for recent data). The material isolated from urine is mainly in the higher molecular weight form and purified extracts of urine have specific activities of up to 161 600 IU/mg. Urokinase from tissue culture is predominantly in the smaller, but still enzymatically active form (specific activity approx. 246,700 IU/mg). The complete amino acid sequence of both forms has recently been published and there is a significant homology between the amino acid sequences of these and other serine proteses such as typsin and prothrombin. Urokinase solutions are relatively stable to brief exposures of pH 1.5 and pH 10 (at 25°C) and to a temperature of 60°C (at neutral pH). By contrast, the t-PAs are not as well characterized and have reported molecular weights in the range of 25,000 to 100,000. Recently a t-PA obtained from a melanoma cell line has been cloned and expressed in Escherichia coli, and its complete amino acid sequence (56 amino acids) deducted from the cDNA fragments. This too has sequence homologies with other serine proteases.

In 1969, Bertuk and Kwaan showed that a fibrinolytic activity was present in cultures of human kidney cells. The greatest yields came from foetal kidney and the fibrinolytic agent produced was immunologically identical t urokinase isolated from human urine. This led to the first commercial production of a PA (urokinase) by cell culture using human embryonic kidney cell.

The large scale culture of animal cells presents greater problems than encountered with microbes. In particular, animal cells grow relatively slowly with generation times of 10-40 h; they are very fragile and sensitive t(infection; their nutritionai requirements are complex and expensive sera are usually necessary; and finally, product yields are generally at ng-ug level which is 10 to 10 lower than most proteins in microbial cultures. The presence of high concentrations of protein in the medium (approx. 5 g/litre) make it difficult to isolate the desired product. Successful scale-ups of animal cell culture techniques have been achieved over the last 10 years, particularly in the production of vaccines and lymphoblast interferon and this has led to a renewal of confidence in this area of technology.

The urgent clinical need for cheap and efficient PAs, and the constant increasing commercial market has stimulated rapid progress over the last few years. Interest in urokinase as a human therapeutic drug is beginning to decline in favour of t-PAs which exhibit greater affinity and specificity for the blood clot itself. Good clinical results have been obtained from the us of t-PAs in a range of thrombotic conditions. The development of acylate PAs is also beginning to show promise. While an acyl group is present PA cannot activate plasminogen, but once inside the body the acyl groups are gradually released from the molecules. It seems that the acyl groups remain on most of the PA molecules until after they have bound to tibrin and thin plasminogen is activated mainly in the area of the clot. Two very important benefits can be expected from acyl PAs: the avoidance of increased plasmin concentrations throughout the body, with the attendant risk of haemorrhage and a more sustained plasminogen activating effect. Progress in the technology of PA production and use will be governed by:

(1) Finding suitable stable cell lines, preferably non-malignant, which can grow in suspension cultures or can be immobilized satisfactorily. Tumour cell lines (e.g. melanoma) are good producers, but whether or not regulator authorities will accept the products of a tumour cell is uncertain.

(2) The discovery of new stimulating agents for PA production and ways of derepressing the genes responsible for the synthesis of PAs.

(3) Reduction or removal of inhibitors which are produced by cells to make PAs.

(4) The efficient scale-up of cell culture to above the 1,000 litre size to give continuous PA production. Considerable biochemical engineering problems still exist, but the groundwork is well-laid.

(5) The use of specific monoclonal antibody affinity columns may aid in the approval of PAs for human use even if they are produced from tumour cell lines or genetically engineered microbial strains. Improvements in cell hybridization technology and genetic engineering could lead to more efficient PA production and thus dramatically lower the price of these enzymes.

Encouraging results have been reported on the production of melanomaderived t-PA cloned in E. coli, by researchers at Genentech and the University of Leuven and further cloning experiments are being undertaken in many other laboratories. However, the commercial viability and biochemical efficacy of genetically engineered PAs remain to be assessed.

Biofuels, Biofilters, Biosensors & Biochips

Although the true definition of Sensors, Fuels and Filters as physical entities and compositions remain constant, the utility of various microorganisms and enzymes is being exploited to mimic these definitions and practicability. In fact, the purpose is to have these properties impregnated in or discharged through the living cell and their components. Recent advances in enzyme electrochemistry have resulted in practical designs for sensors and other commercial devices. Recently, a lot of emphasis has been laid on applications of carbon monoxide utilizing microorganisms as biofilters, biosensors and biofuel cells.

Carbon monoxide (CO) is a colourless, odourless gas with a high toxicity for aerobic organisms because of its high affinity for metal ions, particularly those in the cytochrome oxidase complex involved in respiration. In man, it readily binds to the haemoglobin in red blood cells and can cause rapid arid lethal toxaemia. It is a major atmospheric constituent occurring in the unpolluted troposphere at a concentration of 0.1 ppm and in polluted urban districts at 50-100 ppm. It has been estimated that its atmospheric concentration may increase at a rate of about 6% per annum.

Although man produces small quantities of CO as a result of the oxidation of haemoglobin by microsomal haeme oxygenase, a more substantial source is the incomplete combustion of fossil fuels in domestic and industrial processes: Blast furnace gas contains 25% CO, automobile exhaust gas 0.5 to 12%, cigarette smoke 2 to 5% and cigar and pipe tobacco smoke 5 to 14%. Workers in enclosed conditions in coal mines and those involved in transport and sewage treatment can suffer increased exposure to CO. A number of processes in the chemical industries use synthesis gas and water gas that contain high concentrations of CO.

Three types of devices exploiting CO oxidizing enzymes and organisms have been described-filters, fuel-cells and sensors.

Biofilters

Incomplete combustion of carbonaceous materials results in a build-up A, unwanted CO in a variety of situations: Underground car parks, coal mines and smoke-ridden pubs would all benefit from an effective CO filter. CO is poisonous because it combines with haemoglobin in the blood and prevents oxygen transport. This principle is exploited in a biological filter in which haemoglobin is used to soak up CO. Increasing concern about the hazards of smoking has led manufacturers to be more aware of CO contents, which have been implicated in heart disease. Market pressure to reduce the intake of CO by smokers is liable to result in a reexamination of filter materials. The size restrictions of this application require either a particularly active component or the ability to turnover CO continuously. CO oxidizing enzymes and microbes may be incorporated in cigarette filters to catalyse the continuous conversion of CO to harmless carbon dioxide. A relatively heat stable enzyme with a high affinity for CO has been isolated and the oxygen problem may be alleviated by the use of insoluble non-toxic electron acceptors such as ferrocene. Chemical engineering and long-term stability problems remain to be solved.

Biofuel cells

Fuel cells affect the direct conversion of chemical energy to electricity without the thermal inefficiency associated with conventional generators. Biological fuel cells or biofuel cells have been known since 1910 when Potter, an English botanist showed that an anaerobic culture of yeast or E. coli could generate a negative potential at a platinum electrode against another electrode placed in an aerobic sterile medium. He observed a small current of 0.2 mA. Later on, interest in biofuel cells increased considerably for advancement in space biological research during the 1960s. These have been categorized as (i) Product biofuel cell, (ii) Depolarizer biofuel cell (iii) Regenerative biofuel cell. Biofuel cells exploit the catalytic properties of organisms and their enzymes to broaden the range of fuels amenable to this conversion. Additional advantages offered by biofuel cells include the ability to operate at ambient temperatures under mild chemical conditions and to use fuels present as dilute solutions. The potential of biofuel cells to use waste materials has led to the proposal that they may be valuable for space programmes converting waste into useful products such as power, oxygen or food; the Third World may also benefit from such characteristics. Specialist military needs may be met by noiseless battery rechargers which do not emit infrared radiation. The direct conversion of industrial waste, such as CO, to electricity as part of a detoxification process also has an obvious appeal.

A CO biofuel cell has been produced which uses mediators to shuttle electrons between the enzyme and a platinum or carbon anode. The hydrogen ions produced in the anode pass through the membrane into the cathode where they combine with oxygen to form water. The electrons flow through a circuit from anode to cathode producing electrical power. To date, only very small CO biofuel cells have been constructed, yielding a few milliamperes. Considerable work is required to turn these into economic power sources or detoxification plants. The reverse reaction may also be possible using electricity to generate CO from carbon dioxide. Given the normal relative value of power and CO, however, this would not be economic.

Biosensors

One of the most rapid and substantial returns for research investment in new areas of biotechnology is liable to come from the field of biosensors. Clinical applications presently command the most interest with new products likely to be based largely on either enzymes or antibodies coupled in various ways to control and/or monitoring systems. A smaller but significant market exists for process control and environmental sensors. Detection of CO is required in mines, underground car parks, road tunnels and various industrial situations. Quantification of the gas is important in combustion control systems for furnaces or engines. A CO detector can also provide the basis for a fire alarm. Many of these applications require a greater degree of specificity than is offered by conventional technology in order to minimize false alarms or readings. A further consideration is the potential market that may be available for very low-cost devices. In recent years, however, basic research on biosensors has been growing at a terrific rate because of their enormous applications. The scope includes both sensors that employ biological molecules or systems in the sensing elements and other devices that sense parameters in biological processes. Thus, many sorts of measuring devices are covered including enzymes, whole organisms, immunoelectrodes, piezo-electric crystal detectors, novel chemical sensors, optoelectronic devices, specialist applications of mass spectrometry and NMR (nuclear magnetic resonance), types based on field effect transistors and those which utilize the principles of biological fuel cells. In general, the field of biosensors is multidisciplinary, spanning fundamental and applied aspects of biochemistry, electrochemistry and electronics.

Using microorganisms

Soils and, to a lesser extent, the oceans are major sinks for CO, and about half of the total produce is oxidized by bacteria in these environments. Bacteria that oxidize CO may be subdivided into two groups; utilitarian, where the oxidation of CO provides energy for growth, and non-utilitarian where CO is accepted as a pseudosubstrate for an enzyme system normally used to catalyse a different process, oxidation, which provides no nutritional value to the organism.

There are several biotechnological applications of anaerobic utilitarian and nonutilitarian CO-utilizing bacteria, but much research is required before any of these achieve commercial status. However, it has been proposed that these bacteria may be used to convert industrial waste gases containing CO to methane or equimolar amounts of hydrogen and carbon dioxide, or to acetate.

Methane may be produced by methanogenic bacteria, either by growing *Methanobacterium thermoautotrophicum* anaerobically with CO as the sole energy source, although growth is very slow, or by the oxidation of CO with hydrogen by Methanosarcina barkerii or Methanobacterium formicicum according to the overall equation:

$$CO + 3H_2 \longrightarrow CH_4 + H_2O$$

M. barkerii is also able to oxidize CO in the absence of hydrogen according to the equations:

$$4CO + 4H_2O \longrightarrow 4CO_2 + 4H_2$$
$$CO_2 + 4H_2 \longrightarrow CH_4 + 2H_2O$$
$$4CO + 2H_2O \longrightarrow 3CO_2 + CH_4$$

This reaction can occur in an atmosphere of 100% CO.

The purple non-sulphur photosynthetic bacteria Rhodospirillum rubrum are able to grow anaerobically in the dark, using CO as sole source of carbon and energy and producing equimolar amounts of hydrogen and carbon dioxide according to the equation:

$$CO + H_2O \longrightarrow CO_2 + H_2$$

However, this is not a general property of this group of photosynthetic bacteria since strains of Rhodomicrobium vannielii, Rhodopseudomonas sulphidophila and Rhodopseudomonas palustris are unable to use CO under light or dark growth conditions.

Acetate may be produced by the anaerobic growth of a stable mutant of Butyribacterium methylotrophicum on CO alone, or by Eubacterium limosum grown anaerobically with CO as sole source of energy in a medium supplemented with rumen extract. The acetogenic clostridia, Clostridium pasteurianum, C. formicoacetlcum and C. thermoaceticum oxidize CO under anaerobic conditions. CO stimulates acetate production in these organisms and may replace pyruvate as the C-I donor during synthesis of acetate in C. thermoaceticum.

Biochips

The Japanese electronic specialists, Sharp Corporation, is about to begin basic exploratory work on the development of computers, with biological components. Using Langmuir-Bloodett thin film technology, the company will be attempting to develop peptide electrodes which can be stimulated (turned on, in effect) by enzymes.

Such a molecular switch could function as a basic element in a biocom. It is claimed that a suitable enzyme system could operate at speeds one billion time faster than the best silicon based computer chip. The chemical nature of the interaction is also likely to reduce the amount of heat generated in the chip. It would be necessary, therefore, to use an incomplete enzymatic reaction since all enzyme reactions which go to completion generate a con- amount of thermal energy.

Endorphin

A collaborative venture between researchers at Nagoya University and at Mitsubishi Chemical Industries Limited has produced a significant advance in the biotechnology of E. coli. Scientists have developed a method of producing beta endorphin in E. coli and transporting it from the cell.

Beta endorphin is a 31 amino acid peptide chemical which occurs naturally in the human body and is similar in its activities as a painkiller to morphine. The Nagoya/ Mitsubishi researchers have linked the beta endorphin gene to a gene which codes for a 22-amino acid signal peptide molecule. This signal peptide molecule is derived from an outer membrane protein (OMP- of E. coli which is normally located in the cell membrane of the bacterium. It is the signal peptide which dictates that OMP-F migrates to the exterior of the cell after it is synthesized).

When the signal peptide and the beta endorphin genes are spliced together, 99.9% of the beta endorphin produced is excreted into the culture medium. Because E. coli does not normally secrete proteins externally, the medium will be virtually protein free contrasting significantly with the environment within the cell. This means that purification of the protein should be considerably more straight-forward.

Another Japanese approach to the problem of protein secretion has been adopted by researchers at the Institute of Physical and Chemical Research (RIKEN). This involves making the cell membrane of the bacterium more permeable to protein which can then seep out into the medium. This method, however, is totally non-selective and leads to the excretion of proteins other than the desired one, and other macromoleculas components of the cell.

FIFTEEN

Cell Culture

Recently, it has become increasingly important to achieve a target in a relatively short period of time without impairing the quality. In Plant Biology too, the raising of new plants, manipulation of plants without conventional breeding mechanisms and methods, shortening germination and developmental phase of plants have stimulated the technique of cell culture under controlled and defined conditions. Several leading laboratories have achieved the transfer and expression of bacterial and foreign plant genes in plant cells. Increasing attention is now also been paid to the use of recombinant DNA technology to isolate and transfer genes governing agriculturally important characteristics such as salinity and drought resistance. Like genetic engineering, tissue and cell culture is also a research method. It helps to grow plants as isolated, undifferentiated cells. The objective of performing agricultural genetics (at par with microbial genetics) or plant cell culture in the laboratory is to generate novel phenotypes by single-cell manipulation and selection against a biochemical background while a conventional geneticist can make sexual crosses at the whole plant level. Somatic geneticists work with single cells grown in aseptic culture. Typically, small segments of an organ (a leaf, or root or stem or other) are stimulated to differentiate and divide, generating a disorganized mass of cells or callus. Callus can be placed in liquid shake cultures to disperse and suspend the cell aggregates and grow them as a fine suspension of cells. The development of large scale culture methods for plant cells made it apparent almost 2 decades ago that such cultures could be used to produce secondary metabolites just as in microbial and fungal fermentations. A considerable effort to achieve this has been made in laboratories in Japan and Germany and to a lesser extent in the United States. The achievements have been tremendous, particularly at the laboratory and pilot plant stage, and these are briefly summarized below.

Cell and Plant Tissue Culture

The deliberate culture of living cells is perhaps the oldest type of biotechnology.

Even Brewing, baking, cheese making and the production of other fermented foods such as yoghurt and bean curd all involve microorganisms in a specific media. The techniques of genetic engineering and cell fusion would not be possible without cell culture. The terms 'cell culture' and 'tissue culture' are often used interchangeably. Cell culture, however, covers the growth of any cell such as a microbe or a plant or animal. Tissue culture refers to the cultivation of a plant or mammalian cell which normally forms a multicellular tissue.

In any of these cultivation processes, the cells being cultured may be isolated from nature or may be strains improved by selection, mutation or genetic manipulation. These cells can be grown for several purposes (i) to produce specific useful metabolites such as interferon or penicillin (ii) To convert a sub. into a more useful product such as alcohol or (iii) to convert the substrate medium into more cells ('biomass') thereby either rendering it more useful (food proteins from glucose or methanol or as single cell protein). Aside from the direct combustion of biological material such as wood, cell cultures can produce methane which may prove to be very important sources of energy in the future, as the sources of fossil fuels are depleting. Cell cultures could also be a way of producing a wide range of organic chemicals.

In 1983, the first plant cell culture process was commercialized in Japan. The shikonins, a family of anthraquinones exclusively used in Japan for their antiinflammatory properties, antibacterial activity and as dyes, are found in the roots of Ko'-shikon. (Lithospermum erythrorhizon). Roots make up upto 2% dry weight of Shikonin, but the plant takes several years to reach a commercial size. The Japanese selected the culture method and in a 23-day fermentation period, cells grown in a 750 litre tanks accumulated 23% of their dry weight as Shikonins. Cell culture is now the major commercial source of these compounds. These naphthoquinones are used in Japan for the treatment of burns and skin diseases and for colouring silk.

Plant cell culture can be established from nearly every plant species: Cell cultures have been established from most plants of commercial interest. A large number of compounds are produced by plants (Table).

Ubiquinone-10 is used in Japan for congestive cardiac diseases. A group at the Japan Tobacco and Salt Corporation has selected tobacco cell lines which accumulate higher levels of ubiquinon-10 than any other plant or microorganism. They have been growing tobacco cells in fermenters of upto a 20,000 litre capacity. The production of ajmalcine by Catharanthus roseus cell cultures would also be of commercial interest as this compound is used for the treatment of circulatory diseases. Till some time ago, this industry had been isolating 4000 kg of ajmalicine annually from roots of various Rauwolfia species. Unfortunately, the instability of analytically and screened high-pressuring Catharanthus cell lines has limited use in industry.

Besides, plant cells carry out many complex metabolic processes which yield useful products such as rubber and pain killing opioid alkaloids. These cultured plant cells may provide new sources of pharmaceutical and chemical substances. Some plants are also very difficult to propagate by seeds and the in vitro cloning of single cells from which a whole plant can be raised is now a major area of research. Success in this area has been reported by Unilever using oil palms.

Table

Examples of Products Accumulated in High Levels by Plant Cell Cultures. The Productivity ($g/1\text{-}d^{-1}$) of the Cultures is Variable as Maximal Yields were Achieved After 10-30 Days

Compounds	***Plant Species***	***Yield***	
		g/1	***% dry mass***
Rosmarinic acid	Coleus blumei	3.5	15
Verbascoside	Syringa vulgaris	1.4	16
Cinnamoyl Putrescines	Nicotiana tabacum	1.0	10
Anthraquinones	Morinda citrifolia Galium mollugo	2.0-2.5	15-18
Shikonines	Lithospermum erythrorhizon	1.4	12
Benzylisoquinolines	Berberis stolonifera	2.7	7
	Coptic japonica	1.2	15
Benzophenathridines	Eschscholtzia Californica	0.15	1.7
Nicotine	Nicotiana tabacum	0.1	2.5
Ajmalicine	Catharanthus roseus	0.3	1.0
Serpentine	Catharanthus roseus	0.2	0.8
Diosgenin	Dioscorea deltoidea	0.05	3.8
Ubiquinone-10	Nicotiana tabacum	0.05	0.5

Since 1970, a lot of work has been done on the vegetative multiplication of the oil palm by tissue and cell culture, because of the economic importance of the oil palm. The Indian Government is importing a substantial amount of palmolein at the moment primarily as a cooking medium. Palmolein is obtained from the fleshy mesocarp of the fruits of oil palm tree. On fractionation, palm oil is separated into a liquid fraction, and a semisolid fraction. The former is what is called palmolein, which is then refined to make it suitable for direct consumption. Compared to mustard oil and rape seed oil, which contain a fatty acid (erucic acid) that can cause inflammation of the heart muscle, palmolein is safer. Mustard oil extracted in India contains about 45% of erucic acid. Palm oil is in great demand these days and used in the preparation of edible oils. Plantations of palm trees (Elacis quineensis) are widespread (several million hectares) in the humid tropical zones of Africa, America and South East Asia. This tree bears male and female flowers on the same plant and is always cross pollinated. A lot of work has been done in the Ivory Coast to improve production of oil by the palm tree. However, this is a tedious and time consuming process. Consequently, cell and tissue culture technique has been used to achieve the target. The plant does not grow shoots or suckers naturally and removing small twigs for grafting from its stem is not possible. Therefore, attempts have been made to resort to tissue culture in vitro. Stem fragments have not been successfully used. However, cells from callus have been raised from whatever tissue have been successfully used by workers of Unilever company (U.K.) and Orstom (France). The production of plantlets by the cloning technique is on its way to commercialization in the Ivory Coast. The Ivory Coast is known to be a rich station of oil palm trees in the world, therefore, making it possible to choose the most suitable trees and reproduce them by vegetative propagation. The Ivory Coast has also been providing clones to other countries in Western Africa and Oil-palm plantations. A cooperative agreement had already been made with several institutions in Indonesia and Malaysia for the exchange of plant materials and information on cloning techniques. According to a conservative estimate made by FAO, the average annual consumption of oils and fats was more than 20 kg per person in industrialized countries compared with about 6 kg in developing countries where oil-palm is cultivated to a large extent.

Transformation in Cell Culture

Ever since the discovery and development of the phenomenon of DNA mediated transformation in prokaryotes, this has been considered to be a useful gene transfer process for higher plants also. Earlier, evidence is available for the uptake and integration of bacterial DNA into plants after the application of donor DNA to seeds. In general, most of the foreign DNA is broken down and reutilized, and in some cases, it may escape destruction and persist in the plant tissues. The foreign DNA is either kept free or it binds to the DNA of the recipient cells. Although well established and studied in bacterial and cyanobacterial systems, such genetic manipulation, appeared almost impossible in higher organisms—may be due to the lack of facilities available for sound and more extensive studies. However, with the development of good methods to prepare nucleic acids and identify them in a population of similar molecules in the recipient cell has now made it possible to investigate the process in detail. Besides, as discussed in the following section, at the genetic level the scarcity of mutants with defined biochemical lesions precluded experimentation. Nevertheless, DNA mediated transformation has been reported in animal as well as plant cells, particularly Petunia and Arabidopsis, soyabean and carrot. Good results have been obtained with thiamine deficient mutants of Arabidopsis, and the deficiency corrected using DNA mediated transformation. Among the rare auxotrophs reported in Arabidopsis thaliana, the thiamine requiring ones are the most frequent. These are obtained by X-rays or chemical mutagenesis. The effect of the exogenous DNA (from E. colt) is associated with the concerned gene (thiamine as in mutants). This leads to the concept of the association of the bacterial information with the Arabidopsis genome. Such a close association is also indicated by the absence of significant differences between progenies of reciprocal crosses implying corrected types100% normal plants have been obtained upon crossing either corrected male or female with the mutant type.

The advantage of transformation over protoplast fusion (discussed in the next section) is that for smaller amounts of DNA, even single genes are transferred. Transformation is less likely to result in abnormal products than the mixing of two full genomes. The general requirements for DNA mediated transformation in higher plants are same as that in bacteria. These include (i) easy uptake of DNA (ii) replication of DNA in the recipient cell and (iii) expression of newly-introduced foreign DNA. However, most of the foreign DNA is rapidly degraded.

DNA uptake by pollen grains has also been attempted with sufficient success. It has been possible to introduce exogenous DNA into very young isolated pollen from which haphiod plants with a changed genetic constitution could possibly be raised. Such investigations have already been made with pollens from tobacco plant (Nicotiana glauca) using Rhizobium DNA. With recent developments of cell culture under defined conditions, techniques for obtaining haphoid cell lines, differentiation of whole plants from cultured cells, it is now becoming possible to standardize transformation in cell cultures.

Gene transfer in certain dicotyledonous plants has been simplified by the use of Agrobacterium tumerfaciens as a natural plant-host vector system. (Fig.).

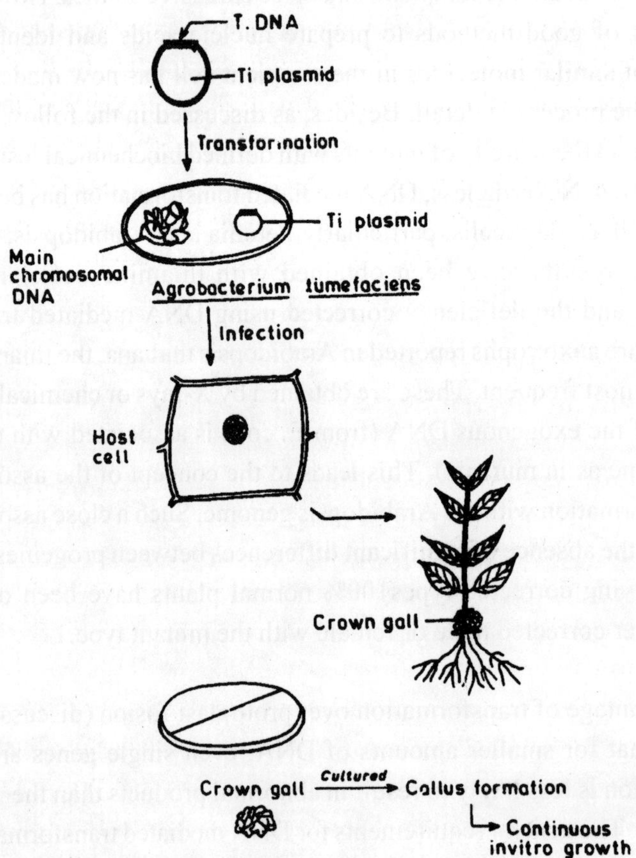

Fig. Agrobacterium tumefaciens—a natural genetic engineers.

Of the various vector systems for transformation of plant cells, the plasmids harboured by Agrobacterium species, and particularly the Ti plasmid of A. tumefaciens are the most promising. However, there is one major drawback with Agrobacterium: It does not infect monocotyledonous plants such as cereals. An obvious aim then for researchers in this field is to find out why Agrobacterium is limited to dicotyledonous species, and one approach to achieving this aim is to elucidate the nature of host-bacterium interactions before and during infection.

Agrobacterium tumefaciens transforms plant cells by transfer of the Tregion (or T-DNA) of the Ti plasmid to the host genome where the genes in the T-DNA are expressed. The successful transformation of a host requires several steps:

recognition of a suitable host, by the Agrobacterium, attraction to and invasion of the host cells, excision of the T-DNA (via the formation of a circular subgenomic molecule) and integration of the T-DNA into a nuclear chromosome. In addition to the T-DNA, a further region of the plasmid is essential for successful transformation. This is the vir (virulence) region which specifies a number of functions in the infection and transformation process. Genetic analysis of the vir region indicates the presence of seven, or possibly eight loci, labelled A-G (the eighth locus, if present, is called the O-locus), and there is some evidence that A and G are regulatory genes, since mutations in these loci prevent expression of any of the vir genes. The A locus is constitutively expressed, while expression of the other loci is induced by the presence of the host.

A detailed study of the role of the host in activating the vir region has recently been made. A series of recombinant plasmids have been constructed in which the gene for ß-galactosidase was placed adjacent to the promoter sequences of the vir genes. Expression of each vir gene could then be assayed simply by assaying for ß-galactosidase. Co-incubation of Agrobacterium with leaf discs, protoplasts or suspension cultures of tobacco resulted in large increases in the levels of expression of all the loci in the vir region of the Ti plasmid, with the exception of the constitutively expressed A locus. Actual contact between the bacterium and the host was not needed, since separation of the Agrobacterium from the plant cells by a dialysis membrane did not abolish the induction. Neither was a signal from the bacterium necessary to elicit the correct signal from the plant: suspension cultures, protoplasts and tissue explants, cultured in the absence of Agrobacterium, all produced a factor (or factors) which was subsequently able to induce expression of the vir genes. Metabolic inhibitors reduced the capacity of the host cells to induce virulence considerably. This data suggests that the inductive factor is a normal product of the metabolism of susceptible host cells (i.e. wounded and/or cultured cells of many dicotyledons). The exact nature of the metabolites is not clear. It has also been shown that Agrobacterium attatches to the host at sites specified by protein molecules. The proteins are definitely on the cell surface, since they are readily removed by brief treatments with proteases. Further, the proteins are not presented in non-susceptible plants. Recognition of the host is partly regulated by genes in the main genome of the Agrobacterium, but the vir region of the plasmid may also be involved, since there is some evidence that some of the vir loci specify or help to specify the host range. Finally, the specific role of one of the vir loci in the transposition of the T-DNA has also been demonstrated recently.

So, there is a steady progress in understanding the interaction between *Agrobacterium tumefaciens* and its hosts. However, we are still a long way from achieving a complete description of the process at the molecular level, and it is this which is required for logical and systematic attempts to modify the make-up of non-host plants, including cereal crops in order to make them susceptible to Agrobacterium-mediated transformation.

Raising Mutants in Plant Cell Cultures

Tissue culture has greatly facilitated the techniques of microbial genetics applicable to higher plant cells. Genetic variability may be induced in a large, homogeneous population of plant cells by exposure to chemical or physical mutagens. It is possible to prepare suspensions of cell cultures of higher plants and keep them in this state for as long as wanted. Consequently, each single cell may be as good as a potential bacterial cell for induction and isolation of mutants and variants.

The importance of cell culture depends upon the development of methods for isolating a wide variety of cultured cell strains with characteristics different from those of cells in the original cultures. To isolate such variant cell strains, well-known techniques in microbial studies will have to be applied to cultured plant cell systems.

One especially important method is the application of mutagens to plant cells to increase the frequency of variant strains in populations of cells so that they may be easily identified and selected. However, work in several laboratories has shown that many chemicals that are potent mutagens in microbial systems are only marginally or not at all effective on cultured plant cells. The search for chemicals effective on a wide variety of plant cells is, therefore, an important aspect of plant somatic cell mutant isolation.

Whether or not a compound is mutagenic depends on the expression of an easily seen characteristic in substantially different in parent cells from the variants derived from them after treatment with the agent. Resistance to a nucleic acid precursor analog called 6-azauracil appears to be a useful characteristic for such work. Parent cells growing in culture are highly sensitive to this compound, but variants that are resistant to it can be found and the difference is easily assayed.

It has been demonstrated that this difference results from an enzyme deficiency in the variant cells. They lack an enzyme (uracil phosphoribosyltransferase) that actually kills the cells. Strains of cells from two different species of plants (diploid Haplopappus gracilis and haploid Datura innoxia) resistant to this analog have been isolated at Riverside California and shown to lack the enzyme.

Some ferns such as *Todea barbara and Osmunda cinnamomea and higher plants including Nicotiana tabacum* have also been helpful for such studies. Many auxotrophic mutants requiring vitamins and amino acids have been raised in T. barbara. Similarly, mutants requiring amino acids, vitamins, and purines have been successfully isolated in N. tabacum. Mutants resistant to antibiotics including Streptomycin and base analogs (such as 5-bromodeoxyuridine and 8-azaguanine) have also been isolated. Mutants of this type have also been isolated in higher plants. Resistant and auxotrophic mutants are important as markers in analyses of genetic linkage, complementation and recombination. Similar results have been obtained using certain metabolic inhibitors. Inhibitors are the substances that interfere with a biological processes, restraining the natural activity of a particular function or metabolic activity of an organism. Inhibitors play an important role in understanding different metabolic processes. By using different metabolic inhibitors, cellular metabolism can be blocked at a specific site and by examining the consequences of such blockages, the nature of the pathway and the factors controlling the process can be studied. With the help of inhibitors, a particular step in DNA, RNA or protein synthesis can be controlled and the metabolic process of an organism thus manipulated. Such mutants may possibly be useful in selecting products of fusion between cells of different genetic backgrounds. In addition to solving several fundamental biological problems, mutants which have been recovered from cultured plant cells are of real biotechnological importance and may be the basis of solving several nutritional, agricultural and industrial problems.

Some interesting results have also been achieved with Pseudomonas tabaci the cause of wildfire disease of Tobacco. Tobacco cells resistant to wildfire toxin have been recovered among a mutagenized haphoid cell population. The successful recovery of tobacco plants resistant to wildfire disease reveals the potential for selection of toxin resistance as a generalized procedure for generating disease resistant varieties. Direct selection for resistance to a pathogen may also be possible in culture.

Phage mediated gene transfer

In recent years, bacteriophages have been employed to mediate gene transfer in higher plant cells. These viruses may serve to protect the donor DNA from nucleases and facilitate its entry into plant cells. Bacterial genes may also be carried by specially constructed transducting phages. The transducing phages X and 0 80, containing bacterial genes for galactose and lactose utilization respectively, have been used to inject callus tissues (unorganized mass of cells joined together). Related experiments have also been conducted on suspension cultures of sycamore cells. This, phenomenon of gene transfer and subsequent phenotypic expression is known as Transgenosis. The use of bacteriophages as vectors to transport selected bacterial genes into higher plant cells appears promising for both fundamental and applied research.

DNA viruses have been successfully adapted as vectors in animal and bacterial systems. However, the plant DNA virus studied so far in detail, the cauliflower mosaic virus does not replicate in a culture system. Although, recent reports indicate that Ti plasmid could be developed as a useful vector for transforming monocotyledonous plants such as maize, no functional system, extended to include other plants, has been developed. However, vector systems have been developed that allow for the introduction of exogenous DNA into protoplasts at very low frequencies.

A remarkably simple genetic system for the study of DNA multiplication and gene expression in plants is provided by DNA plant viruses. These viruses have only about 6 genes that are believed to be regulated in the same way as other plant genes. The DNA replicates in nuclei and may be associated with nuclear proteins (histones) in the same way as plant genetic material. Thus, the virus provides a small-scale, readily manipulated model for gene expression.

One DNA plant virus, the cauliflower mosaic virus, has received more attention than the others. Its biology has been intensively studied during the last few years, and the DNA from two strains of this virus has been completely sequenced. At Davis, for example, an isolate has been found to have 8,031 pairs of nucleotides in its circular chromosome. This sequence of nucleotides and other information has been used to construct a physical map of the virus chromosome. Of the half-dozen genes of this simple virus, one (gene II) is not essential for reproduction, enclothement in protein, or cell-to-cell movement. Recent work at Davis indicates that this non-essential gene is probably involved in insect transmission of the virus in nature. These dispensible regions of the virus chromosome provide sites for insertion of foreign DNA, which is carried into the plant and replicated along with the DNA of the infecting virus.

Another region of the cauliflower mosaic virus chromosome has been identified as being responsible for the severity of disease. This region is gene VI on the physical map. (Fig.). The other five genes appear to have little if any effect on the induction of symptoms. A single change in gene VI can have a profound effect on disease expression: in one case, insertion of 12 base pairs at a particular location in gene VI almost abolished disease. Plants infected with this mutant of the virus show very mild symptoms and grow at the same rate as healthy plant. Information of this sort may enable the investigator to control disease expression, if portions of the viral chromosome are eventually used as a recombinant DNA vector for plants.

Cauliflower mosaic virus has been used as a vehicle to reproduce foreign DNA in plant cells and to carry this foreign DNA from cell-to-cell throughout

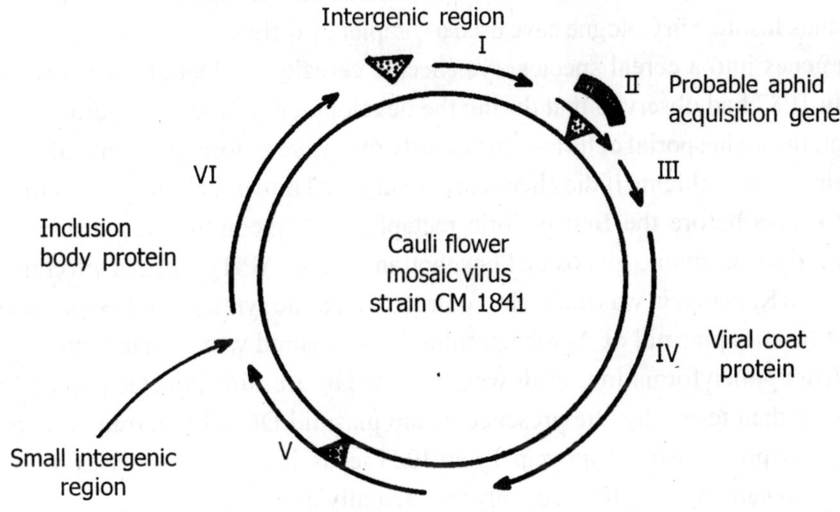

Fig. Physical map of circular choromosmes of CMV

the entire plant. However, not enough foreign DNA to bring about useful transformations of plants can be inserted into the viral chromosome. This limitation appears to be related to the low capacity of the viral particle to accommodate additional DNA. The assembly of DNA and coat protein to form viral particles seems to be necessary before the DNA moves from cell-to-cell in the plant. It does not appear, therefore, that the virus in its present form will be useful as a recombinant DNA vector. However, this virus will probably play an important role in defining the biological activity of those sequences involved in replication and expression of DNA in plants.

Cell Culture, gene transfer and seed banks

Two of the simplest mechanisms for the transfer of genes into plants have been the use of recombinant vectors based on Ti plasmid of Agrobacterium tumefacians (as already described in the previous section) and the direct delivery of foreign DNA into the protoplast, where membranes have been rendered leaky by use of polyethylene glycol. However, this method is unsuitable for cereal species. Delivery of a recombinant Ti plasmid requires infection by Agrobacterium. Under natural conditions, monocots including cereals are not susceptible to Agrobacterium although limited infections of monocots including maize have been achieved under laboratory conditions. Monocot cells have, however, been successfully transformed by DNA delivered directly to protoplasts. But to date, it has been impossible to regenerate transformed whole plants from the transformed cells. Groups of workers at the Max Planck Institute in Cologne have used a completely different approach to transfer foreign genes into a cereal species, rye (Secale cereale L.). Dela Pena, Lorz and Schell in 1987 had observed that during the development of the male germ line of rye plant, the archesporial cells are particularly receptive to foreign chemicals such as caffeine and colchicine if the chemicals are injected into developing floral tillers about 14 days before the first meiotic metaphase. A recombinant plasmid was constructed so that amino glycoside Phosphotransferase (APH) gene, which confers resistance to Kanamycin was under the control of nopatine synthase (NOS) promoter (derived from Ti plasmid of Agrobacterium). The plasmid was injected into floral tillers. After pollen formation, seeds were obtained by crossing injected plants. The seeds were then tested for the presence of the plasmid DNA by germination and growth in the presence of kanamycin. From 3023 seeds, just seven seedlings showed resistance to kanamycin. Of these, only two actually developed.

Information regarding the integration and remittance of foreign DNA within plant cells is limited to the data derived from whole plants. Work on cultured plant cells has not developed enough to allow the application of sophisticated techniques of molecular biology. Three possibilities exist for the mode of inheritance. (i) The exogenous genome is integrated into the eukaryotic genome, thus falling under the host's replication control, (ii) The second alternative involves failure of integration, although the donor genome is maintained as a cytoplasmic plasmid. (iii) The exogenous DNA that enters may be active for a very short time, but due to lack of integration, the autonomous state is lost and vanishes. Presently, many cases of gene transfer in plants are of this type. Perhaps modification of the plant genome may allow integration and subsequent inheritance. In fact, the lack of suitable mutability or the absence of appropriate selection systems for such mutants hampers the future development of more efficient gene transfer systems. The future development of precisely characterized biochemical mutants in cultured plant cells, thus, becomes a major prerequisite for the successful extension of intercellular gene transfers.

The National Institute for Agro-Biological Resources, an organisation associated with the Ministry of Agriculture, Forestry and Fisheries in Japan are backing a research plan to insert the phaseolin gene from the kidney bean into cells from tobacco or petunia plants.

Phaseolin is the major storage protein of the kidney bean and it is hoped that high protein varieties of the hybrid plants can be produced by using an antibiotic resistant Ti plasmid to transfer the gene.

Tobacco and petunia plants are not perhaps, the most likely vehicles for the production of a new type of high protein food. They are, however, among the best understood species from a genetic point of view and the proposed studies are likely to be merely stepping stones to work on more conventional food crops.

The Japanese project is not unique; researchers at the Max Planck Institute in the Federal Republic of Germany have already cultivated high protein hybrid seeds of the type envisaged by the NIABR scientists.

Any seed varieties developed by research could be incorporated into the seed banks which MAFF are currently upgrading with the help of a Y 182 million ($ 0.78 million) grant. MAFF currently holds 99,000 seed varieties in their various laboratory establishments which compares unfavourably with the 316,000 held in the US and the 3,00,000 in the USSR. A computerised catalogue of the seeds is being developed, enabling companies and researchers to readily locate particular plant characteristics.

Among the companies which are likely to take advantage of the new seed catalogue and the sales facility which is also being offered are Kyowa Hakko Kogyo, Kaneko Seeds Company, Nakajima Tenkoen Fruit Nurseries Company and Sumitomo Corporation. The four have recently reached a basic agreement on mutual cooperation in the development of new seed varieties.

While most research and development work in the cooperative project is likely to be done at the Kyowa Hakko laboratories, the other companies all have important roles to play. The two seed companies will be offering stock seeds for development and will be marketing vegetable (Kaneko) and fruit (Nakajima Tenkoen) seeds when they are developed. Sumitomo has a coordinating role and will be responsible for marketing efforts outside Japan, in Korea and China.

Protoplast and Cell Fusion

A protoplast is a "naked" cell or cell without a cell wall and is a very important intermediate in the cell fusion process. A Plant cell wall is decomposed and removed by enzymatic treatment using lytic enzymes. A huge population of protoplasts can be isolated from a single leaf; yields are typically 2-4 million protoplasts per gram of leaf tissue. The development of techniques and procedures causing isolated protoplasts to reform their walls proliferate and eventually regenerate into whole plants is very important facet of plant biotechnology.

Work done on lettuce protoplasts has revealed interesting results. Isolated lettuce protoplasts can be induced to reform their walls and divide to form unorganized clumps of cells which are regenerated under appropriate conditions to form full plants. Theoretically, all plants regenerated from a single lettuce leaf should be identical in physical and chemical properties. Observations of regenerates, however, reveals the interesting result that many of them are different from the source plant and from one another. Similar observations have been .eported in potato protoplast regenerated plants, which have otherwise proved to be stable over many generations.

These variations are not only observed with protoplasts but also with suspension cell cultures. A useful term, 'somaclonal' variation has been coined to describe this phenomenon which is more the rule than the exception ('soma'-refers to somatic tissue and 'clonal' as differences observed within clones). Although the reasons for genetic variation among regenerated plants are not fully understood, it seems to be possible to obtain desirable improvements in crop plants as well as in horticultural characteristics. Attempts are being made to expose the protoplasts to specific selection pressures in the protoplast regeneration medium to isolate all types of desirable regenerates including disease resistant and cold (frost) and heat tolerant plants.

The major difficulty in breeding crops is incompatibility. The use of protoplasts can remove this restriction. Protoplasts from any two plants, regardless of the species, can be fused when protoplasts come into contact with other protoplasts. Fusion of cell membranes is induced and then the cytosols of both protoplasts are mixed. The resulting fused protoplast regenerates the cell wall and retains the genetic properties of both protoplasts. Thus, a 'hybrid cell' is generated by cell fusion. In the entire process of cell fusion described, enzymatic control is very important. For example, insufficient treatment causes a decrease in the fusion frequency. Excessive treatment, however, causes a decrease in the regeneration of the fused cells. Most common fusion inducers include sodium nitrate, polyethylene glycol, high pH and calcium. A conventional method, for protoplast estimation especially in microorganisms is based on the resistivity of osmotic pressure. Protoplasts are swollen and disrupted in a solution of low osmotic pressure, while normal cells grow even under low osmotic conditions. In contrast, both protoplasts and whole cells grow in a solution of high osmotic pressure. There, the difference of the growth rates in low and high osmotic pressure solutions is used as a measure of protoplastization. This method, however, involves a colony counting process and requires more than two days. Recently, Japanese workers have devised an electrochemical method in microbial cells based on the principle that the electric properties change during protoplast formation. They have estimated protoplast concentration and cell fusion using a combined electrode composed of a platinum anode, a platinum cathode and a saturated calomel electrode.

Protoplast and cell and tissue culture have already begun to influence the breeding and vegetative propagation of superior rootstocks and woody perennial trees for an efficient forestry system and urban planting. Recently, hard-to-root biomass species such as Douglas-fir, white spruce and Jack pine have been cloned through micropropagation. Similar methods are being tried with valuable root-stocks of Prunus and Pistacia.

Now that cell suspensions of many woody species can be established, application of recombinant DNA technologies to protoplasts is being evaluated. Attempts are being made to fuse such protoplasts, induce phase changes and improve upon the biosynthetic potential of cells, especially for the products of photosynthetic and the building blocks of proteins. The cell culture is being expanded to cloning redwood trees where time has always been a problem for raising plantlets. Between germination (or planting) and harvest as a renewable source of wood, a redwood must survive and grow in a minimally managed environment for thirty to forty years. Red woods are important now not only in California (in the native range) but also in places like France and New Zealand where red wood is gaining importance as an exotic plantation tree.

Another important and valuable application of cell and protoplast culture is to isolate strains tolerant or resistant to herbicides. Every herbicide is restricted in use by the number of crops it damages or kills. Tolerant mutants of various plant species could broaden the usefulness of currently available herbicides.

In theory, the protoplast fusion technique opens up the potential "gene pool" of a plant to every other living cell. There are, however, a few technical problems. The major problem is that the capacity of a cell culture to regenerate plants normally diminishes over time. This presents difficulties in cell selection techniques, because by the time the selection and testing stages are completed, the selected cells may fail to regenerate. These need to be resolved to use the protoplast and somatic cell fusion technique potentially valuable for biotechnological exploitations such as improved and desirable chemical products and biochemicals from plants.

Thinking about Unusual Plant Biotechnology

Higher plants are still an indispensable source of certain drugs and raw materials despite all the progress made in the synthetic organic chemistry. Several commercial products including Morphinane alkaloids, cardiac glycosides, essential oil, and original rubber are obtained from agriculturally grown plants. Interrupted supplies due to several reasons including political and natural calamities may hamper the needs of the society. Therefore, a continuous production strategy had to be devised and plant cell culture has become an industry in itself. It is quite possible that such cultures would one day replace whole plants as a source of useful products. Unfortunately, the hopes of industry to produce commercially important compounds by large scale fermentation using plant cell cultures have not yet been realized. It will still take somtime to be exploited on a massive scale. In fact, Plant tissue culture is an art and a perfection in an art is never achieved by a short cut. However, it is certainly going to be a very important tool in the area of biotechnology.

The commercially most important plant products have either not yet been found in cell cultures or are only present in a very small concentrations. For example cardiac glycosides are found only in morphologically differentiated cell cultures with concentrations of only 1 mg per litre (suspension). There are also some reports of morphine and codeine being found in cell cultures of Papaver species (about 1.5 mg per gram dry weight). Such concentrations are too low to be commercialized. Although plant cell cultures are able to accumulate sufficiently high concentrations of secondary metabolites to make production feasible, the compounds at these levels are not of great commercial importance.

There are a few technical problems in quantitative recovery of useful products in cell culture.

1. Several metabolites are released in cell vacuoles, instead of in the medium, and are thus locked.
2. The cells must remain viable for long periods to release the product continuously. This problem could be solved to some extent by cell immobilization.

Why have none of the best-known plant products entered the list of compounds produced in high concentrations by plant cell cultures? The cardiac glycosides of Digitalis are found only in morphologically differentiated cell cultures. Even in these rather slowly growing cell cultures levels of only I mg cardiac glycoside I have been found. There are some reports of morphine and codeine being found in cell cultures of Papaver species, but the results could not be reproduced by other leading tissue culture laboratories. Even if the reported yield of 1.5 mg codeine/g dry weight is

Table

Initial Industrial Applications of Immobilized Enzymes and Whole Cells in Japan

Enzyme or Cells	*Carrier*	*Method of attatchment*	*Company*	*Year*
Aminoacylase -	DEAE-Sephadex	Adsorption	Tanabe Seiyaku	1969
E. coli (aspartase)	Polyacrylamide	Entrapment	Tanabe Seiyaku	1973
Brev. ammoni-agenes (fumarase)	Polyacrylamide	Entrapment	Tanabe Seiyaku	1574
Aspartase ammoni-alyase	Duolite A7	Adsorption	Kyowa Hakko	1975
Glucose isomerase	Doulite A7	Adsorption	Kyowa Hakko	1976
Glucose isomerase	Amberlite IRA 904	Adsorption	Mitsubishi Kasei	1976
Streptomyces sp. (glycose isomerase)	Modified anion exchange resin	Adsorption	Denki Kagaku	1976
Penicillin-acylase	Celite	Adsorption	Toyo Zojo	1976

reproducible, the content would have to be increased several fold to become industrially attractive. We have to realize that the commercially most important plant products have either not yet been found in cell cultures or are only present in disappointingly low levels. Published data suggests that plant tissue culturists are ill-advised to forecast the commercial impact of plant cell cultures as sources of pharmaceuticals in the near future. This need for caution is further supported by the following example. Rosmarinic acid is produced by cell cultures of Coleus blumei with a yield of 4 g/l. And this fast-growing, fine suspension should not cause great technical problems for large-scale production. If rosmarinic acid was of any commercial interest, it would seem to be an ideal candidate for production by plant cell cultures. Even so, would industry produce this compound via plant cell cultures? Probably not, as rosmarinic acid is a constituent of several medicinal plants grown by farmers and as such might still be cheaper to isolate from the field-grown material than from high-producing Coleus cells. This would be especially true today as large investments would be necessary if plant cell cultures were employed for the commercial production of useful plant metabolites.

In summary, plant cell cultures are able to accumulate sufficiently high concentrations of secondary metabolites to make production processes feasible, but unfortunately the compounds accumulated at these levels are not of great commercial importance. This raises the question of how these highyielding strains have been established in the first place. Some of them are simply routine cultures that are spontaneously highly productive. The anthraquinone-producing cell lines of Morinda and Galium species (Table 11. 1), as well as the rosmarinic-acid producing Labiatae are examples of these. More often, however, cultures contain only low levels of specific metabolites. By biochemical selection, and/or analytical screening, one can isolate, variants with higher productivity from low-producing wild type cells. The productivity of low producing wild type cells and of the selected variants can often be further enhanced by transferring the cells into a production medium. The claim that the chances of establishing high-producing cell lines are significantly improved if one starts from selected high-producing plant materials has not been convincingly proved. The most disappointing fact, however, is that despite the establishment of hundreds of cell cultures, despite all screening programmes, despite all media variations, many, and unfortunately the majority of the interesting compounds cannot be expressed in cell cultures at present. The difficulty of expressing each compound in an undifferentiated rapidly-growing cell line is evident, if one realizes the complexity of an intact plant. Many of the most important plant compounds are formed or accumulated only in specialized cells. Cardiac glycosides of Digitalis are mainly found in leaf cells; quinine and guinidine are accumulated in the bark of Cinchona trees; tropane alkaloids are synthesized in the roots of some Solanaceae and accumulated mainly in the leaves. During plant growth, the cells become not only morphologically but also chemically specialized. In plant cell cultures, however, the morphological specialization is suppressed and hence most of the chemical specialization is also lost. If one induces morphological differentiation in a cell culture by changing the hormone composition of the medium, the metabolites lost during dedifferentiation would reappear. However, the induction of morphoiogical structures is normally a rather slow process and would, therefore, increase fermentation costs enormously.

If this were the whole story, one should not waste any more time in thinking of plant cell cultures as a source of useful natural products, but fortunately they have much more to offer. For example, plant cell cultures often have not yet been found in the mother plant. This has prompted some research groups to use plant cell cultures not as a source of known compounds but as a source of new pharmacologically active compounds. At the moment, however, there is no indication that such a programme will be of greater commercial success than a screening programme with intact plants. The Schering Company stopped its programme of analyzing plants traditionally used in folk medicine for active principles several years ago. One should, therefore, be very cautious in urging the commercial value of plant cell cultures as a source of new biologically active-compounds.

There are several products and compounds which are normally not produced by whole plants but are produced in cell cultures (Table). Commercially, more important than the observation of new compounds in plant cell cultures is the fact that even nonproducing cultures can be used for biotransformation reactions. It has already been mentioned that cell cultures of Digitalis do not synthesize the cardiac glycosides. However, cell lines of Digitalis lanata are able to perform a specific reaction in this pathway, namely the 12-hydroxylation of digitoxin to digoxin. Digitoxin is a byproduct in the commercial production of digoxin from the leaves of Digitalis lanata, and pharmaceutical companies have tried to find chemical or microbiological ways of transforming digitoxin to digoxin. The group of E. Reinhard and A.W. Alfermann at the University of Tubingen has screened strains of Digitalis lanata cell

Table

Examples of Compounds Hitherto Detected Only in Tissue Cultures and not in the Corresponding Intact Plant.

Compounds	***Cell Cultures***
Pericine	Picralima nitida
Pericalline	
Hinokiol	Thuja occidentalis
Ferruginol	
Pleiocarpamin	
Akuammilin	Catharanthus roseus
Dihydrocondylocarpine	
Vomilenin	Rauwolfia serpentina
Paniculid A	Andrographis particulate
Tarennosid	Gardenia jasminoides
Rutacultin	Ruta graveolens

cultures for high hydroxylation activity. When they used α-methyldigitoxin as a substrate, this compound was completely transformed to α-methyldigoxin, a widely used pharmaceutical. The process has been scaled up successfully, and the participating pharmaceutical company now has to decide the commercial future of this process.

Vinblastine and vincristine, bisindole alkaloids of Catharanthus reseus are used to fight leukaemia. Other bisindole alkaloids are of potential interest. J.P. Kutney and co-workers at Simon Fraser University, Vancouver, have shown that chemically synthesized 2 3, 4-dehydrovinblastine was metabolized to leurosine, catharine, and several other bisindole alkaloids by a cell line of Catharanthus roseus which accumulated no indole alkaloids.

These two examples demonstrate that even non-producing cell cultures may be useful for biotransformation reactions. However, plant cell cultures will have a commercial impact in this field only when they are not competing with processes based on microorganisms. Plant cell cultures should be successful when they are used for plantspecific reactions that are not easily performed by microorganisms. To achieve a commercial breakthrough in this area, it is important to demonstrate that more commercially interesting reactions can be performed by plant cell cultures. For example, the enzymatic conversion of reticuline to salutaridine as a step in the chemical synthesis of morphinane alkaloids, or the epoxidation of hyoscyamine to scopolamine would be reactions of great commercial interest.

A lot of work is going on at GBF in West Germany, in plant cell culture in the hope of obtaining commercially important products. Cell cultures of Peganum harmala, Catharanthus roseus, Nicotiana tabacum have been used for production of serotin, indole alkaloids and cinnamoyl putrescines respectively. As in microbial system, the selection of high-producing cell lines in the case of Nicotiana and increasing the product accumulation by altering the culture media in case of Catharanthus, Peganum, and Nicotiana are possible. It would be important to understand and characterize the biochemical processes responsible for the increased accumulation of secondary metabolites. Increased synthesis may also alter several enzyme systems that directly or indirectly affect the product. For example, the increased formation of indole alkaloids and serotonin was accompanied by a 10-fold increase of tryptophan decarboxylace as observed by German workers. Therefore it may also be possible to increase the product in the cell cultures by enzyme manipulation. it may be possible to transform cell cultures in which the lack of enzyme lysine carboxylase prevents alkaloid formation with the aid of a cloned gene of a bacterial lysine decarboxylase. The future of plant cell culture will ultimately depend on a better understanding of the regulation of secondary pathways. It will be more important and commercially significant to alter the metabolic pathways in cell culture for better yield of products rather than chance selection of high producing cell lines. This might involve the team work of tissue culture experts, microbiologists, geneticists and biochemists.

SIXTEEN

Intellectual Property

An important issue has been raised due to the emergence of modern biotechnology, is the legal characterization and treatment of trade related biotechnological processes and products, popularly described as Intellectual Property. Its protection (Intellectual Property Protection = IPP) and the Rights (Intellectual Property Rights = IPR) available to protect this property have been the subject of discussion in recent years. In this connection, one may like to compare biotechnology with other technologies, the advances which are covered by the patent system and are, therefore, routinely licensed and marketed.

The term property is often found associated with physical objects only, such as household goods or land, for which ownership and associated rights are guaranted and protected by law prevalent in a country. This property is described as tangible. Intellectual property, on the other hand, is intangible and includes 'patents', 'trade secrets', 'copyrights' and 'trademarks'. The rights to protect this property prohibits others from making, copying, using or selling the proprietary subject matter. Under biotechnology, one of the most important examples of intellectual property is the processes and products, which result from the development of genetic engineering techniques through the use of restriction enzymes to create recombinant DNA. The characterization of these research results as intellectual properties encourages industries to allocate labour, research and development (R&D) units and funding to facillitate the production of commercially marketable items. Due to these intellectual properties, many legal and public policies, which are impediments to biotechnological research are also being challenged and are, therefore, undergoing changes. This is understandable, because if public policies do not allow the development and commercial use of an intellectual property, no industry would like to invest funds in this research.

Another example of intellectual property is the development of crop varieties, which are protected through 'plant breeder's rights' or PBRs (PBRs are available in developed countries, not in India). Through PBR, the plant breeder who developed a variety enjoys the exclusive right for marketing the variety, although use of the variety for further breeding or for replantation of seed saved by a farmer (farmer's exemption) is permissible. More recently, however, utility patents for genetic materials, both plants and animals, have also been allowed in some countries, so that the patented material can neither be used for further breeding, nor will the farmers be allowed to save and use the seed for cultivation without paying a fee to the patent holder. Similarly, if patents on superior animal breeds are allowed, a dairy farmer will find that a calf born to his hybrid cow will belong to the company, which sold him the animal. These intellectual property rights affect (favorably as well as unfavourably) the conservation, availability and use of plant and animal genetic resources (PGR and AGR). The impact of IPR on the availability of genetic diversity will also be witnessed. There are also arguments against patenting life forms like transgenic animals and plants, because these patents will work as impediments in free exchange of genetic materials for improvement of crops and livestock. IPRs may also adversely affect (i) food security; (ii) use of evolved agricultural practices; (iii) biological diversity and ecological balance; and (iv) the livelihood of the poor in developing countries.

Since a better understanding of intellectual property by research scientists and university/institute administrators will increase the pace of research for technological development in biotechnology, two separate chapters are devoted to this subject in this book. In the present chapter, we will discuss the various intellectual properties relevant to developments in biotechnology. In the next chapter, we will discuss the specific problem of patenting of biological material, the problem related with deposit of material for obtaining the patent claim and the international conventions in this connection. We will also discuss in these chapters the legal requirements to enforce intellectual property rights, since they will work as incentive for further advances in biotechnology.

Intellectual Property Rights

Patents

The granting of special exclusive rights (for trading new articles) has been a practice to encourage innovations. As an example, monopoly rights (only to inventors) were granted in some countries of Europe, as an incentive to develop new articles that would be of benefit to the society. Under U.S.A. law, a patent means grant of "right to exclude others from making, using or selling" an invention for a 17 year period. Patents are usually allowed for a specified period. In India, The Indian Patents Act of 1970 allows process patents, but no product patents for lands, chemicals, drugs and pharmaceuticals. If no product patent is available for a product, the same product may be manufactured by an alternative process (which is cost effective) without any infringement of the patent granted for the process. The duration of the patent in India is five years from the date of grant of patent or seven years from the date of filing the application, whichever is less. In case of a dispute, the burden of proving that the product has not been manufactured by an alternative process, lies with the original patent holder, who has lodged the complaint.

A patent, however, is subject to some restrictions, since other Federal or State Laws need to be followed. Following are some examples. (i) A patented pharmaceutical, in USA, is subject to regulatory purview of Food and Drug Administration (FDA) before its clinical use is allowed; (ii) Under the U.S. Federal Insecticide, Fungicide and Rodenticide Act, genetically engineered microbial pesticides require Environment Protection Agency (EPA) permits, before they arc released; (iii) Use of excessively noisome genetically engineered inventions might also be curtailled under local nuisance ordinances.

Patents are granted after submission of an application, fulfilling certain statutory requirements. When granted, a patent is also published (in USA, patent applications are published weekly by Patent and Trademark Office = PTO), so that other competitors may try to improve the patented invention, thus further advancing the technology. Progress is actually made through dissemination of ideas and only market ultimately determines which patented inventions are commercially successful.

Patentable subject matter should generally meet three requirements: utility, novelty and statutory subject matter. It is believed that although utility of a new invention (biotechnological or otherwise) can be proved (in some cases, even utility is questionable;), there have been problems in proving novelty and in defining the limits of statutory subject matter as applied to biotechnological products. There have been several court cases in this connection, which were decided both for and against the patents. Infringement of a patent is often found when an accused object is physically different, but legally equivalent to the subject matter claimed. For instance, if a Pseudomanas strain is prepared for remediation of oil spill, then an analogously prepared Bacillus strain will infringe the patent, but if a different mechanism is involved (although for same purpose), the infringement may be avoided. Patents for life forms are discussed in greater details in the next chapter.

Trade secrets

Trade secrets, often include private proprietary information or physical material that allows a definite advantage to the owner. This can be illustrated by the popular example of Coca-Cola brand syrup formula. Trade secrets in the area of biotechnology may include material like (i) hybridisation conditions; (ii) cell lines; (iii) corporate merchandisin plans; or (iv) customer lists. Unlike patents, trade secrets have an unlimited duration and, therefore, may not be required to satisfy the more difficult conditions laid down by law for patent applications. Disclosure of a trade secret and its unauthorized use can be punished by the court and the owner may be allowed compensation. However, if a trade secret becomes public knowledge by independent discovery or other means, it is no longer protectable.

However, due to large degree of research component in biotechnology, there is increased risk in maintaining trade secrets. The results are published and discussed in conferences and disclosed due to exchange of graduate students, thus making it difficult to maintain secrecy. Despite this, in some cases reliance on trade secrets is more prudent than the patents, which sometimes may become outdated before the patent is granted after the application is filed (it takes about 2 years in USA and at least 5 years in India).

Copyright

The best example of copyright is the authored and edited books, or audeo and video cassettes, which can not be reproduced without the permission of the person (author, editor or publisher), who holds the copyright. While patents and trade secrets get the protection for the basic idea, expressed or not expressed in writing, the copyright is possible only on the expressed material (printed, painted, tape recorded, video recorded or expressed in any other form). Copyright rules may be modified from time to time as has been done to allow copy writability to computer software.

In biotechnology, the copyright may cover DNA sequence data which may be published. However, an alternative sequence coding for same protein may be prepared using wobble in the genetic code, so that the copyright is not infringed. Computer data bases and photomicrographs of DNA instruction manuals related to biotechnology can also be copyrighted. The protection to copyright is limited, however, since although one may not be allowed to photocopy the present book on biotechnology due to copyright, but the ideas given in the book can be used for any purpose, whatsoever. This is not so in case of patents and trade secrets. An instruction manual can be copyrighted and also protected as a trade-secret, but can not be patented.

Trademarks

A trademark is a word or symbol "adopted and used by a manufacturer or merchant to identify his goods and distinguish them from those manufactured or sold by others". In biotechnology research, laboratory equipment bears trademarks that are well known to workers in this field. Certain vectors useful in recombinant research are also known by their trademarks. Although ordinarily this intellectual property may not raise any legal issue, it does make a part of intellectual property of any industry.

Choice of Intellectual Property Protection

Biotechnological products can be protected by any one of the four intellectual property rights discussed above. These properties can be bought, sold and licensed like any other property. Selection of the most appropriate mode of protection is a business judgement based on several factors which differ from case to case. These include the following : (i) Pace of technological development: if it is rapid then a trade secret approach may be preferable to patenting; (ii) Associated costs : although the cost of secrecy may exceed the cost of obtaining; a patent, but it may be more expensive to enforce the patent, so that one will have to decide, whether trade secret is a better protection; (iii) Security considerations: is may be impossible to prevent disclosure of a trade secret, so that one may prefer rights other than secrecy; (iv) Need to show patents: patents may have to be shown to investors as a measure of success. The choice of protection also depends on the nature of subject matter that need to be protected. For instance, an instruction manual can be copyrighted or also protected as trade secret, but can not be protected as a patent.

IPR and Plant Genetic Resources (PGR)

In most industrialized countries, crop varieties are subject to intellectual property rights in the form; of PBR (Plant Breeder's Rights). However, the present intellectual property right (IPR) systems do not promote, nor were they designed to promote, the protection of diversity of whole ecosystems or unmodified plants. However, the evaluation, use and mantenance of plants may be encouraged indirectly by the patent system, when a Constituent of the plant has a commercial value. For instance, if PBR is, extended to all plants, as is currently being discussed at the international level, this might help in extending breeding work to more crops, which in turn will contribute to the development of greater crop diversity. In this connection, it is significant that changes are being proposed by some industrialized countries to the

General Agreement on Tariffs and Trade (GATT) and the, World Intellectual Property Organization (WIPO). If these changes are made, the only form of human innovation, that are not patentable will be the informs innovations in developing countries.

At present, we do not understand fully the implications of the extension of intellectual property rights (IPR) to plant genetic materials and crop varieties. This may have far reaching impact on maintenance and preservation of plant genetic resources, because the current IPR systems reinforce the tendency of plant breeders to decrease genetic diversity. In view of this, the impact of extending IPR to plant genetic material is being studied, and discussed at the international level.

GATT (General Agreement of Tariffs and Trade) and

TRIP (Trade Related Intellectual Property)

General Agreement on Tariffs and Trade (GATT) was framed in 1948 and was meant to be a temporary arrangement to settle amicably among countries, disputes regarding who gets what share of world trade. This is achieved by determining both tariff rates and quantitative restrictions on imports and exports globally. Ever since its establishment, GATT remained a provisional treaty—a sort of contractual agreement. The terms of agreements are modified by GATT from time to time and are placed for consent from as many as 102 countries which are contracting parties of GATT, but not its members. GATT meetings are held behind closed doors, which makes it easier to forge and strike deals (which may sometimes be against the public interest) before the public is fully aware of the deal. Although GATT has made the world a better place to do business by allowing more free and fruitful flow of goods and services, this benefit has , unfortunately, gone mainly to developed countries to the disadvantage of the countries in the Thrid World.

It is said that the contracting parties in GATT are equal only in theory, because in practice when developing countries have tried to assert, they have been ignored or overawed by the arguments that the countries with the largest share of the world trade have more at stake, and hence their views should prevail. In addition to exercising its influence through GATT, in 1988 the US Congress enacted a law; 'the Omnibus Trade and Competitiveness Act', which gives the American Government power to 'investigate' trade and investment- related laws of the other countries to check if they are to the detriment of US interests. If after a 'warning' period, the investigated country refuses to change its laws, then the US takes retaliatory action. In this connection, US gave a warning to India in February, 1992 to change its laws relating to IPR, copyright and patents. However, in India, scientists, professionals and political activists argued that "the total package of Trade Related Intellectual Property Rights (TRIPs), when applied, would be a sure recipe for economic and technological subjugation of the country".

In a draft of suggestions for GATT prepared in 1991/92 by its then Director General, Arthur Dunkel (the present Director General is Dr. Sutherland), following are some of the provisions: (i) India will not be able to give any kind of subsidies for the production of oil seeds or pulses, since the international prices of these commodities are lower than the domestic prices; (ii) India will have to alter its patent laws firstly to introduce product patents and for increase in the duration of the patent, secondly to open up its markets to foreign patent holders and patented biotechnology will be opened up to the agriculture sector.

Patenting of Biological Material

Of the different intellectual properties and the rights to protect these properties, particularly when they involve biotechnological processes or products patents are the most important. Even though patenting of inventions arising from basic research (commercially singificant) were earlier considered unfashionable or ethically dubious, it is now viewed more favourably by many. Different countries have different patent laws, which can be modified from time to time. Patents are granted or complaints of infringement of these patents decided by courts in accordance with the patent law of the concerned country. Before 1980, when the discovery of an oileating bacterium (Pseudmonas) by a non-resident Indian scientist (Dr. Chakrabarty), was patented in USA by a multinational corporation, the life forms could not be patented. A later patent issued for 'oncomouse' was another milestone in patenting of life forms.

Recently, the issue of granting patents to life forms has also been discussed in India. Even though there are arguments against patenting life forms particularly in developing world, but the advent of biotechnology has made it necessary that patents of life forms be allowed, because in its absence, commercial firms would not like to invest in biotechnology research. In this connection, there has been a lot of debate in India, because in recent years, USA has been pressurising India through the multinational forum of GATT (General Agreement of 'Trade and Tariffs) to change many of its laws pertaining to patents (as discussed in previous chapter). In view of recent biotechnological developments, the discussion has now progressed beyond merely the question of whether living organisms can be patented, and is now concerned with establishing solid criteria for granting patents. The existing patent laws are also being re-interpreted with regard to biological material.

In India, the patent law is based on Indian Patent Act of 1971, which is being reviewed now for change due to international pressure. As mentioned in the previous chapter, the existing patent law does not allow product patents on foods, chemicals, drugs, and pharmaceuticals. Only process patents are granted in these areas. Therefore, the same product may be produced by an alternative process, which is more cost effective. In case of a dispute, the burden of proving that the product has not been produced by alternative process lies with the original patent - holder, who lodges the complaint. The patent laws in USA and Europe, however, differ from those in India.

International Conventions

In connection with intellectual properties dealing with biological material, there are international conventions, which do not allow patents for processes or products dealing with alleviation of human diseases. For instance, it is not possible to patent techniques for surgery such as by-pass heart surgery, organ transplants, implant of heart valves, artificial limbs, dialysis for renal failure, plastic surgery and other surgical methods used in cancer, appendicitis or neurosurgery. Similarly, one can not patent the following: (i) the use of drugs, antibiotics or vaccines for any form of diagnosis, prevention or cure of diseases; (ii) artificial insemination; and (iii) in vitro fertilization and embryo transfer etc.

In plant biotechnology also, live plants (not the transgenic plants), naturally occurring microorganisms, micropropagation, tissue and organ culture techniques, biological control of pests or hybrid varietics can not be protected using patents. Even techniques for cell fusion, protoplast fusion and gene transfer can not be generally protected through patents. Commericial benefits from hybrid varieties can, however, be secured through control of parent plants, which may be a trade secret. Similarly, although new crop varieties can not be patented, but transgenic animals and plants can be patented, although there may be arguments against it.

In case of microorganisms selected for production of antibiotics, amino acids, enzymes, alcohol etc., patent protection is generally not available, although under specific situations, protection for microorganisms modified for commerical use may be allowed. In these cases (e.g. fermentation), generally commercial benefits are derived by contracts involving transfer of technology through restrictions on further transfer to third parties. Thus, trade secrecy rather than patents in such cases protect the commerical benefit.

International Cooperation

International cooperation has also been sought for application of patent law to solve problems, which are not confined to one or few countries. In a convention called 'Paris Convention of 1988', the basic principles of equal treatment for domestic and foreign inventors were established. The Paris Convention now has 100 member states. The convention allow investors to claim international priority by filing of a patent application initially in one member state and subsequently in others.

The main instrument of international collaboration for intellectual property is the World Intellecutal Property Organization (WIPO) based in Geneva. It administers (but not enforces law) the Paris Convention and all subsequent conventions. WIPO operates by asking member state to ratify a convention and to introduce the agreed basic principles into their national laws. Another international group is the European Patent Convention (EPC) of 1973, which began operation in 1978 and has all member states. EPC has the distinction of being the first patent state to introduce specific provisions for biotechnology inventions. Two of these important provisions include the following : (i) need of culture collections as patent depositories for the placement of microorganisms referred to by patent applications ; this is necessary, because the complete description of living material is difficult and may not be fully reproducible; (ii) exclusion of certain plant related inventions from the list of those which can be patented, e.g. plant and animal varieties bred through classical methods.

First major study of international patent protection for biotechnology was published in 1985 by the Organization for Economic Cooperation and Development (OECD). WIPO also began a wider study through experts on biotechnological inventions with an objective to provide 'suggested solutions' for problems related to patent law. Taking EPC as a model in 1988, the European Commission (EC) formulated a European Commission Directive to help the member states modify their national laws, so that they become uniform in European Countries. This directive has following three major proposals, which will make patent laws more favourable for patent applicants : (i) An invention must not be refused protection simply because it involves living matter (earlier no patents were granted for living forms). Examples of patentability of microbiological processes and patentability given below (ii) The Scope of patent should extend to all progeny produced by multiplication of parental material, provided it retains the characteristics of this patented material. Generally, under patent law, there is provision of 'exhaustion of rights' once the product is sold, so that the purchaser can cultivate unlimited quantity from a small quantity purchased from the patent holder (this is overcome in the directive). Further, the patented invention can be used for experimental use without infringement of the patent. But the experiments may sometimes be designed as a 'way around' the patent and may thus be utilized for commerical purposes. Therefore, in the European Commission Directive (Article 10), the 'experimental use' is deleted. Thus, if this directive is accepted, patented living forms will neither be allowed multiplication by the purchaser, nor will it be allowed to be used for experiments without paying the royalty for the patent; (iii) The Microorganisms to be patented will have to be deposited in culture collections. For convenience, in table earlier only microorganisms are used for illustration, but the categories shown in this table also apply equally to other types of biological material (e.g. cell lines, plant and animal cells).

Table

Some examples of patentable microbiological inventions (processes and products).

Processes	***Products***
1. 'Methods (reproducible)' for producing new microorganisms, e.g. method for (i) reducing pathogenicity; (ii) enhancing biological utility; (iii) isolation (new methods) media or culture conditions, (iv) constructing new organisms by genetic recombination or cell fusion (normal screening methods to isolate naturally occurring microorganisms are not patentable	1. Antibiotics and other products are protected as 'product per se' so, that it is protected even if produced by an alternative process, including a chemical process.
	2. New microorganisms can also be applied for 'product claims', although there are arguments against 'patenting life'.
2. 'Use claim' is also classified as a process, e.g. use of (i) strain to control agricultural pests; (ii) new microorganisms for fermentation, biomass production, extraction of metals, clean up operation or for biotransformation.	3. 'Composition' of microorganisms can be used for product claims' in addition to 'use claim' which is a process patent.

Obligations with Patent Applications

Before a patent can be issued, following specific conditions must be. met (i) the invention must be new, (novelty) and should have utility; (ii) it must be inventive (which means it should not be obvious but should represent a real advance made through the insight of the inventor); (iii) it must be disclosed (disclosure) in a way, which enables a person of normal skill to reproduce it; (iv) the scope of protection to be granted must be in, proportion to the invention; (v) it must relate to' a technology, where patents are permitted (patentable). The application of these conditions can vary to some extent in different countries. A fundamental requirement of patent law, thus, is that in the application, the inventor should describe fully the invention. If the organism is already known and only discovery of a new property is the subject of patent, the microorganisms need not be described and can be simply named. However, if the organism is modified, it needs to be fully described including the method involved in modification. Since, the disclosure in such cases may sometimes be difficult and incomplete, a provision for 'culture collection' has been made in the to application for patent) a sample of the material with an officially approved culture collection, which is equipped to store and handle it. The deposit of material serves the following important functions: (i) it provides a reference material for any dispute about its novelty, or when a third party is using it without a licence from the patentee; (ii) the deposited material provides for a source to enable others to use it, when they are legally free to do so (i.e. when the patent expires at the end of normal term).

Implications of Patenting

The patent application, with its accompanying description and the deposited material, is held secret in the early stages only for a short time, after which the application must be published. In many countries (particularly in Europe, e.g. Germany), the process of granting patent starts 18 months after the publication of the application. However, the application can be withdrawn before publication, if the assessment suggests that the patent may not be granted. The publication gives advantage to applicant's competitors, who can use this information for developing an improved process or product. Publication of application also makes the deposited material available to the public. This law in Europe has serious drawback and has worked as a disincentive for microbiological inventions. In view of this, efforts are underway to change this provision of patent law. In the USA and Japan, however, access of public to deposited material is allowed only after some enforceable right has been granted to the applicant. Following are some of the serious consequences of the easy availablity of the deposited material (microorganisms) to the third parties without any geographical limitations: (i) the loss of effective control on the microorganism and its uses; (ii) irrevocable loss of the option to revert back to trade secrecy, if patent is not granted; (iii) possibility of circumvention of the patent protection, by other competitors of the applicant through genetic modification of the deposited microorganism. Some efforts, are being made to overcome this problem associated with the European Patent Law.

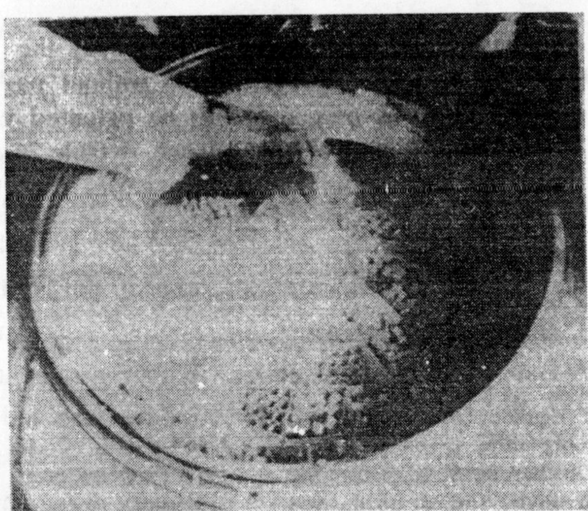

A tank of liquid nitrogen used as specimen repository meant for freezing and preserving living cell cultures, etc. (Fro.. Sci, Amr., March, 1991)

Current Issues: Can Live Forms be Patented ?

Although there was initial resistance to the patents of living organisms, this has now been overcome in most Industrialized countries. In India, there is still some discussion and we are arguing against the patent of live forms, although eventually we may have to agree to the international practice of patenting life forms, if they fulfil the minimum requirement of invention, novelty and utility (see later). This decision of patenting life forms was initially influenced, when in 1981 a US Court allowed the following patent in the famous Chakrabarty case.

"A bacterium from the genus Pseudomonas containing therein at least two stable energy generating plasmids, each of said plasmids providing a separate hydrogen degradative pathway."

The subject of the above claim is an organism, which was made not effective in treating oil spills by manipulating a natural Pseudomonas. Question, therefore, is whether something that already existed in nature can be patented. This is known as the 'Product of Nature' problem in patent law.

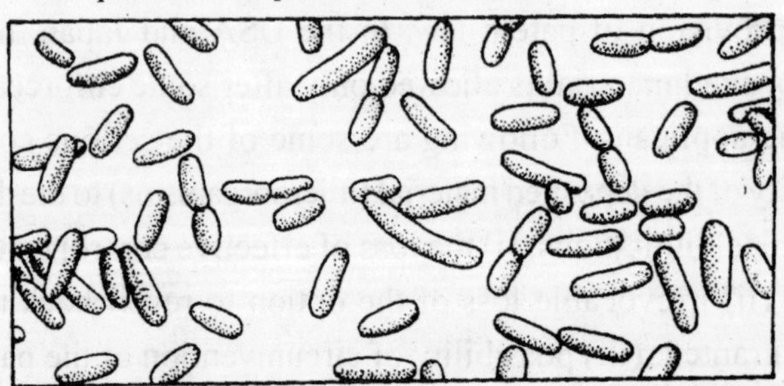

Oil eating bacteria (genetically engineered Pseudomonas), which were the first living organisms to be protected under a standard US patent; these are used for cleaning up oil spills. (From Sci. Amr., March, 1991)

Patent law makes a distinction between invention and discovery and does not allow patent for a discovery. In view of this, if isolation of a substance from nature is mere discovery, it may not he patentable. However, if this substance found in nature has first to be isolated from its surroundings and a process is developed to obtain it from nature, then this process may be patentable as suggested by European Patent Office (EPO). Furthermore, if the substance is characterised and is 'new' having 'no previously recognized existence', then the substance purpose may also be patentable. Antibiotics and other microbial metabolites or a newly isolated microorganism are examples of such patentable substances. Preparations of pure proteins or vitamin B by sophisticated purification techniques or by recombinant DNA methods are other examples. Before 1951, Vitamin B_{12} had been available only as crude liver extracts, which were of no use for therapeutic application, but a patent claim for pure vitamin B_{12} has been allowed in USA. Microorganisms such as E.coli, in which human genes have been incorporated for production of human insulin, human growth hormone, human tissue plasminogen activator (t-PA), etc., have been recognized for patents in the USA. Microbial cells engineered to produce antigens and antibodies also qualified for patents. A protein may be modified by altering the gene sequences responsible for its synthesis. This alteration may or may not amount to infringement of a protein patent. Several cases dealing with patents of products of nature have been recently contested in US courts and a few will be briefly described to illustrate the issue.

Factor VIII, procoagutant activity protein (VIII: C)

Protein 'VIII:C', meant for coagulation, is absent in patients suffering with hereditary sex linked disease hemophilia. The protein is found in plasma complexed with another protein VIII:RP. Scripps Clinic and Research Foundation (USA) were successful in preparing purified VIII:C by removal of VIII:RP from the VIII:C/VIII:RP complex through adsorption on specific monoclonal antibody. Scripps initially obtained a patent for this process and also for the product of such a process. Later they obtained a reissue potent, which covered a purified VIII:C made by any method whatsoever (including VIII:C produced by recombinant method). Genentech (USA) later produced VIII:C by recombinant method and was sued by Scripps for infringement of their reissue patent. Meanwhile, a patent on recombinant method for producing VIII:C was issued to 'Genetics Institute Inc' in July 1988, but the Scripps - Genetech litigation continued.

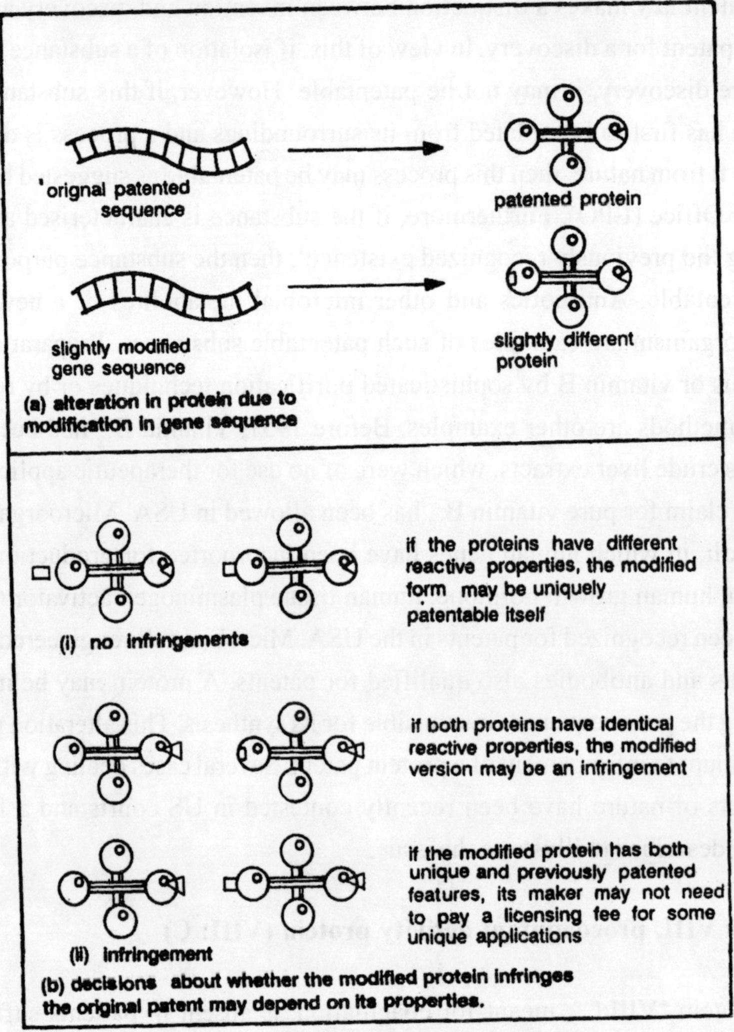

Fig. Diagrammatic representation of conditions which amount to infringement of original patent (From Sci. Amr., March, 1991)

Erythropoietin

Amgen Inc. holds a US patent for preparation of erythropoietin, by recombinant method. In this case, DNA sequences, vector and transformed cells were protected, but not the product, since US patent office usually does not allow patents on the naturally occurring products. However, Genetics Institute Inc. was successful in obtaining a broad product claim for 'erythropoietin' (isolated from urine) like the one obtained by Scripps for VIII:C (see previous section). Are such claims valid and should they aver even the product of recombinant DNA?

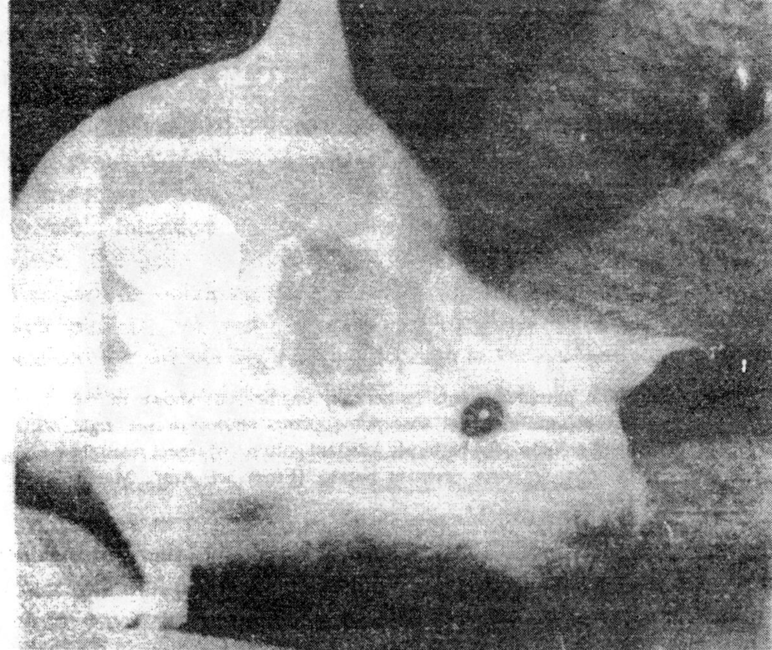

Genetically engineered mouse, also called 'oncomouse' (carries a human cancer gene), that was protected by a US patent in 1988. (From Sci. Amr., March, 1991).

Tissue plasminogen activator (t-PA)

A UK patent claim for 'human tissue plasminogen activator', as produced by 'recombinant DNA technology', was allowed to Genetech. This claim failed on legal grounds (product was known and was produced using known methods). The corresponding LIS patent was 'A DNA isolate consisting essentially of a DNA sequence encoding human tissue plasminogen activator.

Human nerve growth factor (β - NGF),

'US Patent Office Board of Appeals' rejected a claim to recombinant protein 'β-NCF' defined firstly by amino acid sequence and secondly for being free of other proteins of human origin. This rejection was based on the absence of evidence that the recombinant product was significantly different from the natural material or possessed any unexpected properties.

Hybridoma technology

In 1983, 'Wistar Institute' applied to the 'UK Patent Office', for a patent of 'hybridoma technology' for producing viral antibodies. Since the technology was published in 1975 by Milstein and Kohler, the application of this technology was obvious and no patent was allowed in UK, although earlier in 1980, for this broad process claim, a patent was allowed in USA. There are also examples where US patent of monoclonal antibodies was held invalid by District Court but was later allowed by the Higher Court of Appeals for different reasons.

Fig. Some patented p ants (genetically engineered) shown in the left column (normal susceptible plants shown in the right hand column); (a) herbicide resistant cotton, (b) insect resistant tobacco, (c) virus resistant potato. (From Sci, Amr., March, 1991)

Patents for Higher Plants and Higher Animals

Under patent laws of most countries, a patentable invention must be capable of industrial application. Agriculture is also treated as an industry in this connection, so that a wide range of agricultural and horticultural methods and products can be patented, provided these are 'inventions' and represent 'novelties'. Biopesticides and bioinseticides are such examples. Extensive patent literature is also available on the use of Bacillus thuringiensis and michoherbicides (fungi). Other examples include (i) novel techniques of plant micropropagation and (ii) plane-cell and tissue-culture methods to prepare useful metabolises.

Among higher plants, an important example is the patent for a 'tryptophan overproducing maize' obtained through tissue culture. The patent known as 'Hibberd Patent' was issued to 'Molecular Genetics Research and Development'. A similar patent was allowed in US for a transgenic animal, popularly described as 'oncomouse patent'. Patent was also allowed for a polyploid oyster produced by the application of hydrostatic pressure to zygotes. In view of these cases, Patent and Trademark Office now considers claims for patents of 'non-naturally occurring non- human multicellular living organisms'. Japanese patent law is not very different from the US law and patents for animals and plants have been granted in Japan also, provided these were products of specific methods or inventions. In Europe also patents are allowed for such plants and animals, even though plant varieties and animal varieties are excluded. In this connection there have also been dispute regarding what constitutes a variety. Patent claim for a soybean variety applied by Pioneer Hi-bred, was rejected by Supreme Court of Canada, on the ground that no description of the method was available, although the seeds of the variety were deposited. In some cases, hybrid varieties were also considered patentable. It is not certain whether the meaning of a 'variety' will be restricted, or if the scope of a variety will be wider. This is an area which is receiving major attention in several countries and will be resolved in the coming years.

Patenting Transgenic Organisms and Isolated Genes

Transgenic plants and animals can be protected through patent claims in several countries now including USA, Japan, Europe, etc. 'Oncomouse' is one such example, where product claim was initially rejected by EPO, but the deicision has been overruled on appeal. This is despite the fact that animal genetic manipulation may be considered by some to be an ethical question. Among transgenic plants, herbicide resistant cotton, insect resistant tobacco and virus resistant potato have been patented. A boll worm resistant cotton may also be allowed patent protection (consult Sci. Amen, March, 1991). There is no doubt, therefore, that in future patent claims for transgenic plants and animals fulfilling certain conditions will have to be allowed in several countries.

Patenting of Genes and DNA Sequences

The genes that are synthesized artificially, normally fulfil the requirements of patents and, therefore, have no difficulty in getting patent protection. If the protein, that the artificial gene makes, and the organism into which the gene is inserted are also novel having desirable attributes, such a patent may extend to the protein and the organisms also. Thus, the patents involving artificial genes and novel higher organisms (plants or animals) are relatively straight forward.

In recent years, one of the most widely discussed political issue in biotechnology concerns the patenting of naturally occurring useful genes, because they do not fulfil the requirement of novelty (they are only discovered, not invented and for a patent, invention rather than discovery is neccessary). The developing countries have been arguing that patenting of naturally occurring genes will allow the developed countries to make use of a weed from their land, isolate a desirable gene from this weed, transfer it to a commercial crop and then sell it back to the same country (which was source of the gene). There are counterarguments that companies should be rewarded for their investment and labour involved in making the natural genes useful, which may otherwise never be exploited for human welfare.

The law on the above issue is both confusing and changing through judgements on disputes by courts. However, the law in U.S.A allows patent on purified form of a chemical, if in nature it occurs in an unpurified form. In UK, on the other hand, the courts recently held that a naturally occurring gene sequence can not be patented. It is, however, now certain that in U.S.A and some other developed countries, patents will be allowed in future for genes isolated and cloned from nature. An example of such a patent granted involved cloning of a DNA fragment, which originated due to mutation in a microorganism and imparted resistance to the herbicide 'glyphosate'. Intended for integration into plants to confer glyphosate resistance, the gene is patented by Calgene Inc. in USA in terms of a DNA sequence containing the relevant structural gene. Increasing number of such patents arc now appearing particularly in USA, and similar applications for patents are awaiting decision in Europe and Japan. Whether a patent claim for a DNA sequence can protect its expression product (proteins) also needs to be seen. In case of plants, a patent for a DNA sequence will protect the transgenic plants carrying this sequence. Therefore, whenever a patent for a gene is allowed, biotechnology companies would like the protection to extend to the marketed products. This area of IPR in biotechnology will receive further increased attention in coming years.

In 1992, the question of patenting isolated genes or DNA and cDNA sequences has once again been addressed in USA due to an application on behalf of National Institute of Health (NIH), USA, filed for patenting hundreds of cDNA sequences. The main issue for discussion was whether these sequence are 'useful' as required in- patent statute (35 USC 101). This requirement, in 1960 (Brenner vs Manson, US Supreme Court, 148. USPQ 689, 1966) was interpreted to mean that in the process or product to be patentable, there should exist a 'specific benefit in currently available form'. Subsequently the word 'useful' was interpreted to mean 'practical utility'. Many of the isolated genes may not offer specific benefit in its currently available form and may not be of practical utility. This will render them unpatentable, even though it may discourage research in this area. Despite the above requirement, isolated genes, vectors and transformed cells expressing the hormone 'angiogenesis factor (AGF)', which increases vascularization, has been allowed a patent. This patent was allowed, even though at the time of application (1985), AGF did not really have practical utility for therapy or for diagnostic tests. Many other patents of isolated genes are similar in nature and their validity may be questionable if challenged in a law-suit. Such law suits, challenging the validity of patents for isolated genes already granted, will settle this issue in future. Patenting in such cases may be necessary to encourage research leading to isolation of genes, which if patented, will be available to other researchers and if not patented will be kept a secret. Therefore, patent applications for isolated genes may, in future, either face a rigid attitude asking for 'specific practical utility in its currently available form' or may receive a relaxation in view of its utility in immediate research with future potential of practical utility.

Plant Breeder's and Farmer's Rights

In order to understand the full implications of intellectual property rights on agricultural production, one should also understand how 'plant breeder's rights' and patents function. A distinction between plant breeder's right and patents should also be fully appreciated. As discussed earlier, plant varieties are excluded from the list of patentahle articles. If so, how will these be protected, and when (if at all) can these be covered under patent law? These questions are briefly discussed in this section .

Plant varieties are generally protected in several countries (not in India) through plant breeder's right (PBR) or plant variety rights (PVR). Through these rights, further propagation of the variety is restricted. Under the existing convention due to 'International Union for the Protection of New Varieties (UPOV)', the breeder's right does not prohibit the farmer from reuse (plantback) of farm saved seed of a variety from his own harvest for planting another crop. Furthermore, the protected plant variety can be freely used as a plant genetic resource for the purpose of breeding other varieties. Since the revised UPOV convention extends PBR to cover the reuse, countries are now free to extend PBR, to cover the reuse of seed, so that farmers, in such a case, can not use his own seed without paying a royalty to the PBR holder. However, moll countries are expected to limit the PBR with regard to farmer's plant hick, although a farmer can not sell the seed. Enforcement of such rights as above, in case of freely reproducible material, is only possible with large holdings of land, as in case of plantations or in very large farms, and in case of high value cash crops (e.g. cashew, species, medicinal plants, etc.). Even in these cases, infringements are difficult to prove. When 'patent' or 'plant breeder's rights' are not available for true breeding crop varieties, plant breeder (particularly private plant breeder in developed countries like Germany) may feel tempted to focus their efforts on developing hybrid varieties. Because hybrids do not breed true and give higher yields, no one would raise a crop from harvested seed that will give reduced yield. Thus hybrid varieties may give to the plant breeder an advantage, which is equivalent to intellectual property protection.

PBR has analogies to patents, but there are also important differences. Rights are granted for a limited period (usually 21 years) to the breeder. Breeder seeking PBR can not seek exclusive rights for a unique feature of his variety, although under patent this is allowed. For instance, a breeder of the first 'blue rose' can not monopolise blue colour of rose. It will be open to other breeders to breed and protect black roses, which are distinct from the first variety having blue roses. A protected variety must fulfil some requirements. It should be (i) new; (ii) distinct; (iii) uniform and; (iv) stable. 'New' means, the variety should not have been previously exploited commercially. 'Distinct' means, it should be clearly distinguishable from all other varieties known at the date of application for protection. 'Uniform' means that all plants of the variety should be sufficiently uniform. 'Stable' means that the variety can be reproduced and multiplied without losing its characteristics and uniformity.

In India, new crop varieties are bred at state; Agriculture Universities and at state Departments of Agriculture. The seed of new crop varieties flows freely to farmers and to the private companies and no royalty is payable. This really encouraged farmers, in the past, to grow new varieties leading to green revolution. Imposition of PBR in India will lead to the following problems: (i) the cost of seed will increase; (ii) there will be delay in the spread of a new variety to the farmers; (iii) the benefit of new varieties will be restricted to a small segment of tanners.

If multinational companies are allowed in India to invest in seed research and development, they will have monopoly through PBR. The farmer will not be allowed to export or supply seed multiplied by him to other farmers. This may not be in the interest of Indian Agriculture at least for many years to come. In the present circumstances, for India, following options are available: (i) uniform patenting of all inventions including plants; (ii) patenting of plants not obligatory, but plant breeder's rights (PBRs) obligatory; (iii) complete exclusion of plant varieties from any form of IPR. The third option allows the present practice to continue, where complete freedom for introduction and promotion of new varieties is allowed. This option seems to be the best for Indian Agriculture atleast for the time being although there is considerable pressure from developed countries to accept some form of IPR. It is possible that India may have to introduce some sort of PBR for crop varieties. This will encourage private companies to enter plant breeding programmes.

Farmer's rights

Farmer's rights is a concept, which has been developed and adopted in FAO (Food and Agriculture Organization) its a resolution and endorsed by all member countries. It recognizes the fact that farmers and rural communities have greatly contributed to the creation, conservation, exchange and knowledge for the utilization of genetic diversity. Therefore, it is the obligation of world community to help these farmers to carry out this task and also help them in utilizing the genetic diversity available with them. No intellectual property system or any other mechanism exists for compensation or reward to the farmers for this work, Efforts, therefore, have been made to give a concrete legal form to these 'farmer's rights'. In a document published in 1989 (RAFI = Rural Advancement Fund International 'Farmers rights'), it is suggested that a tax should be payable on commercialized biological material derived from developing countries, but used in developed countries. Automatic patenting rights, based on material derived from developing countries, are also suggested. This is an interesting situation, since most accessions in gene banks are primitive cultivars or land races from farmers in developing countries, who will have to be paid royalty. The idea of informal innovation eligible for farmer's rights is based on the following: (i) farmer's land races, plants (used for medicinal purposes and other biological products), processes and innovations are the result of human ingenuity and represented immediate inventions; (ii) most of these inventions arise from informal efforts, which are purposeful and creative; (iii) collectors gather present-day improved germplasm and the knowledge about its breeding and discovery. These aspects in relation to farmer's rights are not covered under TRIPS within the GATT negotiations.

In 1991, an international debate on "Biotechnology and Farmers' Rights: Opportunities and Threats for Small Scale Farmers in Developing Countries" was organised in Amsterdam, the Netherlands. Following are some of the recommendations made towards the concept of farmer's rights: (i) Plant breeder's rights should be maintained; (ii) Patent legislation should not be extended to genetic material of plants and animals and there should be legal sanctions prescribed in this connection; (iii) An 'International Gene Fund' be established, and used to compensate the local communities in the Third World including farmers and farm communities (not the individual farmers) for their past and present contributions. Payment towards this fund by member countries will be obigatory and the amount will be based on the degree to which the users of the genetic material profit from it. The United Nations Conference on Environment and Development (UNCED) held in 1992 took further steps to establish an international fund, which gave substance to the 'principle of farmer's rights' and also to the genetic resource as a 'heritage of mankind'.

SEVENTEEN

International Market

Biotechnology, to succeed, depends on more than just its technology. Changes in the legal, ethical and political guidelines which define the com- in which we live often trail pitifully behind advances in our ability to apply our knowledge of the natural world. While the level of debate amongst those involved in biotechnology is of a high quality, it remains essëntially private, confined to individual companies or laboratories.

Biotechnology basically relies on the principle that "why trouble to make compounds yourself when a bug will do it for you." Apparently, it would seem that pure chemists are of no more use to society. However, in actual practice biotechnology actually emphasizes the increased importance of traditional chemists in expanding their experimentation from the laboratories to that of society. In particular, we have learned a great deal more about the capabilities of microorganisms and how they can be put to work to provide us with an enormous range of useful substances. This increased knowledge, allied with plant and animal cell culture techniques, genetic engineering, Recombinant DNA research, enzyme engineering, hybridoma production, fermenter design, bioreactors including perfusion bioreactors, downstream processing and large number of other aspects have aroused remarkable international interest.

The biotechnology business has two main parts: small specialist hightechnology companies and multinational manufacturing companies. In the 1940s, antibiotics were quickly adopted by all the major pharmaceutical companies, some of which achieved their present position largely as a result of their successful marketing of these products. In the early 1970s, many large chemical companies needed to diversify and, especially as petroleum prices began to rise sharply, speciality chemicals and feedstuffs from microbial cultures were evaluated and, in some cases, developed. However, many of these companies, expert in chemical engineering and organic chemistry, were unfamiliar with biological systems. Thus, it was not surprising that around 1975, when the enormous potential of genetic engineering was just beginning to be recognized and monoclonal antibodies had been discovered, there was a sudden flurry of commercial activity. This was mainly characterized by academic researchers (chiefly molecular biologists, biochemists and immunologists) realizing that their research could have significant industrial uses in the immediate future. In the USA where a lot of this research was taking place, (the 'entrepreneurial spirit' is a strong cultural force) many scientists realized that there was a very real chance that they could see their own discoveries through commercial application and also obtain considerable financial rewards for their efforts. 'Venture capital' was not particularly hard to come by in the early days and so partners could be found to provide financial backing. Arising from these circumstances, such companies as Cetus, Genex and Genetech were born, which in the last ten years have become well-known 'biotechnology majors'. Each now has a staff of several hundred and several subsidiaries have been formed for geographical and financial reasons. Biogen, which began in Switzerland during the same period, now also has a major laboratory in the USA. Several hundred other companies have been established. Some have only a handful of employees who mainly act as consultants as a side-line to their academic research; others are fullfledged companies with 20 to 200 employees. Until about 1980, most of these specialist biotechnology companies obtained revenue mainly by undertaking contract research, particularly for large pharmaceutical, chemical and oil companies. This period gave a valuable start to the older firms, but those formed later found that most of the obvious targets had been tackled and the results patented. Many of the large companies have now had time to set up their own departments in-house and some who had unhappy experiences because of unsuccessful projects, or over-optimistic targets, have become less enthusiastic about biotechnology. It is fair to say that the extensive media coverage given to biotechnology in this period may have been somewhat harmful in that it encouraged a 'boom and bust' phenomenon where unrealistic rewards were often promised.

In the early 1980s, there has been a sobering of attitudes. Contract research is still done on a considerable scale but with more carefully defined terms. There has been a general swing towards health-care products which offer large profits, and a growing awareness among sponsors and investors that biotechnology may not bring quick results but will still be an important new business area. A few badly managed or underfunded companies have failed, some over-extended research programmes have been cut back drastically and a few mergers and acquisitions have begun. Many of the initial entrants were privately funded but, at least in the USA, most have now had to make public stock offerings to obtain additional financial resources. In most cases, these have been quite successful and some of the major public companies still obtain a major part of their revenue from interest on share capital. In Europe, venture capital is less available and biotechnology has taken off more slowly but still along broadly similar lines. New companies are still being formed all over the world but many now start with one or more well-developed products ready to be launched. One of the earliest examples of this was Celltech, set up in the UK in late 1980. Celltech remains privately funded but has grown to employ about 150 people in two years and now ranks among the world leaders in recombinant DNA and monoclonal antibody techniques. This growth was made possible by a judicious combination of substantial equity funding and the early launch of a range of antiinterferons which were available during the company's first year of trading.

In Japan, where fermentation was already well-developed, almost all commercial biotechnology development has been done by large companies. World-wide, the biomass industry is quite separate from immunology and recombinant DNA. Biomass projects may be large-scale and often involve considerable government funding. Nevertheless, some small biotechnology companies have also been formed.

Biotechnology takes research science into the commercial domain on an astonishingly short time-scale. This phenomenon is already known in chemical engineering and computering, but is relatively new in biology. There have, therefore, been strains on the structure of many academic establishments, worries about 'brain drains' and concern about the safety of genetic manipulation. These have not all been resolved but useful discussion continues and considerable progress is being made, particularly, in the area of safety regulations. All these have been the growing pains of an industry, not really a new one, but one suddenly given new opportunities by technological advances. Only the next 20 years can tell where this opportunity may lead. Meanwhile, information is vital to sustain innovation and to allow informed decisions to be taken about this technology, whether for business investment or for legislation.

Universities versus Private Enterprises

Biotechnology is perhaps the single unparalleled discipline which has established a network of intimate relationships and contacts between Universities—the desks of basic research—and various public and private sectors. In fact here is a branch of academics in which basically nonacademicians have also suddenly become seriously interested. The interest may be more for their business lust and profiteering attitude, 'but it does involve basic academics somewhere. Consequently, the question now is whether the discipline should be confined to Universities or to research

Institutions or should it be left entirely to the industrial entrepreneurs to blaze the trail into the brave new world? During the last 10 years particularly in Japan, the United States and the United Kingdom, many private enterprises have been established to conduct basic research, reap the harvest, and get them patented for commercial uses. Several private institutions of basic research and process also endeavour to ensure the liaison between research and industry. Interestingly, these private enterprises have also been successful in drifting talent and brains from Universities to their private pool to a large extent. Consequently, biotechnology has entered a phase of stiff competition at a national and international level arousing the interest of Governments, and there is a brisk race between various cotuntries as in several other regular competitive fields such as space and nuclear programmes. Although such bifurcations of trifurcation have taken place depending upon financial status and individual business trends, it certainly would be unwise to allow such division without any mutual contact. No other technology includes such a wide diversity of activity from sewage treatment to interferon production; from the traditional fermentation process of citric acid and antibiotic production to the new single cell protein projects. Therefore, coordination between a whole host of industries, Government laboratories and teaching centres are essential to avoid duplication and miss links of efforts. There is every reason for asserting that biotechnology and the benefits it can bring would be best served by the creation of one or more specialized centres where knowledge could be gained, stored and disseminated without hinderance.

The University-Industry relationship with respect to Biotechnology research can not be established on a uniform basis in all the countrics because of different political and financial backgrounds. For example, in the United States, there are two types of relationship which exist:

(i) Industries directly setting up the research departments in the University and taking up the patent rights besides provision for training company's personnel. A well-known case is that of Hoechst AG-German Pharmaceutical Company, for making an important direct contribution towards establishing a department of Molecular Biology at Massachusetts General Hospital (Harvard Medical School & University). Such a relationship is quite binding because patents, although obtained in the name of Massachusetts General Hospital, can be exploited commercially only by Hoechst AG with royalties to the Laboratory workers. Several other companies and Universities in the United States have entered into such contracts which are renewable.

(ii) Industries could support research in the particular University. Such relationship could be profitable to a large extent, particularly, to prevent the brain drain. In such a case the University staff could work on commercially oriented research in addition to maintaining close link to research in their own laboratories.

These relationships may be possible for highly developed countries such as the United States where the private sector is becoming relatively more interested in biological research than the State Government.

In the United Kingdom also, several private companies including Beecham, Glaxo,. Shell United Kingdom, Unilever, Wellcome as well as Swiss company Sandoz have been funding University research and even setting up independent research laboratories with some research staff on the company's payroll. Even funding such as CASE (Cooperative Awards in Science and Engineering) has been instituted by private companies in collaboration with science and Engineering Research Council to support postgraduate students in Universities.

The major emphasis everywhere is that professional relationships should not interfere with teaching and research responsibilities. It is also argued that research contracts between Universities and Industry require the minimum amount of survey and that scientists in Universities utilizing the industry funds should retain full rights of disclosure and publications on their research reports. A large number of other European countries as well as Japan are entering into close-knit University-industry coopted biotechnology research.

In India, biotechnology, in the true sense, has begun quite late and the industrial sector has yet to realise the potential benefit of such a relationship. Much, however, will depend on private interests also.

It would be meaningless to treat biotechnology like Chemistry, Physics or Biology and to assert that Universities could provide the basic knowledge. Although a large number of universities abroad and a few in India are attempting to teach biotechnology, specialized training centres or enterprises, with a practical approach and aim to inculcate the biotechnology cult, are essential because biotechnology is inter-disciplinary. This is essential because in universities the classical view of physical and natural sciences has led to the neglect of subjects which hold the key to modern biotechnology.

Copyrights of Genetically Engineered Cells

While this has yet to begin in India, Genetic engineers may be surprised to learn that manipulations of genetic information within cells can also be works of authorship which copyright law protects from unauthorized reproduction.

The idea that genetic engineers can make use of copyright law may be startling, but a close examination of the US Copyright Act of 1976 indicates that this law could have an immense impact on biotechnological research, development and production. The central arguments of this act are that virtually all original works of a genetic engineer are copyrighted automatically when they are created; that scientists generally can enforce copyrights; and that those copyrights may provide more effective protection than patents or trade secrets.

Advantages of copyright Protection

What is the value in copyrighting genetic information? Under certain circumstances, from a practical and legal viewpoint, copyright protection may be the only or the most effective way an author can protect a valuable genetic work. Furthermore, copyright protection offers unusually attractive remedies.

The range of benefits open to a copyright owner is truly impressive. The Justice Department may prosecute anyone who wilfully infringes a copyright for commercial advantage or private gain. In addition, upon conviction, a court may order the destruction of all 'implements, devices, or equipment used in the manufacture' of infringing copies. The convicted copyright infringer is not only out of the copying business, but also out of any microbiological business for want of an industrial plant. A copyright owner can recover both his own actual damages and the infringer's profits.

While patent protection exists for seventeen years and trade secrets last for as long as they can be kept secret, copyright exists from the time of the work's creation to years after the death of the author. Most importantly, the copyright of works made for hire, e.g. by an employee of a corporation, endures for one hundred years after the creation of the work or seventy-five years after publication of the work, whichever is earlier. Thus, copyrights last much longer than patents and impose none of the problems or expense of safeguarding a trade secret.

Copyright protection under the 1976 Act

We now come to the central questions: Can engineered organisms be copyrighted and, if so, can the copyright be enforced? The Act of 1976 provides copyright protection from the moment a 'literary work' is created. A work is created when it is 'fixed in a tangible medium of expression'. In turn, a work can be fixed in any 'form, manner, or medium as long as it is sufficiently permanent or stable to permit it to be perceived, reproduced, or otherwise communicated for a period of more than a transitory duration'. The act of creation may take place in the quiet and secrecy of a private laboratory. Property rights immediately accrue, which may be enforced once the copyright is registered.

Libraries of spliced DNA fragments and cultures of engineered cells with a foreign DNA sequence are certainly fixed in a permanent and tangible media of expression. Moreover, genetic works are stable enough to be perceived and reproduced. Reproduction, in the cell, is the entire point of genetic engineering and perception is feasible by DNA sequencing. Under the 1976 Act, perception may be direct or 'with the aid of a machine or device'.

Some may argue that recombinant DNA work is not sufficiently original to be copyrightable. Every DNA fragment plasmid introduced into a host tissue is only distantly related to that of the growing cells in intact tissues.

Calluses or cell aggregates without support matrices may have similar productivity of metabolites and longevity to immobilized cell systems. Therefore, before the superiority of immobilized techniques can be used to justify immobilizing plant cells, detailed comparisons should be made of the physiology and biochemistry of immobilized plant cells and callus, and of cells in suspension.

By June 1980, more than 100 applications had been filed for patents at the United States Patent and Trade Mark office. The first patent to be officially approved was that of Ananda Chakrabarty (General Electric Company, GEC) for the development of a bacterial strain of the Pseudomonas genus having the ability to degrade hydrocarbons especially against oil spills in the oceans. It took him more than eight years to obtain this patent. However, in history of science, this was not the first patent to cover a living cell because in 1873 Louis Pasteur had also patented brewer's yeast. This decision upheld the progress in biotechnology. It may be emphasized that the bacterium produced by Chakrabarty was not produced by recombinant DNA techniques but by employing conventional genetic techniques. This was immediately followed by granting of a patent to Stanley Cohen (Stanford) and Herbert Boyer (California) primarily for introducing basic techniques of recombinant DNA and introduction of foreign genes in cells. Later, under terms and conditions, these techniques could be used by research workers in Universities freely, but commercial establishments and bioindustries were required to pay a fixed royalty for using these techniques for marketing substances. There has been a rapid increase in the number of patents following the 1980 Supreme Court decision permitting patenting of live microorganisms. Patent battles are becoming frequent in this highly competitive field. The ability to obtain separate patents for a gene sequence and its product is leading to crosslicensing arrangements. According to Budapest treaty among 19 countries (India excluded), one deposit in any depository will satisfy patent office requirements, whereas previously deposits in multiple countries were necessary. Non reproducibility of deposited material could lead to future patent applications, it is essential to deposit the cultures in individual countries in which a patent is sought (except in those cases where countries have reached international agreements recognizing each other's depositories). To facilitate the entire process, the World Intellectual Property Organization (WIPO), deals with international agreements on patents and copyrights with headquarters in Geneva.

There have been many conflicting cases and legitimate solutions concerning several discoveries especially plasmids and their patenting. Besides, there have also been reports of patents being challenged. Therefore, scientific research in this field has lots in common with any public or private court hearing and follow up of decisions.

At present, patent laws are fairly practiced in the United States, Japan, Australia, and several European countries. In the United States, a patent, by definition, should be for any new invention, discovery of a machine or a new and useful manufacturing process. It also includes new varieties of cultivated plants. However, there have been some conflicting reports regarding characterizing a microorganisms to be patented. Should it be recognized by (i) its culture conditions, (ii) metabolic activity or (iii) products formed and released. Similar problems also existed in france, where laws say that the manufacturing process and the synthesized product can be patented easily. However, there is difficulty in patenting the microbe releasing the product.

In Europe in general, according to the convention held at Munich in 1973, animal races and plant varieties are not patentable. However, microorganisms and the products obtained from microorganisms can be patented including man-made or genetically engineered microbes. In Canada also, genetically engineered microbes are patentable.

Some research at the Department of Biochemistry, the Agricultural University, Wageningen, The Netherlands, has recently become the subject of an international patent, a license to which has been sold to the Dutch pharmaceutical company, Duphar, for Rs. 35,000. The research concerns 'Enzymatic reactions in reversed micelles' and has been conducted by Dr. Maria H. Hilhorst.

Reversed micelles are structures that are formed if water is dispersed in an organic solvent containing detergent molecules with a polar 'head' and an apolar 'tail'. Because of their amphiphilic nature, the detergent molecules will be located between the water phase and the organic phase with their polar heads directed towards the waterdroplets and their tails sticking out in the organic phase.

During the preparation of reversed micelles, enzymes can be contained in the aqueous centre. The resulting system enables the use of enzymes in reactions of compounds which do not readily dissolve in water, but only in organic solvents. The layer of detergent molecules protects the enzymes from inactivation by the organic solvent.

This research provides some important clues to the effect of the composition of the micellar medium on the activity of the enclosed enzymes. In the research, the enzyme 20-beta-hydroxysteroid dehydrogenase was used as a model. This enzyme catalyzes the reduction of the hormone progesterone and can be assayed rather easily. The systems that were used in the study contained the enzyme in the water phase, the substrate (progesterone) in the organic phase, the detergent CTAB (cetyltrimethylammoniumbromide) in the interphase, various solvents as organic phase, and various amounts of hexanol to reduce the repelling activity of the head groups of the detergent as to produce stable micelles.

The composition of the system appeared to influence the enzyme activity considerably. The most active system (with dodecane as organic phase and 100% hexanol) showed forty times as much activity as the least active system (with toluene and 30% hexanol). For industrial applications of the micellar system, it is important to be able to predict the enzyme activity from the composition of the system. Dr. Hilhorst developed a method to do this, on the idea that differences in enzyme activity in the examined systems are caused by the varying solubility of the substrate in the interphase.

Particularly in the pharmaceutical industry, many compounds insoluble in water were also made, and that production with enzymes was out of the question until now. The results described by Dr. Hilhorst bring the use of enzymes for these processes near, since it is now possible to predict the optimal composition of the micellar media.

The presentations on new biotechnology and patent law and other legal issues 'allowed by open forum discussions, seems to have been aimed at bringing together the biological scientist and the patent lawer so that each could learn from the other in an exchange of philosophies (which may be conflicting). Indeed, the very idea of patenting inventions in this field soon comes under examination and reflects the lingering unease of some academic scientists with the constraints they believe it places on academic freedom. The real opposites are, of course, not publications verus patents but total secrecy versus the combination of patenting and a compatible publication policy.

Resource Data Bank

More than 20 years ago, Professor Skerman from the University of Queensland in Australia took up the mamoth task of cataloguing a computerized list of the various microbial strain maintained by institutions all over the world. It has been expanded later on and took the shape of World Data Centre which has also taken up the task of documenting viruses (as a special programme of the international Committee on the Taxonomy of Viruses). In 1974, the concept of Microbiological Resource Centres (MIRCEN) was proposed by the Executive Director of United Nations Environment Programme (UNEP) with funding by UNESCO. The major programme of the MIRCEN is the management, distribution and utilization of microbial gene pools. This was facilitated by correlating World Data Center with MIRCEN in collaboration with World Federation of Culture Collections. Now, MIRCEN serves as a pivotal unit for the formation of culture collections (including Rhizobium species for legume inoculations) in several developing countries (Table) and also for providing data services to the centers acting in close contact with World Data Centre.

A Representative List of Major Culture Collections

Australia

Fungi, yeasts, and actinomycetes
Australian National Reference
Laboratory in Medical Mycology
The Institute of Medical Research
The Royal North Shore Hospital of Sydney
St. Leonard, N.S.W., Australia

Bulgaria

Pathogenic bacteria and fungi

Bulgarian Type Culture Collection
Institute for State Control of Medical Preparations Sofia, Bulgaria

Czechoslovakia

Bacteria, mycoplasmas, viruses, and fungi
Czechoslovak Collection of Microorganisms
U.E. purkyne University Frida
OBRANCU
Miru 10,66243
Brno, Czechoslovakia

France

Bacteria
Collection of the Institut Pasteur
25, rue du Docteur Roux
Paris, France 75015

Germany

Bacteria, Fungi, yeasts, and plants viruses
Deutsche Sammlung van Mikroorganismen
D-3400 Gottigen
Grisebachstrase 8, West Germany

Huangary

Bacteria, fungi, and virus
Microbiological Gene Bank
Microbiological Department Group of
the Department of Food Technology
and Microbiology
University of Horticulture
1064 Budapest, Izabella U.46,
Hungary.

India

Bacteria and fungi
Indian Type Culture Collection
Indian Agricultural Research Institute
New Delhi 12, India

Japan

Bacteria, fungi, viruses, bacterfophages, algae, protozoa, and rickettsia
Japanese Federation of Culture
Collections of Microorganisms
c/o Institute of Applied Microbiology
University of Tokyo, Bunkyo-Ku
Tokyo, Japan

Fungi, yeasts. bacteria, and bacteriophoges
Institute for Fermentation
4-54 Juso-tushnocho
Higashiyodogawaku
Osaka, Japan

Bacteria and fungi
Laboratory of Culture Collection of Microorganisms
Faculty of Agriculture
Hokkaide University
Sapporo, Japan

Actinomycetes
Kaken Chemical Company, Ltd.
6-42, Jujodal I-chome, Kita-Ku
Tokyo 114, Japan

The Netherlands

Centraalburesu voor Schimmel Culture
Oosterstraat 1
Baarn, The Netherlands

United Kingdom

Algae and protozoa
Culture Centre of Algae and Protozoa
Institute of Terrestrial Ecology
36 Storey's Way
Cambridge, CB3 ODT, England

Bacteria of industrial importance
National Collection of Industrial Bacteria
Tarry Research Station
PO Box 31, 135 Abbey Road
Aberdeen AB9 SDG, Scotland

Bacteria of medical and veterinary importance
National Collection of Type Cultures
Central Public Health Laboratory
Colindale Avenue
London NW9 5HT, England

Bacteria pathogelnic for plants
National Collection of Plant Pathogenic Bacteria
Plant Pathology Laboratory
Hatching Green
Harprenden, Hertfordshire
AL5 2 BD, England

Fungi (other than animal pathogens and wood-rotting fungi)

Collection of Fungus Cultures
Commonwealth Mycological Institute Ferry Lane
Kew, Surrey TW9 3 AF, England

United States of America
Bacteria, fungi, algae, protozoa, bocteriophages cell culture, viruses, antiserum, rickettsiae and chlamyciae

American Type Culture Collection
12301 Parklawn Drive
Rockville, MD 20852 U.S.A.

Bacteria and fiingi (no catalogue)

Northern Regional Research Laboratory
U.S. Department of Agriculture, Science and
Education Administration
1815 North University Street
Peoria, Il, 61604, U.S.A.

WORLD LIST
Rhizobium Collections
International Biological Program
World Catalogue of Rhizobium Collections
7 Marylebone Road
London N.W.L, England.

Marcen is an acronym for Microbiological Resources Centres that in reality are actually existing academic and/or research institutes in the developed and developing countries. These centres participate in a global collaborative network effort in the promotion of international scientific cooperation and harnessing the beneficial applications of the microbial world for human progress. Eight centres i.e. those located in Australia (Now relocated in Japan (See Table). Brazil, Egypt, Guatemala, Kenya,

Table

Some Major International Culture Collections Useful to Biotechnologists

Culture collection	***Address***
American Type Culture Collection	12301 Parklawan Drive, Rockville, Maryland 20852, USA
Deutsche Sanmlung von Microorganismen	Grisebachstrasse 8, Gottingen 3400, FRG
Centraalbureau voor Schimmelcultures	Oosterstraat 1, PO Box 273, 3740 AG Baarn, The Netherlands
I	
Czechoslovak Collection of Microorganisms	J.E. Purkyne University, 662 43 Brno, Czechoslovakia
Collection Nationale de Cultures de Microorganismes	Institut Pasteur, 28 Rue du Docteur Roux, 75724 Paris Cedex 15, France
Culture Collection of the Institute for	Institute for Fermentation, 17-85 Jugo Hohmachi, 2, chome, Yodogawa-Ku, Osaka, Japan
USSR All-Union Collection of Microorganisms	Institute of Microbiology, USSR Academy of Sciences, Profsojuznaja 7, Moscow B-133, USSR

Senegal, Sweden and Thailand were identified and established within the framework of the now completed (1975-1985) UNESCO/UNEP project.

Today, the global MIRCEN network programme, which embodies activities that are carried out within the framework of UNESCO's regular programme activities in co-operation with the international scientific community and as a follow-up to earlier cornerstone activities, is comprised of 15 such centres (see Table 13.3) *and from 1987* onward also include centres in Australia and France.

The world-wide programme focussing on efforts in preserving microbial gene pools and making them accessible to the developing countries and launched through the establishment of a world network of Microbiological Resources Centres MIRCENsC attempts to:

(a) Provide the infrastructure for a world network which would incorporate regional and interregional co-operating laboratories geared to the management, distribution and utilization of the microbial gene pool;

(b) Reinforce efforts relating to the conservation of microorganisms, wit, emphasis on Rhizobium gene pools, in developing countries, with an agrarian base;

(c) foster the development of new inexpensive technologies native to the region;

(d) promote the applications of microbiology in order to strengthen rural economies; and

(e) serve as focal centres for the training of manpower and the diffusion of microbiological knowledge.

Table

Global Mircen Network

MICRCEN Speciality	*Location*	*Place and country*
Rhizobium MIRCEN	University of Nairobi	Nairobi, Kenya
Rhizobium MIRCEN	Instituto de Pesquisas Agronomicas	Porto Alegre, Brazil
Fermentation, Food and Waste Recycling MIRCEN	Thailand Institute of Scientific and Technological Research	Bangkok, Thailand
Biotechnology MIRCEN	Ain Shams University	Cairo, Arab Rep. of Egypt
Biotechnology MIRCEN	Central American Research Institute for Industry	Guatemala City Guatemala
Rhizobium MIRCEN	University of Hawaii	Hawaii, U.S.A.
Biotechnology MIRCEN	Karolinska Institute	Stockholm, Sweden
World Data Centre MIRCEN	Life Science Research Information Section, RIKEN	Saitama, Japan
Rhizobium MIRCEN	Centre National de Racherches Agronomiques	Bambey, Senegal
Biotechnology MIRCEN	Planta Piloto de Proce SOS Industriales Microbiologicos (PROIMI)	Tucuman, Argentina

MICRCEN Speciality	*Location*	*Place and country*
Rhizobium MIRCEN	Cell Culture and Nitrogen-Fixation Laboratory	Maryland, U.S.A.
Fermentation Technology MIRCEN	University of Osaka	Osaka, Japan
Biotechnology MIRCEN	University of Kent and Centerbury	Kent, U.K.
Biotechnology and Agriculture MIRCEN	University of Waterloo/ University of Guelph	Waterloo/Guelph, Canada
Marine Biotechnology MIRCEN	University of Maryland	Maryland, U.S.A,

In 1986, the third edition of the World Directory of Collections of Cultures of Microorganisms supported by UNEP and UNESCO enunciated a time-honoured concept that "culture collections occupy a central position in microbiology. Today culture collections also occupy a central position in life sciences and biotechnology as they are conceived by man as treasurehouses for the economic and social progress of mankind, be it on a national, regional or international scale. In this context, the World Data Centre of Microorganisms (WDC) is a tool for culture collections and their users to realize quick and smooth exchanges of information on microorganisms to support research and development in those fields.

In the summer 1986, the WDC was relocated from University of Queensland in Australia to RIKEN in Japan.

In 1966, a survey of culture collections was recommended in a meeting of specialists, convened by UNESCO, to discuss the coordination of activities of culture collections. Subsequently, the IAMS Section on Culture Collections (later to become the World Federation for Culture Collections, WFCC) undertook the preparation of a World Directory of Collections of Cultures of Microorganisms. Consequently, a survey initiated by Dr. S. Martin of the National Reseach Council of Canada and continued by Professor, V.B.D. Skerman at the University of Queensland, Brisbane, Australia, resulted, with support from UNESCO, in 1972, in the first edition of the World Directory. The Directory was published by John Wiley and Sons. Inc., New York for the World Federation for Culture Collections. The nucleus at Brisbane became the World Data Centre, and in 1975 acquired MIRCEN status to maintain and update the Directory on computer. In 1982, the second edition of the World Directory was published by the WDC with the support of UNESCO, FAO, WHO, UNU, UNIDO, UNEP and the EEC.

In 1984, at the fifth International Congress of Culture Collections in Bangkok, on the occasion of the imminent retirement of Prof. Skerman, the World Federation for Culture Collections officially reaffirmed the WDC as a component of WFCC and thus the WFCC's responsibility for the operation and management of the WDC.

A steering group was set up by the Executive Board of WFCC to ensure a smooth transfer of the WDC to a new host institution to be chosen by open competition. Following the evaluation of a number of proposals, the WFCC adopted the group's recommendation that the WDC be transferred to the Institute of Physical and Chemical Research (RIKEN), Saitama, Japan.

To this effect, WFCC and RIKEN exchanged a letter of agreement for the relocation of WFC to RIKEN in May, 1986. In August, the data was transferred from Brisbane to RIKEN and is now stored in the Life Science Research Information Section (LSRIS) minicomputer. The WDC at RIKEN has since been accorded MIRCEN status.

The WDC is an information centre which supports the culture collections and their users. Its primary tasks are:

(1) Publication of the description of culture collections.
(2) Production of a species-oriented directory.

In addition, other tasks will be performed based on necessity and available resources.

The primary data sources for the WDC are the culture collections. WDC encourages the affiliated culture collections to send updated data at any time through a variety of mechanisms such as catalogues or newsletters that are accompanied with instructions for data updating. In response, the WDC will send printouts from the databases the culture collections if required. Unlisted, non affiliated culture collections, sources for WDC are international and local information systems, federations and institutions. The WDC is envisaged to use and to co-operate with the Microbial Strain Data Network (MSDN)-a joint initiative of CODATA, WFCC and IUMS, as it becomes operational.

In due course, online access and retrieval with terminals through a network of VENUS-P is envisaged.

Two levels of user guides are prepared by the WDC. These are:

(1) a menu of services indicating how to access the WDC;
(2) detailed user guides of protocols to disseminate the data of WDC.

As the co-operation of culture collections is essential to WDC, it is necessary to reinforce a revitalized communication network between WDC and other organizations, such as WFCC, culture collections, federations, regional and/or national data centres, the MIRCENS and international data activities such as MSDN. For this purpose, the WFCC newsletter and the MIRCEN Journal, published by Oxford University Press in association with UNESCO, play an important role. The WDC prepares its own brochures and posters for mailing and presentation at meetings.

LSRIS is responsible for the daily operation of WDC. in addition, the Japan collection of Microorganisms (JCM) in RIKEN also plays a cooperative role. WDC operations at RIKEN are under the direction of the Director, Dr. Kazuo Komagata, with supporting staff contributed by LSRIS; the Japan Collection of Microorganisms; RIKEN and advisory services from Universities and Research Institutes in Japan.

The WDC currently holds information on 327 culture collections from 56 countries. The core data of the WDC is on culture collections and their holdings. Users of the WDC can locate a culture collection and/or category of microorganisms by use of the core data. They may find a specific strain through the WDC by consulting a list of species preserved in culture collections.

In the database, a list of species with a collection number is prepared as an index.

The data received from Brisbane in three volumes of magnetic tapes was successfully transferred onto disks of the LSRIS minicomputer, PANAFACOM 53300. The core has been installed in a database management system. Other data in the magnetic tapes is currently under analysis.

WDC brochures and posters were presented at the XIV International Congress of Microbiology, Manchester, U.K. in September 1986.

Within the framework of UNESCO'S global network of Microbiological Resource Centres, the World Data Centres on Microorganisms had organized a number of training courses, primarily for researchers from the developing countries, in 1975,

1977 and 1980 at Brisbane, Australia. A training course on "Bioinformatics: Data management and computer usage in culture collections" was organized at the new venue within the rubric of UNESCO'S approved programme: Research, training and international cooperation in key areas in Science and Technology. Consequently, the activity derived support from the organization's programme actions in two key areas i.e. (i) informatics and (ii) applied microbiology and biotechnology. The course was organized by the Life Sciences Research Information Centre (RIKEN) and was supported by the International Centre for Cooperative Research in Biotechnology (OSAKA MIRCEN) and the MIRCEN for South East Asia at Bangkok, Thailand.

This training course was sponsored by UNESCO, Japan Federation for Culture Collection (JFCC), Institute for Fermentation, Osaka (IFO), Japan Collection of Microorganisms (JCM) and other funding agencies.

Culture Collections

There is now a growing awareness that culture collections provide essential resources and services to all aspects of Biotechnology. The collection, maintenance and characterization of microorganisms form part of the infrastructure of microbiology and together with advisory identification, preservation and patent services, make an important contribution to the development of biotechnology. Among the millions of microbial, plant and animal cells stored in the world's culture collections, it is often possible to find one that fits exactly with a biotechnologist's particular requirements or one that can be developed to perform a specific task.

Organization of culture collections

The World Data Centre on Microorganisms lists 566 culture collections in its directory. While many of these collections provide culture only by special arrangement and do not publish catalogues, others not only provide cultures, advice and ancillary services on demand to researchers in industry and elsewhere but also publish catalogues. The number of these 'service collections' is small, but many countries are now seeking to establish national microbiological resources centres. In addition, a global network of eleven Microbiological Resource Centres has been set up by the International Cell Research Organization. Each is concerned with developing collections of microorganisms, providing training and advisory centres and promoting regional collaboration.

Service collections have developed and are administered in several different ways. The US and the FRG, for example, have single collections which maintain all types of microorganisms; other countries, such as the UK and Czechoslovakia, have a decentralized system where each type of microorganism is housed in a separate institute and the individual collections are coordinated to form a national network. The United Kingdom supports eleven service collections housed in separate institutes and maintaining bacteria, fungi (including yeasts), algae, protozoa and animal cells. Some have specialist interests in medical, veterinary, industrial or genetical strains; others maintain taxonomic collections of a more general interest. The services of these collections are available to scientists throughout the world. Table lists some of the major culture collections in other countries which are particularly useful to biotechnologists. National federations have been established in several countries and the World Federation for Culture Collections (WFCC) provides international coordination. The federations hold scientific meetings, arrange training courses and prepare books, documents and newsletters. In addition, they provide essential links with government and funding bodies.

Clearly, the prime function of culture collections must be to provide a comprehensive range of pure, authenticated microorganisms that are of past, present or potential interest.

Curators have learned the most reliable preservation methods for the strains in their collection, through experience and empirical investigations, although for particularly sensitive cells—leptospita, for example—reliable methods have yet to be developed. There is an urgent need for research on the molecular events that occur during preservation so that the genetic stability of strains can be assured. The problem becomes more urgent with the increasing need to maintain genetically manipulated microorganisms that are often unstable; moreover, it is important that cultures deposited for patent purposes should be maintained in such a way that strain drift is minimized.

Maintenance methods are very diverse and the choice of the most suitable for the preservation of a particular strain is often difficult. In general, methods fall into four classes: *subculturing* (active transfer), *drying, freeze-drying and cryopreservation.* Subculturing is convenient and cheap and requires little in the way of equipment and skills. Moreover, many 'cultures' remain alive after storage for months or even years. However, the proportion of cells surviving within the original population may be very low and there is growing evidence that substantial changes take place. To retain strain stability, subculturing must be used in conjunction with selection procedures and this diminishes the simplicity of the method. Culture collections, while recognizing the short term convenience of the method favour other procedures.

The three other types of preservation methods arrest cell multiplication and thus halt strain drift. Cells can be dried on soil, sand, silica gel or paper, but changes in strain characteristics may occur. Freeze-drying is widely practised because it is convenient to distribute cells in this form and makes it possible to prepare large batches of microorganisms. However, while this method is very useful for most bacteria, it is unsuitable, or gives very low survival levels, for a number of other microorganisms. In addition, it has been implicated in several reports of genetic change. Nevertheless, freezedrying provides substantially greater stability than drying or subculturing for a wide range of microorganisms and cells that withstand the process survive storage for long periods.

Cryopreservation at temperatures between -70°C and -196°C is generally accepted as the preservation method conferring maximum stability to cells. A wide range of microorganisms and cell types can be stored successfully at low temperatures and, in general, survival levels are significantly higher than those obtained with other methods. Preliminary experimentation is needed to establish the best growth parameters, cryoprotectants and freezing and cooling rates for particular cells. Internationally, the computerization of catalogue and strain data has been under consideration by a number of agencies (the World Federation for Culture Collection the EEC; national organizations in Scandinavia, Japan, Europe; MIRCENS in the developing countries) and it is clear that great progress will be made over the next decade.

Another important requirement of biotechnology is access to collections with deposit facilities for patent strains. It is a legal requirement that if a process involving the use of a microorganism is to be patented, a culture of the microorganism must be deposited with a recognized culture collection. Different countries have different patent laws and in the past the regulations governing the deposit and release of patent strains have confused depositors. In addition, patents taken out in more than one country required the culture to be deposited in each country. The Budapest Treaty, which came into effect in 1977, simplifies proceedings by allowing a single deposit made in an accepted International Depository Authority (IDA) to serve for patents taken out in any number of member states. To qualify as an IDA, a collection must meet requirements of stability, facilities and scientific competence. Many service collections have now been accepted.

Culture collections exist to support all microbiology—the future requirements of education and research as well as industry must be safeguarded. There is clearly a need for a more extensive examination of present holdings for useful properties, for improved access to existing information by the development of computer links and for the development of research into the area of strain stability. In addition, there is a need to develop culture collections as banks not only for microorganisms but for animal and plant cells, viruses and hybridomas, genetically marked strains, genetically manipulated strains, plasmid-bearing strains and perhaps even plasmids and gene fragments.

It is clear that culture collections provide many services that support the infrastructure of biotechnology. It is vital that the infrastructure of the collections themselves is protected; the fundamental work of microbiology the comparative studies, the examination and identification of new isolates, the basic taxonomy from which other microbiological studies develop-must be fostered. It is encouraging that there are many international developments with this objective.

International Biotech Scenario

Since the mid 1970's, biotechnology and one of its scientific precursors, recombinant DNA research have been the focus of appreciations and controversy. Issues such a work-place safety, University-industry relationship, changes in patent laws, gene conservation and the environmental impact of various projects have been vigorously debated. In some cases, it has been taken as a symbol of certain kind of innovation and resisted. This is true at the moment in the Federal Republic of Germany where some supporters of Green Politics oppose commercial biotechnology as a continuation of an oppressive industrial system.

Although biotechnology is quite old and traditional discipline covering areas such as farming and brewing, yet over the past one and a half decade, the advances in the 'New Biotechnology' would certainly affect wide range of activities. Microbial and other types of cells will be used to make several organic substances on a large scale at an economic cost. The aim is to renew the resources and output of existing industries and creation of new ones. (Table).

Table

Selected Biotechnology and Genetic Engineering Companies in Different Countries

Agrigenetics	USA
Amgene	USA
Bethesda Research Laboratories	USA
Biologicals	USA

Bio-response	USA
Bio-tech General	USA
Calgene, Inc.	USA
Centocor	USA
Centaur Genetics Corporation	USA
Cetus	USA
Cetus Immune Corporation	USA
Clonal Research	USA
Collaborative Genetics	USA
Cytogen	USA
Damon Biotech	USA
DNA Science, Inc.	USA
DNAX Ltd.	USA
Enzo Biochem	USA
Genetech	USA
Genetics Institute	USA
Genetics Replication Technologies, Inc.	USA
Molecular Genetics	USA
Neo Bionics	USA
New England Bio Labs	USA
Phytogen	USA
Transgene	USA
Unigene	USA
University Genetics	USA
Eli Lilly & Company	USA
Gen Probe Inc.	USA
Integrated Genetics, Inc.	USA
Sigma cell culture Reagents	USA
Traders Protein	USA
Whittker M.A. Bioproducts	USA
Genetic systems	USA
Diagenetics Laboratories Inc.	USA
Xenogen	USA
Endotronics	USA
Woodland	USA
Seed Tec	USA
Syntex	USA

Antibioticos S.A.	USA
Abbott Diagnostic	USA
E.1. du Pont de Nemours and Company	USA
Haldene Consultants Limited	UK
Bioprocessing Limited	UK
Bio Enzyme	UK
Bio-Legicals	Canada
Biogen	Switzerland
Jenssen	Belgium
Gentross	Sweden
Nitto Chemical Industry	Japan
Kirin	Japan
Kirin-Am Gen	U.S.-Japan
Lyonnasie des Eaux	France
Rhone-Poulenc Agrochemie	France
Kemira Oy,	Finland
Novo Industri A/s	Denmark

Currently, biotechnology has spread to almost all parts of the world (Table) is well documented with the UNESCO sponsored International Centre for Genetic Engineering and Biotechnology now located in India (with industrial microbiology section at Triesta in Italy). In 1983, South American Biotechnology MIRCEN started to function after a long consulting period with many of the laboratories now within the network. At the moment, there are more than 40 co-operating laboratories in Bolivia, Chile, Brazil, Ecuador, Colombia, Paraguay, Peru, Uruguay and Argentina with different strengths and capabilities.

The International Center of Cooperative Research in Biotechnology, Japan (ICBiotech) was founded in April 1985. It has resulted from the International Center of Cooperative Research and Development in Microbial Engineering, Japan (ICME), that was established in April 1978 in the Faculty of Engineering, Osaka University.

Activities of the Center include:

1. Conducting studies on microbial engineering, with special emphasis on studies dealing with the microbial utilization of natural resources, microbial ecosystems and microbial reactions engineering.

2. Educating graduate students with the aim of providing a pool of manpower to contribute to international cooperative research and development in biotechnology.

Table

International Organizations with Activities in Biotechnology

Agency	***Activities***
Bio-Energy Council	Non-profit organization monitoring biomass projects worldwide. Publishes a directory.
Commission of the European Union	Organized in Directorate-Generals (DGs). Communities (CEC) DG7 coversagri-culture, fisheries and food, DG12 education and research and DG13 includes the FAST (Forecasting and Assessment in Science and Technology) programme, an extensive research programme on 'biomolecular engineering' and biomass projects.
Commonwealth Agricultural Bureaux (CAB)	An association of research institute concerned with agriculture. Publish several abstracts journals and databases.
Commonwealth Science Council (CSC).	Funds alternative energy projects.
European Molecular Biology	Operates the European Molecular Biology Organization (EMBO). Laboratory in Heidelberg, which, among other projects, has major nucleotide sequence.
Food and Agriculture Organiza-tion (FAO).	Interests in biomass fuels, single-cell protein and biological control.
Inter-American Development Bank (IDB).	Funds a central American biotuets project.
International Cell Research Organization (ICRO).	Microbiological research and training.
International Development Research Centre (IDRC).	Funds various research programmes.
International organization for	An international network of laboratories

Agency	Activities
Biotechnology and Bioengineering (IOBB).	who collaborate.
Organization for Economic Cooperation and Development (OECD)	Has commissioned several studies on the prospects and effects of biotechnology on member states
United Nations Industrial Development Organization (UNIDO)	Is establishing an International Centre for GeneticEngineering and Biotechnology, probably in India,and has produced a number of papers considering the effects of biotechnology on Developing countries. It also publishes a news letter.
World Data Centre	Helps to maintain and assist communications between culture collections.
World Federation for Culture Collections	Helps to maintain and assist communications between culture collections.
World Health Organization (WHO) (United Nations)	Establishes standards and reference preparations for biological substances such as hormones and vaccines. Is considering similar activities for monoclonal antibodies and materials used in recombinant DNA research. Promotes research and national collaboration on all aspects of human health care and pest and disease control. Convenes various expert groups whose deliberations are published.
World Intellectual Property Organization (WIPO)	A United Nations body concerned with patent cooperation and classification.
United Nations Educational Scientific and Cultural Organization (UNESCO)	Funds research and teaching projects particularly in the Gulf States and Africa.
United Nations Environmental Programme (UNEP).	Assists in culture collection development and maintenance.

3. Organizing a Japanese Universities' group to cooperate in making workplans and executing programmes of scientific exchange in biotechnology (exchange ol scientists, cooperative research, and scientific seminar) under the Japan Society for the Promotion of Science (JSPS).

4. Cooperating in "International Post-Graduate University Course in Microbiology" by UNESCO at the Department of Fermentation Technology, Faculty of Engineering, Osaka University.
5. Collecting specific information in the field of biotechnology from all parts of the world and disseminating up-to-date information on scientists in developing countries.
6. Working as the national point of contact (Japan) of Southeast Asian Regional Network of Microbiology in UNESCO Programme on Regional Cooperation in Basic Sciences; cooperative research, workshop and seminar.
7. Issuing annual scientific reports and newsletters periodically.

Cooperative Programme

In the past eight years, since the establishment of ICME, the Center has implemented the programme of the Japan Society for the Promotion of Science in Biotechnology with Southeast Asian countries, by means of the exchange of scientists, cooperative research, and scientific seminars. It is operated by University groups in Japan and Southeast Asian countries in the framework of the "core University system". So far, Thailand, Singapore and Philippines, participate in the programme.

In Japan, Osaka University acts as the core university with participation of cooperating universities and coresearchers recognized by JSPS, NRCT (Thailand), NUS (Singapore) and NSTA (Philippines).

(i) JAPAN-THAILAND COOPERATIVE PROGRAMME IN BIOTECHNOLOGY

Cooperative research is classified into six subjects as follows:

1. Enzyme technology
2. Industrial microbiology and genetics
3. Food microbiology
4. Utilization of agricultural products and wastes
5. Development of novel products
6. Cell technology

(ii) JAPAN-SINGAPORE COOPERATIVE PROGRAMME IN BIOTECHNOLOGY

The counterpart implementary agency in Singapore is National University of Singapore (NUS).

Cooperative research is classified into the following three subjects:

1. Fermentation technology
2. Cell technology
3. Food microbiology

(iii) JAPAN-PHILIPPINES COOPERATIVE PROGRAMME

The counterpart implementing agency in the Philippines is the National Science and Technology Authority (NSTA).

The National institute of Biotechnology and Applied Microbiology **(BIOTECH)** in University of the Philippines at Los Banos, as the core university, and the National Institute of Science and Technology are the Institutes involved in the programme. Cooperative research is classified into:

1. Genetic engineering
2. Enzyme technology
3. Microbial engineering

(iv) JAPAN-INDONESIA COOPERATIVE PROGRAMME

The cooperative programme with Indonesia in the field of Biotechnology in the general exchange system is handled jointly by JSPS and its counterpart agency in Indonesia, LIPI (Indonesian Institute of Sciences).

(v) JAPAN-MALAYSIA COOPERATIVE PROGRAMME

The Center takes part in a cooperating programme with Malaysia in the field of Integrated Engineering as a cooperating body to Kyoto University (corn university).

UNESCO Promotion of Microbiological and Biotechnological Research

Country	*Item*	*Source*
Belgium	Joint Research study with Cote d'Ivoire on N-fixation by Cyanobacteria in Tropical Soils	UNESCO
Argentina	Regional MIRCEN research activities	UNESCO
Uruguay	Research studies on Rhizobium	UNESCO
Kampuchea	Establishment of microbiology laboratory	UNESCO
Egypt, A.R.	Regional MIRCEN research activities	UNESCO
ALEC*	Regional Training Course on Bioenergy	UNESCO

*Arab League Educational Cultural and Scientific Organization.

In 1971, the Council for Science and Technology, an advisory organ to the Japanese Prime Minister, recognized the importance of life sciences for human welfare. In 1980, the Council in reiterating the importance of life sciences and the recent progress of research being made in this area globally, proposed the following:

1. To select important subjects from the disciplines of life sciences and provide them with long-range support.
2. To establish and support research services for the life sciences, especially through the establishment, maintenance and supply of laboratory organisms, experimental materials and information relating to these.
3. To organize a research system and a research service system coordinated to ensure efficient progress in research.

In 1972, an office for Life Science Promotion was organized in RIKEN and six goal-oriented research projects were formulated. These projects are to be carried out over periods of ten years by research groups made up of Scientists affiliated with Universities and research institutes throughout Japan. The office promotes research services (for the establishment and supply of laboratory organisms) and the development of an information management system for the organisms and related research. The office also promotes recombinant DNA research, an important aspect of the life sciences.

The office promotes the life sciences not only by providing research funds but also by building the required facilities. In 1980, the office built the Culture Collection of Microorganisms, a facility whose main purpose is to collect, maintain, and supply microorganisms, as one of the programs in the establishment of the research services. In 1986, the RIKEN Gene Bank was established at Tsukuba Science City which supplies cell cultures and genes.

1. DEVELOPMENT OF AN INFORMATION SYSTEM FOR LABORATORY ORGANISMS

Recognising the growing need for nationally unified information on laboratory organisms for life sciences, the office in 1976 initiated the development of a National Information System for Laboratory Organisms (NISLO). The LSRIS is now developing NISLO in collaboration with six study groups which specialize in laboratory animals, microorganisms, plants, algae, animal tissues and cell cultures, and plant tissue and cell cultures. NISLO processes accession information and characteristic data on laboratory organisms using the data base technique.

2. JAPAN COLLECTION OF MICROORGANISMS

In 1980, the Culture Collection of Microorganisms was established at the Office to promote research in life sciences related to microbiology. The Culture Collection will be a center for culture collections and represents a milestone achievement that has long been awaited by the Japan Federation for Culture Collections and the microbiological societies in Japan.

The Culture Collection is a national organization dedicated to the collection, maintenance, and distribution of authentic cultures of microorganisms. Research efforts at the Culture Collection will be focused on comparative microbiology, microbial systematics and improvement of methods in the characterization and preservation of the cultures. The culture collection will also provide a data bank of information on the cultures working jointly with the Research Information Section of the Office.

The microorganisms to be maintained in the Culture Collections include bacteria, actinomycetes, molds and yeasts. These will be arranged in a cooperative network with the various culture collections throughout the country.

WORLD FEDERATION FOR CULTURE COLLECTIONS (WFCC)

Few microbiology courses include information about the support available to working scientists from service culture collections. Students often start their professional life with no knowledge of the whereabouts of these microbial resource centres or the kinds of services provided by them.

To remedy this situation, the WFCC has produced a 15-minute video film "BACKUP" to show the ways in which culture collections support microbiology. The video introduces students to the importance of using authenticated material and of maintaining it properly for future use. In addition, it describes the services provided by culture collections-information, preservation, identification, safe-deposit, patent depository and training. Students are encouraged to make contact with their local culture collections and to make good use of them.

The teaching video for University Department and Training Centers is available for sale in several video systems and with different language commentaries.

Ethical and Value Judgements

Besides, other aspects as mentioned in this section, the ethical and moral questions in biotechnology also pertains to the risk of development of biological weapons. Growing apprehensions about the possible military use of genetic engineering and other new biotechnologies are threatening to undermine the international ban on biological weapons. An issue is the use of biology, like chemistry and physics before it, to produce new weapons of mass destruction. At a conference held in Geneva about two years ago to review the workings of the Biological Weapons Convention, one of the conference leaders, asked: *"Why is it that scientists look upon opposition to weapons as 'political', but work on the development of weapons as 'patriotic'?"*

The Biological Weapons Convention of 1972 prohibits the development and possession of biological and toxin weapons; their use was previously outlawed by the Geneva Protocol of 1925. The Convention now has 103 signatories, including all the major nations outside the Middle East. Perhaps more worrisome is the fact that there is an ambiguity in the Biological Weapons Convention that the advance of biotechnology has turned into a serious problem. The convention permits the development and possession of biological agents and toxins of types and in quantities that are justified for "prophylactic, protective or other peaceful purposes", including defense. However, although there is genuine concern for the protection of troops and civilians, most scientists agree that defensive and offensive research activities are largely indistinguishable.

This could suggest that novel agents will have to be designed and produced in order to assess and defend against the possible threat of their development by an opponent. It is this loophole in the treaty that is arousing worldwide suspicions and concern that defensive programmes, which are usually carried out in secret, might be used to mask the development of "designer weapons" that could evade opposing defenses.

A variety of previously unimaginable biological weapons can now be conceived. New toxin or antibiotic resistance genes can be inserted into naturally infectious agents, thereby making them more dangerous and more resistant to treatment. Moreover, by altering surface antigens, pathogens could be rendered insensitive to standard vaccines and to the immunological defenses of the host. Antigen alteration would also interfere with diagnosis and identification of the agent, which is often based on serological methods. The stability of an agent, in principle, can be increased by splicing in genes for resistance to inactivation by ultraviolet light, drying, aerosol formation, and other stresses encountered in use as a weapon.

The very existence of a new pathogen would be a threat to life, because no degree of security can fully guarantee against its eventual escape or release by careless or irrational individuals or terrorists, nor can logic assure that, if an unrealiable weapon is at hand, hazards such as the possibility of mutation after release will not someday be ignored in the heat of battle.

Even defensive activities like the production of vaccines against exotic agents tend to raise suspicions, for some form of the agent is needed to make the vaccine, and an aggressor would need the vaccine to immunize before using the agent as a weapon. Regardless of intention, any successful specific defense carries with it the potential for offense and is, therefore, actually inconsistent with the Convention's objective.

Outside the United States, variations and inconsistencies in biotechnology regulation abound, according to a survey of Western European countries and Japan recently completed. Unsettled jurisdictional questions among government agencies within several countries, similar to uncertainties in the United States, threaten to affect development of coherent international policies. Although most of the surveyed countries have liberalized their recombinant DNA research guidelines, France and the Federal Republic of Germany are considering whether to restrict biotechnology because of ethical questions associated with gene therapy protocols.

In France, as in most of the countries, industry has been involved in formulating biotechnology guidelines and regulations. For example, France's recombinant DNA guidelines were eased considerably five years ago in December 1984 when jurisdiction over such research was shifted primarily to local institutions. Although there is some legal ambiguity about the deliberate release of genetically engineered organisms, it is generally believed that current laws and regulations are adequate. However, this notion has not yet been challenged by any actual proposal for such an experiment.

In the Federal Republic of Germany, compliance of researchers with recombinant DNA guidelines is mandatory if they are funded by the government but voluntary if they receive funding from the private sector. However, pending revisions of the guidelines are expected to extend coverage to commercial operations but also to shift jurisdiction over most such research to local management in both the private and public sectors. Deliberate release of genetically engineered organisms currently is expressly prohibited, both by the guidelines and by environmental laws. Interest aroused by the Green Party over possible misuses of gene therapy could seriously curtail such research in Germany.

Although the European community in general is trying to harmonize the way it treats the regulation of biotechnology progress towards developing community-wide laws has been "quite limited." The main effort so far has been to expedite premarket approval of human and veterinary drugs. The need for new laws to deal with the deliberate release of genetically engineered organisms and similar matters is under review.

Within the group of countries considered in the survey, Japan's biotechnology regulations are considered the "most extensive." Considerable effort is being devoted in Japan to make changes that will enhance commercial developments. However, deliberate release experiments are still prohibited, and guidelines for changing them remain to be developed.

Looking Forward

Despite the antiquity of mankind's use of microorganisms and other cells, it is only relatively recently that we have been able to envision new largescale biological processes created by the skills of many different types of highly trained scientists and technologists. There is nothing like *a complete or perfect biotechnologist* because it derives information from Genetics, Biochemistry, Microbiology (a combination resulting in Molecular Biology) and Engineering. No one can be expert in all the arts of microbiology, Genetics and Biochemistry, Chemical and process engineering and the design of computer control systems and several other related areas. It is important and essential that those involved in biotechnology do not retreat into narrow specializations although it is difficult keeping up to date with the major developments and ideas across the whole range of biotechnological endeavours. Biotechnologists, in general, are in an unusually fortunate position. They are now gaining access to funds and facilities that are the envy of many other scientists and technologists. But the *success of biotechnology depends on more than just its technology.* Changes in the legal, ethical and political guidelines in which we live often trail pitifully behind advances in our ability to apply our knowledge of the natural world. While the level of debate amongst those involved in biotechnology is of a high quality, it remains essentially private, confined to individual companies or laboratories. The present benign attitude of investors, industry and even many Governments may diminish or fade away in due course unless many of the promises of todays biotechnology are fulfilled with the development of new, profitable industrial processes, novel drugs and so forth. Basic knowledge must be corporately shared among all who need it. This is essential so that those industries already involved can take immediate advantage of new developments. Besides, those companies not yet using biotechnology can see how the subject relates to and could contribute to their own activities. Those companies, especially in India, need guidance and a point on which they can focus. The need will continue for as long as biotechnology continues. The onus is, therefore, on governments to provide the essential foundations for future industrial expansion (Table) and fundamental research relating to that.

In the context of possible future developments, there is great concern about emissions of sulphur oxides from the combustion of coal in thermal power stations. In future, high and medium sulphur coals will require desulphurization before use. Desulphurization, by controlled microbial action, offers a possible biotechnological solution.

Biotechnology is linked with the energy and agricultural sectors of regional and natural economies.

Table

Government Agencies in the UK, USA and India with Activities in Biotechnology

Agency	*Activities*
U.K.	
National Institute for Biological Standards and Control (NIBSC)	Concerned with establishing standards and reference preparations for biological materials.
Natural Environmental Research (NERC)	Funds research, mainly on subjects of ecological importance, such as pollution control.
Science and Engineering Research Council (SERC) (including the SERC Biotechnology Directorate Management Committee (BTMC).	Fundus cooperative training awards with industry and has established a Directorate of Biotechnology to coordinate its programme.
USA	
Department of Agriculture (USDA)	Funds research.
Department of Commerce (DOC) Office of Technology Assessment and Forecast (OTAF)	Concerned with influencing congressional policy decisions on new technologies. Prepared an excellent report Impacts of Applied Genetics and is working on a follow-up concerned with biotechnology commercialization and international competitiveness.

Agency	***Activities***
Department of Energy	Funds research aimed at biomass energy sources and other alternative fuels, especially through the Office of Alcohol Fuels.
Environmental Protection Agency (EPA)	Has a general control over environmental matters which extend from the composition of "gasohol" to the possible regulation of genetic engineering under the Toxic Substances Control Act (TSCA).
Food and Drug Administration	Oversees the safety of food and pharmaceutical products, including in vitro diagnostics through various Bureaus.
National Bureau of Standards (NBS)	Develops standards, some of which will relate to biotechnology.
National Institutes of health (ineluding the Recombinant DNA Advisory Group RAC)	Major site of federally funded biomedical research conducted by a number of subsidiary Institutes, Includes the RAC.
National Science Foundation	Major source of grants funding research and training.
National Technical Information Service (NTIS)	NTIS supply research reports from many US government projects.
INDIA	
Department of Biotechnology	The major Government body to develop a Biotechnology programme in India and establish a link between Institutions, Industry and Universities.
Department of Non-Conventional Energy Sources.	To explore and extend various sources of energy harvesting and utilization.

In India, which faces periodic drought and floods (and thus strongly affecting agriculture and Industrial set-up), the introduction of new technologies required a social revolution in the age-old tradition bound society. The importance of agriculture to Indian development is shown by the fact that today it contributed 50 per cent of total employment and 33 per cent of exports. But for this crucial sector, the country spent only 0.17 per cent of the GNP for research and development as against 20 per cent spend by Japan. The other major sector is industry which provides 30 per cent of employment and one third of the country's exports depended on the small-scale industry.

India is traditionally the biggest buyer of fertilizers from Qutar, importing some 2,00,000 tonnes of urea and ammonia each annually. Recently, India and Qutar began a review of bilateral economic relations for setting up a joint gas-based multi-million dollar fertilizer project here. It would be most economical to exploit the biotechnological resources for ammonia production as described in earlier chapters.

There are some basic questions and choices to be exercised. In any discussion of the status and future prospects of several biotechnological productions mentioned in the preceding chapters, the major question must concern the relative consistency of the concepts. The existence of processes designed to produce methane (biogas) from biomass and others to produce biomass from methane (natural gas) must initially seem inconsistent. Even though the plant biomass used for biogas production is carbohydrate rich and the microbial biomass produced from natural gas is protein rich, the question whether it might be more sensible to produce protein-rich plant biomass by either agriculture and aquaculture must be posed. Similarly, there are questions regarding the production of ethanol or biopolymers from sugar. Biopolymers are produced extracellularly by various microbes that can use carbohydrates and/or petrochemicals as their carbon energy substrates. In biopolymer production processes, the fermentation step is the key operation from the point of view of overall process economics. This is largely because most microbes that produce biopolymers are mesophilic resulting in major problems with respect to effective process cooling. Hence, for bulk polymer production, feedstocks that can be converted into product with the minimum of heat production must be used. In this respect, carbohydrates are greatly superior to either hydrocarbons or methanol.

The biotechnological operations have proved extremely successful for both the manufacture of traditional products and for fine chemicals, particularly those exhibiting unique biological properties. Several processes which produce existing and novel products will form the supporting base for extensive programming of biotechnology. A case in point is SCP and Biopolymer production. For developing countries brain drain is not something new or which has appeared suddenly since a better brain is always attracted towards a better scientific environment. This does not, however, imply that such scientists as do not force brain drain are not better scientists. Even a developed country such as United Kingdom is now seriously facing this problem as scientists are moving to the United States, Australia and Continental Europe. The establishment of major biotechnological processing ventures is not so quick. Most ventures will require not only design, construction, start-up and market development phases, but also significant product and process development phases. Hence, the time between initiating a venture and manufacturing a specified product at the design capacity of the plant will probably be 5-10 years (with fully profitable operation several years after project initiation and expansion). Therefore, biotechnology which is now going to be a way of life instead of merely a scientific discipline, will have to be developed with utmost planning, prudence and caution.

Index

□□□